WATER
& WASTEWATER
INFRASTRUCTURE

Energy Efficiency and Sustainability

Frank R. Spellman

CRC Press
Taylor & Francis Group
Boca Raton London New York

CRC Press is an imprint of the
Taylor & Francis Group, an **informa** business

CRC Press
Taylor & Francis Group
6000 Broken Sound Parkway NW, Suite 300
Boca Raton, FL 33487-2742

First issued in paperback 2019

© 2013 by Taylor & Francis Group, LLC
CRC Press is an imprint of Taylor & Francis Group, an Informa business

No claim to original U.S. Government works

ISBN-13: 978-1-4665-1785-1 (hbk)
ISBN-13: 978-1-138-38221-3 (pbk)

Library of Congress Cataloging-in-Publication Data

Spellman, Frank R.
    Water & wastewater infrastructure : energy efficiency and sustainability / Frank R. Spellman.
        p. cm.
    Includes bibliographical references and index.
    ISBN 978-1-4665-1785-1 (hardback)
    1. Waterworks--Energy conservation. I. Title. II. Title: Water and wastewater infrastructure.

TJ163.5.W36S64 2013
628.1028'6--dc23   2012034119

Visit the Taylor & Francis Web site at
http://www.taylorandfrancis.com

and the CRC Press Web site at
http://www.crcpress.com

For

Melissa L. Stoudt

Contents

Section II   Energy-Efficient Equipment, Technology, and Operating Strategies

Section III   Energy-Efficient Technology

Section IV   Biomass Power and Heat Generation

Section V   Sustainability Using Renewable Energy

Section VI   Appendices

Preface

At its core, this book is about facing reality—a pressing reality. That is, this book is about energy use today and future energy availability for the treatment of the water we drink and the water we foul through our use or abuse (or both). The pressing reality? Have you noticed the current prices of gasoline, heating oil, and diesel fuel in the United States? The pressing reality of the high cost of hydrocarbon fuels is just one facet of the growing problem. How about our growing population and the corresponding need to increase our use of energy to support, maintain, and sustain the growing population? Moreover, we also need to consider worldwide growth in energy needs, now and in the future. China and India, for example, are expanding their populations, their economies, and thus their need for energy.

With regard to the focus of *Water & Wastewater Infrastructure: Energy Efficiency and Sustainability*, keep in mind that water and wastewater treatment facilities across the country are facing many common challenges, including rising costs, aging infrastructure, increasingly stringent regulatory requirements, population changes, and a rapidly changing workforce. Accordingly, to address these issues and to focus managerial energy toward mitigating these issues, water and wastewater facilities must strive to operate in a manner that provides economic and environmental benefits, saves money, reduces environmental impacts, and leads to sustainability. This can best be accomplished by improving energy efficiency and water efficiency.

The bottom line: This book not only details processes that can assist facilities to become more energy efficient but also provides guidance to ensure their operational sustainability.

Author

Frank R. Spellman, PhD, is a retired U.S. naval officer with 26 years of active duty, a retired environmental safety and health manager for a large wastewater sanitation district in Virginia, and a retired assistant professor of environmental health at Old Dominion University, Norfolk, Virginia. He is the author or co-author of 75 books, with more soon to be published. Dr. Spellman consults on environmental matters with the U.S. Department of Justice and various law firms and environmental entities around the globe. He holds a BA in public administration, a BS in business management, and a MBA, MS, and PhD in environmental engineering. In 2011, he traced and documented the ancient water distribution system at Machu Pichu, Peru, and surveyed several drinking water resources in Amazonia, Ecuador. Dr. Spellman also studied and surveyed two separate potable water supplies in the Galapagos Islands.

Acronyms and Abbreviations

°C	Degrees Centigrade or Celsius
°F	Degrees Fahrenheit
μ	Micron
μg	Microgram
μm	Micrometer
A/O	Anoxic/oxic
A^2/O	Anaerobic/anaerobic/oxic
AC	Alternating current
ACEEE	American Council for an Energy Efficient Economy
Al_3	Aluminum sulfate (or alum)
Amp	Amperes
Anammox	Anaerobic ammonia oxidation
APPA	American Public Power Association
AS	Activated sludge
ASCE	American Society of Civil Engineers
ASE	Alliance to Save Energy
ATM	Atmosphere
AWWA	American Water Works Association
BABE	Bio-augmentation batch enhanced
BAF	Biological aerated filter
BAR	Bioaugmentation reaeration
BASIN	Biofilm activated sludge innovative nitrification
BEP	Best efficiency point
bhp	Brake horsepower
BNR	Biological nutrient removal
BOD	Biochemical oxygen demand
BOD-to-TKN	Biochemical oxygen demand-to-total Kjeldahl nitrogen ratio
BOD-to-TP	Biochemical oxygen demand-to-total phosphorus ratio
BPR	Biological phosphorus removal
CANON	Completely autotrophic nitrogen removal over nitrate
CAS	Cyclic activated sludge
CBOD	Carbonaceous biochemical oxygen demand
CCCSD	Central Contra Costa Sanitary District
CEC	California Energy Commission
CEE	Consortium for Energy Efficiency
cfm	Cubic feet per minute
CFO	Cost flow opportunity
cfs	Cubic feet per second
CHP	Combined heat and power
Ci	Curie
COD	Chemical oxygen demand
CP	Central plant
CV	Coefficient of variation
CWSRF	Clean Water State Revolving Fund

DAF	Dissolved-air flotation unit
DCS	Distributed control system
DO	Dissolved oxygen
DOE	Department of Energy
DON	Dissolved organic nitrogen
DSIRE	Database of State Incentives for Renewables and Efficiency
EBPR	Enhanced biological phosphorus removal
ECM	Energy conservation measure
ENR	Enhanced nitrogen removal
EPA	Environmental Protection Agency
EPACT	Energy Policy Act
EPC	Energy performance contracting
EPRI	Electric Power Research Institute
ESCO	Energy Services Company
$FeCl_3$	Ferric chloride
FFS	Fixed-film system
GAO	Glycogen accumulating organism
GBMSD	Green Bay, WI, Metropolitan Sewerage District
GPD	Gallons per day
GPM	Gallons per minute
H_2CO_3	Carbonic acid
HCO_3^-	Bicarbonate
HDWK	Headworks
hp	Horsepower
HRT	Hydraulic retention time
Hz	Hertz
I&C	Instrumentation and control
I&I	Inflow and infiltration
IFAS	Integrated fixed-film activated sludge
IOA	International Ozone Association
IUVA	International Ultraviolet Association
kW	Kilowatt hour
kWh/year	Kilowatt-hours per year
LPHO	Low-pressure, high-output
M	Mega
M	Million
MBR	Membrane bioreactor
MG	Million gallons
mg/L	Milligrams per liter (equivalent to parts per million)
MGD	Million gallons per day
MLE	Modified Ludzack–Ettinger process
MLSS	Mixed liquor suspended solids
MPN	Most probable number
MW	Molecular weight
N	Nitrogen
NAESCO	National Association of Energy Service Companies
NEMA	National Electrical Manufacturers Association
NH_4	Ammonium
NH_4-N	Ammonia nitrogen

NL	No limit
NPDES	National Pollutant Discharge Elimination System
NYSERDA	New York State Energy Research and Development Authority
O&M	Operation and maintenance
ORP	Oxidation–reduction potential
Pa	Pascal
PAO	Phosphate accumulating organisms
PG&E	Pacific Gas and Electric
PID	Phased isolation ditch
PLC	Programmable logic controller
PO_4^{3-}	Phosphate
POTWs	Publicly owned treatment works
PSAT	Pump system assessment tool
psi	Pounds per square inch
psig	Pounds per square inch gauge
RAS	Return activated sludge
rpm	Revolutions per minute
SBR	Sequencing batch reactor
SCADA	Supervisory control and data acquisition
scfm	Standard cubic feet per minute
SRT	Solids retention time
TDH	Total dynamic head
TKL	Total Kjeldahl nitrogen
TMDL	Total maximum daily load
TN	Total nitrogen
TP	Total phosphorus
TSS	Total suspended solids
TVA	Tennessee Valley Authority
UV	Ultraviolet light
UVT	UV transmittance
VFD	Variable-frequency drive
VSS	Volatile suspended solids
W	Watt
WAS	Waste activated sludge
WEF	Water Environment Federation
WEFTEC	Water Environment Federation Technical Exhibition & Conference
WERF	Water Environment Research Foundation
WMARSS	Waco Metropolitan Area Regional Sewer System
WPCP	Water Pollution Control Plant
WRF	Water Research Foundation
WSU	Washington State University
WWTP	Wastewater treatment plant

Section I

The Basics

1

Introduction

The US economy, because it's so energy wasteful, is much less efficient than either the European or Japanese economies. It takes us twice as much energy to produce a unit of GDP as it does in Europe and Japan. So, we're fundamentally less efficient and therefore less competitive, and the sooner we begin to tighten up, the better it will be for our economy and society.

—**Hazel Henderson (on** *ENN Radio***)**

1.1 Setting the Stage

Several long-term economic, social, and environmental trends—the so-called *triple bottom line* (Elkington, 1999)—are evolving around us. Many of these long-term trends are developing because of us and specifically for us or simply to sustain us. Many of these long-term trends follow general courses and can be described by the jargon of the day; that is, they can be alluded to or specified by specific buzzwords commonly used today. We frequently hear these buzzwords used in general conversation (especially in abbreviated texting form). Buzzwords such as *empowerment, outside the box, streamline, wellness, synergy, generation X, face time, exit strategy, clear goal,* and so on and so forth are just part of our daily vernacular.

In this book, the popular buzzword we are concerned with, *sustainability,* is often used in business. In water and wastewater treatment, however, sustainability is much more than just a buzzword; it is a way of life (or should be). Many of the numerous definitions of sustainability are overwhelming, vague, or indistinct. For our purposes, there is a long definition and short definition of sustainability. The long definition is *ensuring that water and wastewater treatment operations occur indefinitely without negative impact.* The short definition is *the capacity of water and wastewater operations to endure.* Whether we define long or short fashion, what does sustainability really mean in the real world of water and wastewater treatment operations?

We have defined sustainability in both long and short form. Note, however, that sustainability in water and wastewater treatment operations can be characterized in broader or all-encompassing terms than these simple definitions. We can use the triple bottom line scenario, with regard to sustainability, the environmental aspects, economic aspects, and social aspects of water and wastewater treatment operations, to define today's and tomorrow's needs more specifically.

Infrastructure is another term used in this text; it can be used to describe water and wastewater operation in the whole, or it can identify several individual or separate elements of water and wastewater treatment operations. In wastewater operations, for example, we

TABLE 1.1

2009 Report Card for American Infrastructure

Infrastructure	Grade
Bridges	C
Dams	D
Drinking water	D–
Energy	D+
Hazardous waste	D
Rail	C–
Roads	D–
Schools	D
Wastewater	D–
America's infrastructure GPA:	D

Source: Adapted from ASCE, *2009 Report Card for America's Infrastructure*, American Society of Civil Engineers, Reston, VA, 2009 (http://www.infrastructure reportcard. org/report-cards).

can focus extensively on wastewater collection and interceptor systems, lift or pumping stations, influent screening, grit removal, primary clarification, aeration, secondary clarification, disinfection, outfalling, and a wide range of solids handling unit processes. On an individual basis, each of these unit processes can be described as an integral infrastructure component of the process. Or, holistically, we simply could group each unit process as one, as a whole, combining all wastewater treatment plant unit processes as "the" operational infrastructure. We could do the same for water treatment operations; for example, for individual water treatment infrastructure components, fundamental systems, or unit processes, we could list source water intake, pretreatment, screening, coagulation and mixing, flocculation, settling and biosolids processing, filtering, disinfection, and storage and distribution systems. Otherwise, we could simply describe water treatment plant operations as the infrastructure.

How one chooses to define infrastructure is not important. What is important is to maintain and manage infrastructure in the most efficient and economical manner possible to ensure its sustainability. This is no easy task. Consider, for example, the 2009 Report Card for American Infrastructure produced by the American Society of Civil Engineers, shown in Table 1.1.

Not only must water and wastewater treatment managers maintain and operate aging and often underfunded infrastructure, but they must also comply with stringent environmental regulations and must keep stakeholders and ratepayers satisfied with operations and with rates. Moreover, in line with these considerations, managers must incorporate economic considerations into every decision; for example, they must meet regulatory standards for quality of treated drinking water and outfalled wastewater effluent. They must also plan for future upgrades or retrofits that will enable the water or wastewater facility to meet future water quality and future effluent regulatory standards. Finally, and most importantly, managers must optimize the use of manpower, chemicals, and electricity.

1.2 Sustainable Water/Wastewater Infrastructure

The U.S. Environmental Protection Agency (USEPA, 2012) has defined *sustainable development* as that which meets the needs of the present generation without compromising the ability of future generations to meet their needs. The current U.S. population benefits from the investments that were made over the past several decades to build our nation's water/wastewater infrastructure.

Practices that encourage water and wastewater sector utilities and their customers to address existing needs so that future generations will not be left to address the approaching wave of needs resulting from aging water and wastewater infrastructure must continuously be promoted by sector professionals. To be on a sustainable path, investments need to result in efficient infrastructure and infrastructure systems and be at a pace and level that allow the water and wastewater sectors to provide the desired levels of service over the long term.

Sounds easy enough: The water/wastewater manager simply needs to put his or her operation on a sustainable path; moreover, he or she can simply accomplish this by investing. Right? Well, investing what? Investing in what? Investing how much? These are questions that require answers, obviously. Before moving on with this discussion it is important first to discuss plant infrastructure basics (focusing primarily on wastewater infrastructure and in particular on piping systems) and then to discuss funding (the cash cow vs. cash dog syndrome).

1.2.1 Maintaining Sustainable Infrastructure

During the 1950s and 1960s, the U.S. government encouraged the prevention of pollution by providing funds for the construction of municipal wastewater treatment plants, water pollution research, and technical training and assistance. New processes were developed to treat sewage, analyze wastewater, and evaluate the effects of pollution on the environment. In spite of these efforts, however, the expanding population and industrial and economic growth caused pollution and health issues to increase.

In response to the need to make a coordinated effort to protect the environment, the National Environmental Policy Act (NEPA) was signed into law on January 1, 1970. In December of that year, a new independent body, the USEPA, was created to bring under one roof all of the pollution control programs related to air, water, and solid wastes. In 1972, the Water Pollution Control Act Amendments expanded the role of the federal government in water pollution control and significantly increased federal funding for construction of wastewater treatment plants.

DID YOU KNOW?

Looking at water distribution piping only, the USEPA's 2000 survey on community water systems found that in systems that serve more than 100,000 people, about 40% of drinking water pipes were greater than 40 years old. It is important to remember, though, that age, in and of itself, does not necessarily indicate problems. If a system is well maintained, it can operate over a long time period (USEPA, 2012).

Many of the wastewater treatment plants in operation today are the result of federal grants made over the years; for example, the 1977 Clean Water Act Amendment to the Federal Water Pollution Control Act of 1972 and the 1987 Clean Water Act reauthorization bill provided funding for wastewater treatment plants. Many large sanitation districts, with their multiple plant operations, and even a larger number of single plant operations in smaller communities in operation today, are a result of these early environmental laws. Because of these laws, the federal government provided grants of several hundred million dollars to finance construction of wastewater treatment facilities throughout the country.

Many of these locally or federally funded treatment plants are aging; based on experience, we can rate some as dinosaurs. The point is that many facilities are facing problems caused by aging equipment, facilities, and infrastructure. Complicating the problems associated with natural aging is the increasing pressure on inadequate older systems to meet demands of increased population and urban growth. Facilities built in the 1960s and 1970s are now 40 to over 50 years old, and not only are they showing signs of wear and tear but they also simply were not designed to handle the level of growth that has occurred in many municipalities.

Regulations often necessitate a need to upgrade. By receiving matching funds or otherwise being provided federal money to cover some of the costs, municipalities can take advantage of a window of opportunity to improve their facilities at a lower direct cost to their communities. Those federal dollars, of course, do come with strings attached, as they are to be spent on specific projects in specific areas. On the other hand, many times new regulatory requirements are put in place without the financial assistance needed to implement them. When this occurs, either the local community ignores the new requirements (until caught and forced to comply) or it faces the situation and implements local tax hikes or rate-payer hikes to cover the cost of compliance.

Note: Changes resulting because of regulatory pressure sometimes mean replacing or changing existing equipment, result in increased chemical costs (e.g., substituting hypochlorite for chlorine typically increases costs threefold), and could easily involve increased energy and personnel costs. Equipment condition, new technology, and financial concerns are all considerations when upgrades or new processes are chosen. In addition, the safety of the process must be considered, of course, because of the demands made by USEPA and OSHA. The potential of harm to workers, the community, and the environment are all under study, as are the possible long-term effects of chlorination on the human population.

An example of how a change in regulations can force the issue is demonstrated by the demands made by the Occupational Safety and Health Administration (OSHA) and the USEPA in their Process Safety Management (PSM)/Risk Management Planning (RMP) regulations. These regulations put the use of elemental chlorine (and other listed hazardous materials) under close scrutiny. Moreover, because of these regulations, plant managers throughout the country are forced to choose which side of a double-edged sword cuts their way the most. One edge calls for full compliance with the regulations (analogous to stuffing the regulation through the eye of a needle); the other edge calls for substitution—that is, replacing elemental chlorine (probably the USEPA's motive in the first place; see the following note) with a non-listed hazardous chemical (e.g., hypochlorite) or a physical (ultraviolet irradiation) disinfectant. Either way, it is a very costly undertaking (Spellman, 2008).

Note: Many of us who have worked in water and wastewater treatment for years characterize PSM and RMP regulations as the elemental chlorine "killers." You have probably heard the old saying: "If you can't do away with something, then regulate it to death."

Water and wastewater treatment plants typically have a useful life of 20 to 50 years before they require expansion or rehabilitation. Collection, interceptor, and distribution pipes have life cycles that can range from 15 to 100 years, depending on the type of material and where they are laid. Long-term corrosion reduces the carrying capacity of a pipe, thus requiring increasing investments in power and pumping. When water or wastewater pipes age to that point of failure, the result can be contamination of drinking water, the release of wastewater into our surface waters or basements, and high costs to replace the pipes and repair any resulting damage. With pipes, the material used and how the pipe was installed can be a greater indicator of failure than age.

1.2.2 Cash Cows or Cash Dogs?

Maintaining the sustainable operations of water and wastewater treatment facilities is expensive. If funding is not available from federal, state, or local governmental entities, then the facilities must be funded by ratepayers. Water and wastewater treatment plants are usually owned, operated, and managed by the community (the municipality) where they are located. Although many of these facilities are privately owned, the majority of water treatment plants (WTPs) and wastewater treatment plants (WWTPs) are publicly owned treatment works (POTW) (i.e., owned by local government agencies).

These publicly owned facilities are managed and operated onsite by professionals in the field. Onsite management, however, is usually controlled by a board of elected, appointed, or hired directors or commissioners, who set policy, determine budget, plan for expansion or upgrading, hold decision-making power for large purchases, set rates for ratepayers, and in general control the overall direction of the operation.

When final decisions on matters that affect plant performance are in the hands of, for example, a board of directors comprised of elected and appointed city officials, their knowledge of the science, the engineering, and the hands-on problems that those who are onsite must solve can range from comprehensive to nothing. Matters that are of critical importance to those in onsite management may mean little to those on the board. The board of directors may also be responsible for other city services and have an agenda that encompasses more than just the water or wastewater facility. Thus, decisions that affect onsite management can be affected by political and financial concerns that have little to do with the successful operation of a WTP or POTW.

Finances and funding are always of concern, no matter how small or large, well-supported or underfunded, the municipality. Publicly owned treatment works are generally funded from a combination of sources. These include local taxes, state and federal monies (including grants and matching funds for upgrades), and usage fees for water and wastewater customers. In smaller communities, in fact, their water/wastewater (W/WW) plants may be the only city services that actually generate income. This is especially true in water treatment and delivery, which are commonly looked upon as the cash cows of city services. As a cash cow, the water treatment works generates cash in excess of the amount of cash necessary to maintain the treatment works. These treatment works are "milked" continuously with as little investment as possible, and funds generated by the facility do

DID YOU KNOW?

More than 50% of Americans drink bottled water occasionally or rely upon it as their major source of drinking water—an astounding fact given the high quality and low cost of U.S. tap water.

not always stay with the facility. Funds can be reassigned to support other city services, so when facility upgrade time rolls around funding for renovations can be problematic. On the other end of the spectrum, spent water (wastewater) treated in a POTW is often looked upon as one of the cash dogs of city services. Typically, these units make only enough money to sustain operations. This is the case, of course, because managers and oversight boards or commissions are fearful, for political reasons, of charging ratepayers too much for treatment services. Some progress has been made, however, in marketing and selling treated wastewater for reuse in industrial cooling applications and some irrigation projects. Moreover, wastewater solids have been reused as soil amendments; also, ash from incinerated biosolids has been used as a major ingredient in forming cement revetment blocks used in areas susceptible to heavy erosion from river and sea inlets and outlets (Drinan and Spellman, 2012).

Planning is essential for funding, for controlling expenses, and for ensuring water and wastewater infrastructure sustainability. The infrastructure we build today will be with us for a long time and, therefore, must be efficient to operate, offer the best solution in meeting the needs of a community, and be coordinated with infrastructure investments in other sectors such as transportation and housing. It is both important and challenging to ensure that a plan is in place to renew and replace it at the right time, which may be years away. Replacing an infrastructure asset too soon means not benefiting from the remaining useful life of that asset. Replacing an asset too late can lead to emergency repairs that are significantly more expensive than those that are planned (USEPA, 2012). Additionally, making retrofits to newly constructed infrastructure that was not designed or constructed correctly is expensive. Doing the job correctly the first time requires planning and a certain amount of competence.

1.3 Water/Wastewater Infrastructure Gap

A 2002 USEPA report referenced a water infrastructure gap analysis that compared current spending trends at the nation's drinking water and wastewater treatment facilities to the expenses that they can expect to incur for both capital and operations and maintenance costs. The gap is the difference between projected and needed spending and was found to be over $500 billion over a 20-year period. This important gap analysis study is just as pertinent today as it was 10 years ago. Moreover, this text draws upon tenets presented in the USEPA analysis in formulating many of the basic points and ideas presented here.

1.4 Energy Efficiency: Water/Wastewater Treatment Operations

Obviously, as the title of this text implies, we are concerned with water and wastewater infrastructure. This could mean we are concerned with the pipes, treatment plants, and other critical components that deliver safe drinking water to our taps and remove wastewater (sewage) from our homes and other buildings. Although any component or system that makes up water and wastewater infrastructure is important, remember that no water-related infrastructure can function without the aid of some motive force. This motive force (energy source) can be provided by gravitational pull, mechanical means, or electrical energy. We simply cannot sustain the operation of water and wastewater infrastructure without energy. As a case in point, consider that drinking water and wastewater systems account for approximately 3 to 4% of energy use in the United States and contribute over 45 million tons of greenhouse gases annually. Further, drinking water and wastewater plants are typically the largest energy consumers of municipal governments, accounting for 30 to 40% of total energy consumed. As a percent of operating costs for drinking water systems, energy can represent as much as 40% of those costs and is expected to increase 20% over the next 15 years due to population growth and tightening drinking water regulations.

Not all the news is bad, however. Studies estimate potential savings of 15 to 30% that are "readily achievable" in water and wastewater treatment plants, with substantial financial returns in the thousands of dollars and within payback periods of only a few months to a few years.

In the chapters that follow, we begin our discussion of energy efficiency for sustainable infrastructure in water and wastewater treatment plant operations with brief characterizations of the water and wastewater treatment industries. We then move on to a brief discussion of the basics of energy. We follow this with a discussion on determining energy usage, cutting energy usage and costs, and renewable energy options.

References and Recommended Reading

Drinan, J.E. and Spellman, F.R. (2012). *Water and Wastewater Treatment: A Guide for the Nonengineering Professional*, 2nd ed., CRC Press, Boca Raton, FL.

Elkington, J. (1999). *Cannibals with Forks*, Wiley, New York.

Spellman, F.R. (2008). *Handbook of Water and Wastewater Treatment Plant Operations*, 2nd ed., CRC Press, Boca Raton, FL.

USEPA. (2002). *The Clean Water and Drinking Water Infrastructure Gap Analysis*, EPA-816-R-02-020, U.S. Environmental Protection Agency, Washington, DC.

USEPA. (2012). *Frequently Asked Questions: Water Infrastructure and Sustainability*, U.S. Environmental Protection Agency, Washington, DC, http://water.epa.gov/infrastructure/sustain/si_faqs.cfm.

2

Characteristics of the Wastewater and Drinking Water Industries

Wastewater Treatment

According to the Code of Federal Regulations (CFR) 40 CFR Part 403, regulations were established in the late 1970's and early 1980's to help Publicly Owned Treatment Works (POTW) control industrial discharges to sewers. These regulations were designed to prevent pass-through and interference at the treatment plants and interference in the collection and transmission systems.

Pass-through occurs when pollutants literally "pass through" a POTW without being properly treated, and cause the POTW to have an effluent violation or increase the magnitude or duration of a violation.

Interference occurs when a pollutant discharge causes a POTW to violate its permit by inhibiting or disrupting treatment processes, treatment operations, or processes related to sludge use or disposal.

Drinking Water Treatment

Municipal water treatment operations and associated treatment unit processes are designed to provide reliable, high quality water service for customers, and to preserve and protect the environment for future generations.

Water management officials and treatment plant operators are tasked with exercising responsible financial management, ensuring fair rates and charges, providing responsive customer service, providing a consistent supply of safe potable water for consumption by the user, and promoting environmental responsibility.

The Honeymoon Is Over

The modern public water supply industry has come into being over the course of the last century. From the period known as the "Great Sanitary Awakening," that eliminated waterborne epidemics of diseases such as cholera and typhoid fever at the turn of the last century, we have built elaborate utility enterprises consisting of vast pipe networks and amazing high-tech treatment systems. Virtually all of this progress has been financed through local revenues. But in all this time, there has seldom been a need to provide for more than modest amounts of pipe replacement, because the pipes last so very long. We have been on an extended honeymoon made possible by the long life of the pipes and the fact that our water systems are relatively young. Now the honeymoon is over.

—AWWA (2001)

2.1 Introduction

In this chapter, a discussion of the characteristics of the wastewater and drinking water industries provides a useful context for understanding the differences between the industries and how these differences necessitate the use of different methods for estimating needs and costs and for instituting energy efficiency procedures to ensure sustainability (USEPA, 2002).

2.1.1 Wastewater and Drinking Water Terminology

To study any aspect of wastewater and drinking water treatment operations, it is necessary to master the language associated with the technology. Each technology has its own terms with its own accompanying definitions. Many of the terms used in water/wastewater treatment are unique; others combine words from many different technologies and professions. One thing is certain—water/wastewater operators without a clear understanding of the terms related to their profession are ill equipped to perform their duties in the manner required. Although this text includes a glossary of terms at the end, we list and define many of the terms used right up front. Experience has shown that an early introduction to keywords is a benefit to readers. An upfront introduction to key terms facilitates a more orderly, logical, systematic learning activity. Those terms not defined in this section are defined as they appear in the text.

Absorb—To take in. Many things absorb water.

Acid rain—The acidic rainfall that results when rain combines with sulfur oxides emissions from combustion of fossil fuels (coal, for example).

Acre-feet (acre-foot)—An expression of water quantity. One acre-foot will cover 1 acre of ground 1 foot deep. An acre-foot contains 43,560 cubic feet, 1233 cubic meters, or 325,829 gallons (U.S). Abbreviated as ac-ft.

Activated carbon—Derived from vegetable or animal materials by roasting in a vacuum furnace. Its porous nature gives it a very high surface area per unit mass, as much as 1000 square meters per gram, which is 10 million times the surface area of 1 gram of water in an open container. Used in adsorption (see definition), activated carbon adsorbs substances that are not or are only slightly adsorbed by other methods.

Activated sludge—The solids formed when microorganisms are used to treat wastewater using the activated sludge treatment process. It includes organisms, accumulated food materials, and waste products from the aerobic decomposition process.

Adsorption—The adhesion of a substance to the surface of a solid or liquid. Adsorption is often used to extract pollutants by causing them to attach to such adsorbents as activated carbon or silica gel. Hydrophobic (water-repulsing) adsorbents are used to extract oil from waterways in oil spills.

Advanced wastewater treatment—Treatment technology to produce an extremely high-quality discharge.

Aeration—The process of bubbling air through a solution, sometimes cleaning water of impurities by exposure to air.

Aerobic—Conditions in which free, elemental oxygen is present. Also used to describe organisms, biological activity, or treatment processes that require free oxygen.

Agglomeration—Floc particles colliding and gathering into a larger settleable mass.

Air gap—The air space between the free-flowing discharge end of a supply pipe and an unpressurized receiving vessel.

Algae bloom—A phenomenon whereby excessive nutrients within a river, stream, or lake cause an explosion of plant life that results in depletion of the oxygen in the water needed by fish and other aquatic life. Algae bloom is usually the result of urban runoff (of lawn fertilizers, etc.). The potential tragedy is that of a "fish kill," where the stream life dies in one mass execution.

Alum—Aluminum sulfate; a standard coagulant used in water treatment.

Ambient—The expected natural conditions that occur in water unaffected or uninfluenced by human activities.

Anaerobic—Conditions in which no oxygen (free or combined) is available. Also used to describe organisms, biological activity, or treatment processes that function in the absence of oxygen.

Anoxic—Conditions in which no free, elemental oxygen is present. The only source of oxygen is combined oxygen, such as that found in nitrate compounds. Also used to describe biological activity of treatment processes that function only in the presence of combined oxygen.

Aquifer—A water-bearing stratum of permeable rock, sand, or gravel.

Aquifer system—A heterogeneous body of introduced permeable and less permeable material that acts as a water-yielding hydraulic unit of regional extent.

Artesian water—A well tapping a confined or artesian aquifer in which the static water level stands above the top of the aquifer. The term is sometimes used to include all wells tapping confined water. Wells with water levels above the water table are said to have positive artesian head (pressure), and those with water levels below the water table have negative artesian head.

Average monthly discharge limitation—The highest allowable discharge over a calendar month.

Average weekly discharge limitation—The highest allowable discharge over a calendar week.

Backflow—Reversal of flow when pressure in a service connection exceeds the pressure in the distribution main.

Backwash—Fluidizing filter media with water, air, or a combination of the two so that individual grains can be cleaned of the material that has accumulated during the filter run.

Bacteria—Any of a number of one-celled organisms, some of which cause disease.

Bar screen—A series of bars formed into a grid used to screen out large debris from influent flow.

Base—A substance that has a pH value between 7 and 14.

Basin—A groundwater reservoir defined by the overlying land surface and underlying aquifers that contain water stored in the reservoir.

Beneficial use of water—The use of water for any beneficial purpose. Such uses include domestic use, irrigation, recreation, fish and wildlife, fire protection, navigation, power, industrial use, etc. The benefit varies from one location to another and by custom. What constitutes beneficial use is often defined by statute or court decisions.

Biochemical oxygen demand (BOD$_5$)—The oxygen used in meeting the metabolic needs of aerobic microorganisms in water rich in organic matter.

Biosolids—Solid organic matter recovered from a sewage treatment process and used especially as fertilizer or soil amendment; usually referred to in the plural (*Merriam-Webster's Collegiate Dictionary*, 10th ed., 1998).

Note: In this text, *biosolids* is generally used to replace the standard term *sludge*. It is the author's view that sludge is an ugly four-letter word inappropriate for describing biosolids. Biosolids can be reused; they have some value. Because biosolids have value, they certainly should not be classified as a waste product, and when the topic of biosolids for beneficial reuse is addressed, it is made clear that they are not a waste product.

Biota—All the species of plants and animals indigenous to a certain area.

Boiling point—The temperature at which a liquid boils. The temperature at which the vapor pressure of a liquid equals the pressure on its surface. If the pressure of the liquid varies, the actual boiling point varies. The boiling point of water is 212°F or 100°C.

Breakpoint—Point at which chlorine dosage satisfies chlorine demand.

Breakthrough—In filtering, when unwanted materials start to pass through the filter.

Buffer—A substance or solution that resists changes in pH.

Calcium carbonate—Compound principally responsible for hardness.

Calcium hardness—Portion of total hardness caused by calcium compounds.

Carbonaceous biochemical oxygen demand (CBOD$_5$)—The amount of biochemical oxygen demand that can be attributed to carbonaceous material.

Carbonate hardness—Caused primarily by compounds containing carbonate.

Chemical oxygen demand (COD)—The amount of chemically oxidizable materials present in the wastewater.

Chlorination—Disinfection of water using chlorine as the oxidizing agent.

Clarifier—A device designed to permit solids to settle or rise and be separated from the flow. Also known as a settling tank or sedimentation basin.

Coagulation—The neutralization of the charges of colloidal matter.

Coliform—A type of bacteria used to indicate possible human or animal contamination of water.

Combined sewer—A collection system that carries both wastewater and stormwater flows.

Comminution—A process to shred solids into smaller, less harmful particles.

Composite sample—A combination of individual samples taken in proportion to flow.

Connate water—Pressurized water trapped in the pore spaces of sedimentary rock at the time it was deposited. It is usually highly mineralized.

Consumptive use—(1) The quantity of water absorbed by crops and transpired or used directly in the building of plant tissue, together with the water evaporated from the cropped area. (2) The quantity of water transpired and evaporated from a cropped area or the normal loss of water from the soil by evaporation and plant transpiration. (3) The quantity of water discharged to the atmosphere or incorporated in the products of the process in connection with vegetative growth, food processing, or an industrial process.

Contamination (water)—Damage to the quality of water sources by sewage, industrial waste, or other material.

Cross-connection—A connection between a storm-drain system and a sanitary collection system, a connection between two sections of a collection system to handle anticipated overloads of one system, or a connection between drinking (potable) water and an unsafe water supply or sanitary collection system.

Daily discharge—The discharge of a pollutant measured during a calendar day or any 24-hour period that reasonably represents a calendar day for the purposes of sampling. Limitations expressed as weight are total mass (weight) discharged over the day. Limitations expressed in other units are average measurements of the day.

Daily maximum discharge—The highest allowable values for a daily discharge.

Darcy's law—An equation for the computation of the quantity of water flowing through porous media. Darcy's law assumes that the flow is laminar and that inertia can be neglected. The law states that the rate of viscous flow of homogeneous fluids through isotropic porous media is proportional to, and in the direction of, the hydraulic gradient.

Detention time—The theoretical time water remains in a tank at a given flow rate.

Dewatering—The removal or separation of a portion of water present in a sludge or slurry.

Diffusion—The process by which both ionic and molecular species dissolved in water move from areas of higher concentration to areas of lower concentration.

Discharge monitoring report (DMR)—The monthly report required by the treatment plant's National Pollutant Discharge Elimination System (NPDES) discharge permit.

Disinfection—Water treatment process that kills pathogenic organisms.

Disinfection byproducts (DBPs)—Chemical compounds formed by the reaction of disinfectant with organic compounds in water.

Dissolved oxygen (DO)—The amount of oxygen dissolved in water or sewage. Concentrations of less than five parts per million (ppm) can limit aquatic life or cause offensive odors. Excessive organic matter present in water because of inadequate waste treatment and runoff from agricultural or urban land generally causes low DO.

Dissolved solids—The total amount of dissolved inorganic material contained in water or wastes. Excessive dissolved solids make water unsuitable for drinking or industrial uses.

Domestic consumption (use)—Water used for household purposes such as washing, food preparation, and showers. The quantity (or quantity per capita) of water consumed in a municipality or district for domestic uses or purposes during a given period, it sometimes encompasses all uses, including the quantity wasted, lost, or otherwise unaccounted for.

Drawdown—Lowering the water level by pumping. It is measured in feet for a given quantity of water pumped during a specified period, or after the pumping level has become constant.

Drinking water standards—Established by state agencies, the U.S. Public Health Service, and the U.S. Environmental Protection Agency (USEPA) for drinking water in the United States.

Effluent—Something that flows out, usually a polluting gas or liquid discharge.

Effluent limitation—Any restriction imposed by the regulatory agency on quantities, discharge rates, or concentrations of pollutants discharged from point sources into state waters.

Energy—In scientific terms, the ability or capacity of doing work. Various forms of energy include kinetic, potential, thermal, nuclear, rotational, and electromagnetic. One form of energy may be changed to another, as when coal is burned to produce steam to drive a turbine, which produces electric energy.

Erosion—The wearing away of the land surface by wind, water, ice, or other geologic agents. Erosion occurs naturally from weather or runoff but is often intensified by human land use practices.

Eutrophication—The process of enrichment of water bodies by nutrients. Eutrophication of a lake normally contributes to its slow evolution into a bog or marsh and ultimately to dry land. Eutrophication may be accelerated by human activities, thereby speeding up the aging process.

Evaporation—The process by which water becomes a vapor at a temperature below the boiling point.

Facultative—Organisms that can survive and function in the presence or absence of free, elemental oxygen.

Fecal coliform—The portion of the coliform bacteria group that is present in the intestinal tracts and feces of warm-blooded animals.

Field capacity—The capacity of soil to hold water. It is measured as the ratio of the weight of water retained by the soil to the weight of the dry soil.

Filtration—The mechanical process that removes particulate matter by separating water from solid material, usually by passing it through sand.

Floc—Solids that join to form larger particles that will settle better.

Flocculation—Slow mixing process in which particles are brought into contact, with the intent of promoting their agglomeration.

Flume—A flow rate measurement device.

Fluoridation—Chemical addition to water to reduce incidence of dental caries in children.

Food-to-microorganisms ratio (F/M)—An activated sludge process control calculation based on the amount of food (BOD_5 or COD) available per pound of mixed liquor volatile suspended solids.

Force main—A pipe that carries wastewater under pressure from the discharge side of a pump to a point of gravity flow downstream.

Grab sample—An individual sample collected at a randomly selected time.

Graywater—Water that has been used for showering, clothes washing, and faucet uses. Kitchen sink and toilet water is excluded. This water has excellent potential for reuse as irrigation for yards.

Grit—Heavy inorganic solids, such as sand, gravel, eggshells, or metal filings.

Groundwater—The supply of fresh water found beneath the surface of the Earth (usually in aquifers) often used for supplying wells and springs. Because groundwater is a major source of drinking water, concern is growing over areas where leaching agricultural or industrial pollutants or substances from leaking underground storage tanks (USTs) are contaminating groundwater.

Groundwater hydrology—The branch of hydrology that deals with groundwater: its occurrence and movements, its replenishment and depletion, the properties of rocks that control groundwater movement and storage, and the methods of investigation and use of groundwater.

Groundwater recharge—The inflow to a groundwater reservoir.

Groundwater runoff—A portion of runoff that has passed into the ground, has become groundwater, and has been discharged into a stream channel as spring or seepage water.

Hardness—The concentration of calcium and magnesium salts in water.

Head loss—Amount of energy used by water in moving from one point to another.

Heavy metals—Metallic elements with high atomic weights, such as mercury, chromium, cadmium, arsenic, and lead. They can damage living things at low concentrations and tend to accumulate in the food chain.

Holding pond—A small basin or pond designed to hold sediment-laden or contaminated water until it can be treated to meet water quality standards or used in some other way.

Hydraulic cleaning—Cleaning pipe with water under enough pressure to produce high water velocities.

Hydraulic gradient—A measure of the change in groundwater head over a given distance.

Hydraulic head—The height above a specific datum (generally sea level) that water will rise in a well.

Hydrologic cycle (water cycle)—The cycle of water movement from the atmosphere to the Earth and back to the atmosphere through various processes. These processes include precipitation, infiltration, percolation, storage, evaporation, transpiration, and condensation.

Hydrology—The science dealing with the properties, distribution, and circulation of water.

Impoundment—A body of water such as a pond, confined by a dam, dike, floodgate, or other barrier, and used to collect and store water for future use.

Industrial wastewater—Wastes associated with industrial manufacturing processes.

Infiltration—The gradual downward flow of water from the surface into soil material.

Infiltration/inflow—Extraneous flows in sewers; simply, inflow is water discharged into sewer pipes or service connections from such sources as foundation drains, roof leaders, cellar and yard area drains, cooling water from air conditioners, and other clean-water discharges from commercial and industrial establishments. Defined by Metcalf & Eddy as follows:

- *Infiltration*—Water entering the collection system through cracks, joints, or breaks.

- *Steady inflow*—Water discharged from cellar and foundation drains, cooling water discharges, and drains from springs and swampy areas. This type of inflow is steady and is identified and measured along with infiltration.

- *Direct flow*—Those types of inflow that have a direct stormwater runoff connection to the sanitary sewer and cause an almost immediate increase in wastewater flows. Possible sources are roof leaders, yard and areaway drains, manhole covers, cross-connections from storm drains and catch basins, and combined sewers.

- *Total inflow*—The sum of the direct inflow at any point in the system plus any flow discharged from the system upstream through overflows, pumping station bypasses, and the like.

- *Delayed inflow*—Stormwater that may require several days or more to drain through the sewer system. This category can include the discharge of sump pumps from cellar drainage as well as the slowed entry of surface water through manholes in ponded areas.

Influent—Wastewater entering a tank, channel, or treatment process.

Inorganic chemical/compounds—Chemical substances of mineral origin, not of carbon structure. These include metals such as lead, iron (ferric chloride), and cadmium.

Ion exchange process—Used to remove hardness from water.

Jar test—Laboratory procedure used to estimate proper coagulant dosage.

Langelier saturation index (LI)—A numerical index that indicates whether calcium carbonate will be deposited or dissolved in a distribution system.

Leaching—The process by which soluble materials in the soil such as nutrients, pesticide chemicals, or contaminants are washed into a lower layer of soil or are dissolved and carried away by water.

License—A certificate issued by the State Board of Waterworks/Wastewater Works Operators authorizing the holder to perform the duties of a wastewater treatment plant operator.

Lift station—A wastewater pumping station designed to lift the wastewater to a higher elevation. A lift station normally employs pumps or other mechanical devices to pump the wastewater and discharges into a pressure pipe called a *force main*.

Maximum contaminant level (MCL)—An enforceable standard for protection of human health.

Mean cell residence time (MCRT)—The average length of time a mixed liquor suspended solids particle remains in the activated sludge process. May also be known as *sludge retention time*.

Mechanical cleaning—Clearing pipe by using equipment (bucket machines, power rodders, or hand rods) that scrapes, cuts, pulls, or pushes the material out of the pipe.

Membrane process—A process that draws a measured volume of water through a filter membrane with small enough openings to take out contaminants.

Metering pump—A chemical solution feed pump that adds a measured amount of solution with each stroke or rotation of the pump.

Milligrams/liter (mg/L)—A measure of concentration equivalent to parts per million (ppm).

Mixed liquor—The suspended solids concentration of the mixed liquor.

Mixed liquor volatile suspended solids (MLVSS)—The concentration of organic matter in the mixed liquor suspended solids.

Nephelometric turbidity unit (NTU)—Indicates amount of turbidity in a water sample.

Nitrogenous oxygen demand (NOD)—A measure of the amount of oxygen required to biologically oxidize nitrogen compounds under specified conditions of time and temperature.

Nonpoint-source (NPS) pollution—Forms of pollution caused by sediment, nutrients, and organic and toxic substances originating from land use activities that are carried to lakes and streams by surface runoff. Nonpoint-source pollution occurs when the rate of materials entering these waterbodies exceeds natural levels.

NPDES permit—A National Pollutant Discharge Elimination System permit authorizes the discharge of treated wastes and specifies the conditions that must be met for discharge.

Nutrients—Substances required to support living organisms. Usually refers to nitrogen, phosphorus, iron, and other trace metals.

Organic chemicals/compounds—Animal- or plant-produced substances containing mainly carbon, hydrogen, and oxygen, such as benzene and toluene.

Parts per million (ppm)—The number of parts by weight of a substance per million parts of water. This unit is commonly used to represent pollutant concentrations. Large concentrations are expressed in percentages.

Pathogenic—Disease causing; a pathogenic organism is capable of causing illness.

Percolation—The movement of water through the subsurface soil layers, usually continuing downward to the groundwater or water table reservoirs.

pH—A way of expressing both acidity and alkalinity on a scale of 0 to 14, with 7 representing neutrality, numbers less than 7 indicating increasing acidity, and numbers greater than 7 indicating increasing alkalinity.

Photosynthesis—A process in green plants in which water, carbon dioxide, and sunlight combine to form sugar.

Piezometric surface—An imaginary surface that coincides with the hydrostatic pressure level of water in an aquifer.

Point source pollution—A type of water pollution resulting from discharges into receiving waters from easily identifiable points. Common point sources of pollution are discharges from factories and municipal sewage treatment plants.

Pollution—Alteration of the physical, thermal, chemical, or biological quality of, or the contamination of, any water in the state that renders the water harmful, detrimental, or injurious to humans, animal life, vegetation, property, or to public health, safety, or welfare, or impairs the usefulness or the public enjoyment of the water for any lawful or reasonable purpose.

Porosity—That part of a rock that contains pore spaces without regard to size, shape, interconnection, or arrangement of openings. It is expressed as a percentage of total volume occupied by spaces.

Potable water—Water satisfactorily safe for drinking purposes from the standpoint of its chemical, physical, and biological characteristics.

Precipitate—A deposit on the Earth of hail, rain, mist, sleet, or snow. The common process by which atmospheric water becomes surface or subsurface water. The term *precipitation* is also commonly used to designate the quantity of water precipitated.

Preventive maintenance (PM)—Regularly scheduled servicing of machinery or other equipment using appropriate tools, tests, and lubricants. This type of maintenance can prolong the useful life of equipment and machinery and increase its efficiency by detecting and correcting problems before they cause a breakdown of the equipment.

Purveyor—An agency or person that supplies potable water.

Radon—A radioactive, colorless, odorless gas that occurs naturally in the earth. When trapped in buildings, concentrations build up and can cause health hazards such as lung cancer.

Recharge—The addition of water into a groundwater system.

Reservoir—A pond, lake, tank, or basin (natural or human made) where water is collected and used for storage. Large bodies of groundwater are called *groundwater reservoirs*; water behind a dam is also called a *reservoir of water*.

Return activated sludge solids (RASS)—The concentration of suspended solids in the sludge flow being returned from the settling tank to the head of the aeration tank.

Reverse osmosis—Process in which almost pure water is passed through a semipermeable membrane.

River basin—A term used to designate the area drained by a river and its tributaries.

Sanitary wastewater—Wastes discharged from residences and from commercial, institutional, and similar facilities that include both sewage and industrial wastes.

Schmutzdecke—Layer of solids and biological growth that forms on top of a slow sand filter, allowing the filter to remove turbidity effectively without chemical coagulation.

Scum—The mixture of floatable solids and water removed from the surface of the settling tank.

Sediment—Transported and deposited particles derived from rocks, soil, or biological material.

Sedimentation—A process that reduces the velocity of water in basins so that suspended material can settle out by gravity.

Seepage—The appearance and disappearance of water at the ground surface. Seepage designates movement of water in saturated material. It differs from percolation, which is predominantly the movement of water in unsaturated material.

Septic tanks—Used to hold domestic wastes when a sewer line is not available to carry them to a treatment plant. The wastes are piped to underground tanks directly from a home or homes. Bacteria in the wastes decompose some of the organic matter, the sludge settles on the bottom of the tank, and the effluent flows out of the tank into the ground through drains.

Settleability—A process control test used to evaluate the settling characteristics of the activated sludge. Readings taken at 30 to 60 minutes are used to calculate the settled sludge volume (SSV) and the sludge volume index (SVI).

Settled sludge volume (SSV)—The volume (in percent) occupied by an activated sludge sample after 30 to 60 minutes of settling. Normally written as SSV with a subscript to indicate the time of the reading used for calculation (SSV_{60} or SSV_{30}).

Sludge—The mixture of settleable solids and water removed from the bottom of the settling tank.

Sludge retention time (SRT)—See mean cell residence time.

Sludge volume index (SVI)—A process control calculation used to evaluate the settling quality of the activated sludge. It requires the SSV_{30} and mixed liquor suspended solids test results to calculate.

Soil moisture (soil water)—Water diffused in the soil. It is found in the upper part of the zone of aeration from which water is discharged by transpiration from plants or by soil evaporation.

Specific heat—The heat capacity of a material per unit mass. The amount of heat (in calories) required to raise the temperature of 1 gram of a substance 1°C; the specific heat of water is 1 calorie.

Storm sewer—A collection system designed to carry only stormwater runoff.

Stormwater—Runoff resulting from rainfall and snowmelt.

Stream—A general term for a body of flowing water. In hydrology, the term is generally applied to the water flowing in a natural channel as distinct from a canal. More generally, it is applied to the water flowing in any channel, natural or artificial. Some types of streams include: (1) *ephemeral*, a stream that flows only in direct response to precipitation, and whose channel is at all times above the water table; (2) *intermittent* or *seasonal*, a stream that flows only at certain times of the year when it receives water from springs, rainfall, or from surface sources such as melting snow; (3) *perennial*, a stream that flows continuously; (4) *gaining*, an effluent stream or reach of a stream that receives water from the zone of saturation; (5) *insulated*, a stream or reach of a stream that is separated from the zones of saturation by an impermeable bed and neither contributes water to the zone of saturation nor receives water from it; (6) *losing*, an influent stream or reach of a stream that contributes water to the zone of saturation; and (7) *perched*, a perched stream is either a losing stream or an insulated stream that is separated from the underlying groundwater by a zone of aeration.

Supernatant—The liquid standing above a sediment or precipitate.

Surface tension—The free energy produced in a liquid surface by the unbalanced inward pull exerted by molecules underlying the layer of surface molecules.

Surface water—Lakes, bays, ponds, impounding reservoirs, springs, rivers, streams, creeks, estuaries, wetlands, marshes, inlets, canals, gulfs inside the territorial limits of the state, and all other bodies of surface water, natural or artificial, inland or coastal, fresh or salt, navigable or nonnavigable, and including the beds and banks of all watercourses and bodies of surface water, that are wholly or partially inside or bordering the state or subject to the jurisdiction of the state. Waters in treatment systems that are authorized by state or federal law, regulation, or permit and which are created for the purpose of water treatment are not considered to be waters in the state.

Thermal pollution—The degradation of water quality by the introduction of a heated effluent. Primarily the result of the discharge of cooling waters from industrial processes (particularly from electrical power generation); waste heat eventually results from virtually every energy conversion.

Titrant—A solution of known strength of concentration; used in titration.

Titration—A process whereby a solution of known strength (titrant) is added to a certain volume of treated sample containing an indicator. A color change shows when the reaction is complete.

Titrator—An instrument, usually a calibrated cylinder (tube-form), used in titration to measure the amount of titrant being added to the sample.

Total dissolved solids—The amount of material (inorganic salts and small amounts of organic material) dissolved in water and commonly expressed as a concentration in terms of milligrams per liter.

Total suspended solids (TSS)—Total suspended solids in water, commonly expressed as a concentration in terms of milligrams per liter.

Toxicity—The occurrence of lethal or sublethal adverse affects on representative sensitive organisms due to exposure to toxic materials. Adverse effects caused by conditions of temperature, dissolved oxygen, or nontoxic dissolved substances are excluded from the definition of toxicity.

Transpiration—The process by which water vapor escapes from the living plant, principally the leaves, and enters the atmosphere.

Vaporization—The change of a substance from a liquid or solid state to a gaseous state.

Volatile organic compound (VOC)—Any organic compound that participates in atmospheric photochemical reactions except for those designated by the USEPA Administrator as having negligible photochemical reactivity.

Waste activated sludge solids (WASS)—The concentration of suspended solids in the sludge being removed from the activated sludge process.

Wastewater—The water supply of a community after it has been soiled by use.

Water cycle—The process by which water travels in a sequence from the air (condensation) to the Earth (precipitation) and returns to the atmosphere (evaporation). It is also referred to as the *hydrologic cycle*.

Water quality—A term used to describe the chemical, physical, and biological characteristics of water with respect to its suitability for a particular use.

Water quality standard—A plan for water quality management containing four major elements: water use, criteria to protect those uses, implementation plans, and enforcement plans. An antidegradation statement is sometimes prepared to protect existing high-quality waters.

Water supply—Any quantity of available water.

Waterborne disease—A disease caused by a microorganism that is carried from one person or animal to another by water.

Watershed—The area of land that contributes surface runoff to a given point in a drainage system.

Weir—A device used to measure wastewater flow.

Zone of aeration—A region in the earth above the water table. Water in the zone of aeration is under atmospheric pressure and would not flow into a well.

Zoogleal slime—The biological slime that forms on fixed-film treatment devices. It contains a wide variety of organisms essential to the treatment process.

2.2 Characteristics of the Wastewater Industry

Wastewater treatment takes effluent (spent water) from water users (*consumers*, whether from private homes, business, or industrial sources) as influent to wastewater treatment facilities. The wastestream is treated in a series of steps (*unit processes*, some similar to those used in treating raw water and others that are more involved), then discharged (out-falled) to a receiving body, usually a river or stream. In the United States, 16,024 publicly owned treatment works (POTWs) treat municipal wastewater. Although there are also some privately owned wastewater treatment works, most of the industry (98%) is in fact municipally owned. POTWs provide service to 190 million people, representing 73% of the total population (USEPA, 2008). Of the facilities, 71% serve populations of less than 10,000 people. Furthermore, approximately 25% of households in the nation are not connected to centralized treatment, instead using onsite systems (e.g., septic tanks). Although many of these systems are aging or improperly functioning, this text is restricted to centralized collection and treatment systems.

2.2.1 Wastewater Treatment Process: The Model

Figure 2.1 shows a basic schematic of a centralized conventional wastewater treatment process providing primary and secondary treatment using the *activated sludge process*. This is the model, the prototype, the paradigm used in this book. Secondary treatment provides biochemical oxygen demand (BOD) removal beyond what is achievable by simple sedimentation through such techniques as *trickling filter, activated sludge,* and *oxidation ponds*. This book, for instructive and illustrative purposes, focuses primarily on the activated sludge process. The purpose of Figure 2.1 is to allow the reader to follow the treatment process step-by-step as it is mentioned and to assist in demonstrating how all the various unit processes sequentially follow and tie into each other.

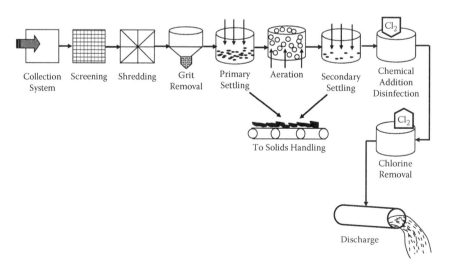

FIGURE 2.1
Unit processes for wastewater treatment.

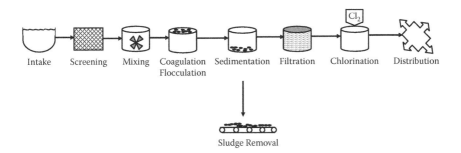

FIGURE 2.2
Water treatment unit process.

2.3 Characteristics of the Drinking Water Industry

Water treatment brings raw water up to drinking water quality. The process this entails depends on the quality of the water source. Surface water sources (lakes, rivers, reservoirs, and impoundments) generally require higher levels of treatment than groundwater sources. Groundwater sources may incur higher operating costs from machinery but may require only simple disinfection (see Figure 2.2).

In this text, we define water treatment as any unit process that changes or alters the chemical, physical, or bacteriological quality of water with the purpose of making it safe for human consumption and appealing to the customer. Treatment also is used to protect the water distribution system components from corrosion. A summary of basic water treatment processes (many of which are discussed in this chapter) is presented in Table 2.1.

The drinking water industry has over ten times the number of systems as the wastewater industry. Of the almost 170,000 public water systems, 54,000 systems are community water systems that collectively serve more than 264 million people. A community water system serves more than 25 people a day all year round. The remaining 114,000 water systems are transient noncommunity water systems (e.g., camp grounds) or nontransient, noncommunity water systems (e.g., schools). The scope of discussion in this text is largely confined to community water systems, as these systems serve most of the population. Small systems serving fewer than 10,000 people comprise 93% of all community water systems in the nation; however, most of the population (81%) receives drinking water from larger systems.

DID YOU KNOW?

The USEPA estimates the nationwide capital investment needs for wastewater pollution control at $298.1 billion. This figure represents documented needs for up to a 20-year period. The estimate includes $193.3 billion for wastewater treatment and collection systems, $63.6 billion for combined sewer overflow correction, and $42.3 billion for stormwater management (USEPA, 2008).

TABLE 2.1

Basic Water Treatment Processes

Process or Step	Purpose
Screening	Removes large debris (leaves, sticks, fish) that can foul or damage plant equipment
Chemical pretreatment	Conditions the water for removal of algae and other aquatic nuisances
Presedimentation	Removes gravel, sand, silt, and other gritty materials
Microstraining	Removes algae, aquatic plants, and small debris
Chemical feed and rapid mix	Adds chemicals (e.g., coagulants, pH, adjusters) to water
Coagulation/flocculation	Converts nonsettleable or settable particles
Sedimentation	Removes settleable particles
Softening	Removes hardness-causing chemicals from water
Filtration	Removes particles of solid matter which can include biological contamination and turbidity
Disinfection	Kills disease-causing organisms
Adsorption using granular activated carbon	Removes radon and many organic chemicals such as pesticides, solvents, and trihalomethanes
Aeration	Removes volatile organic chemicals (VOCs), radon, H_2S, and other dissolved gases; oxidizes iron and manganese
Corrosion control	Prevents scaling and corrosion
Reverse osmosis, electrodialysis	Removes nearly all inorganic contaminants
Ion exchange	Removes some inorganic contaminants including hardness-causing chemicals
Activated alumina	Removes some inorganic contamination
Oxidation filtration	Removes some inorganic contaminants (e.g., iron, manganese, radium)

Source: Adapted from Spellman, F.R., *Handbook of Water and Wastewater Treatment Plant Operations*, 2nd ed., CRC Press, Boca Raton, FL, 2008.

2.4 Capital Stock and Impact on Operations and Maintenance[*]

The different components of capital stock (total physical capitol) that make up our nation's wastewater and drinking water systems vary in complexity, materials, and the degree to which they are subjected to wear and tear. The expenditures that utilities must make to address the maintenance of systems are largely driven by the condition and age of the components of infrastructure.

2.4.1 Useful Life of Assets

The life of an asset can be estimated based on the material, but many other factors related to environment and maintenance can affect the useful life of a component of infrastructure. It is not feasible to conduct a condition assessment of all wastewater and drinking infrastructure systems throughout the United States; however, approximation tools can be used to estimate the useful life of these infrastructure systems.

---

[*] Based on USEPA, *The Clean Water and Drinking Water Infrastructure Gap Analysis*, EPA-816-R-02-020, U.S. Environmental Protection Agency, Washington, DC, 2002.

DID YOU KNOW?

In contrast to the wastewater industry, only about 43% of community drinking water systems are publicly owned. Most of these systems are under the authority of local governments. Ownership type varies by system size; almost 90% of systems serving more than 10,000 people are under public ownership.

One approximation tool that can be used to estimate is the *useful life matrix*. This matrix can serve as a tool for developing initial cost estimates and for long-range planning and evaluating programmatic scenarios. Table 2.2 shows an example of a matrix developed as an industry guide in Australia. Although the useful life of a component will vary according to the materials, environment, and maintenance, matrices such as that shown in Table 2.2 can be used at the local level as a starting point for repair and replacement, strategic planning, and cost projects. The United States as well as other industrialized countries have engineering and design manuals that instruct professional designers on the accepted standards of practice for design life considerations. The U.S. Army Corps of Engineers, American Society for Testing and Materials, Water Environment Federation, American Society of Civil Engineers, and several associates maintain data that provide guidance on design and construction of conduits, culverts, and pipes and related design procedures.

Most of the assets of both wastewater and drinking water treatment systems are comprised of pipe. The useful life of pipe varies considerably based on a number of factors. Some of these factors include the material of which the pipe is made, the conditions of the soil in which it is buried, and the character of the water or wastewater flowing through it. In addition, pipes do not deteriorate at a constant rate. During the initial period following

TABLE 2.2

Useful Life Matrix

Years	Component
	Wastewater
80–100	Collections
50	Treatment Plants—Concrete Structures
15–25	Treatment Plants—Mechanical and Electrical
25	Force Mains
50	Pumping Stations—Concrete Structures
15	Pumping Stations—Mechanical and Electrical
90–100	Interceptors
	Drinking water
50–80	Reservoirs and Dams
60–70	Treatment Plants—Concrete Structures
15–25	Treatment Plants—Mechanical and Electrical
65–95	Trunk Mains
60–70	Pumping Stations—Concrete Structures
25	Pumping Stations—Mechanical and Electrical
65–95	Distribution

Source: Adapted from IPWEA, *International Infrastructure Management Manual*, Version 1.0, Institute of Public Works Engineering Australia, Sydney, 2000.

installation, the deterioration rate is likely to be slow and repair and upkeep expenses low. For pipe, this initial period may last several decades. Later in the life cycle, pipe will deteriorate more rapidly. The best way to determine remaining useful life of a system is to conduct periodic condition assessments. At the local level, it is essential for local service providers to complete periodic condition assessments in order to make the best life-cycle decisions regarding maintenance and replacement.

2.4.2 Operating and Maintaining Capital Stock

Since 1970, spending in constant dollars on operations and maintenance (O&M) for wastewater treatment operations and drinking water treatment operations has grown significantly. In 1994, for example, 63% of the total spending for wastewater operations and 70% of the total spending for drinking water operations were for O&M (CBO, 1999). Likely explanations for the increase in wastewater and drinking water O&M costs include the following:

- Expansion and improvement of services, which translated into an increase in capital stock and a related increase in operations and maintenance costs
- Aging infrastructure, which requires increasing repairs and increasing maintenance costs

Additionally, increases in wastewater operations and maintenance have been driven, in large part, by a large number of solids handling (biosolids) facilities coming online. The installation of these facilities has increased O&M costs beginning in the mid-1980s.

Over the next 20 years, O&M expenses are likely to increase in response to the aging of the capital stock; that is, as infrastructure begins to deteriorate, the costs of maintaining and operating the equipment will increase. An American Water Works Association (AWWA) study found that projected expenditures for deteriorating infrastructure would increase steadily over the next 30 years (AWWA, 2001). The projected increase in O&M costs finds support in the historical spending data, which indicate an upward trend for O&M.

Increasing O&M needs will present a significant challenge to the financial resources of wastewater and drinking water systems. As the nation's water infrastructure ages, systems should expect to spend more on O&M. Some systems might even postpone capital investments to meet the rising costs of O&M, assuming that their total level of spending remains constant. The majority of systems likely would increase spending to ensure that both capital and O&M needs are fulfilled; thus, total spending would increase significantly. Many systems would recognize that delaying new capital investments would only increase expenditures on O&M, as old and deteriorated infrastructure would need to be maintained at increasingly higher costs.

DID YOU KNOW?

Pipes are expensive, but invisible. Most people do not realize the huge magnitude of the capital investment that has been made to develop the vast network of distribution mains and pipes—the infrastructure—that makes clean and safe water available at the turn of a tap. Water is by far the most capital intensive of all utility services, mostly due to the cost of these pipes. The water infrastructure is literally a buried treasure beneath our streets (AWWA, 2001).

2.5 Wastewater Capital Stock

The basic components of wastewater treatment infrastructure are collection/interceptor systems and treatment works. Systems vary across the clean water industry as a function of the demographic and topographic characteristics of the service area, the unique characteristics of the particular wastestream, and the operating requirements dictated in the permit conditions. The type of treatment is largely controlled by discharge limitations and performance specified through state or federal permits.

Pipe networks represent the primary component of a wastewater treatment system. During the last century, as populations grew and spread out from urban centers, the amount of pipe increased as homes were connected to centralized treatment. Although there is not an actual inventory of the total amount of sewer pipe associated with wastewater collections systems in the United States, the American Society of Civil Engineers (ASCE) has developed an estimate based on feet of sewer per capita, with the average length being estimated at 21 feet of sewer per capita. The range varied from 18 to 23 feet per capita. The resulting estimate is about 600,000 miles of publicly owned pipe (ASCE, 1999).

Because there is no nationwide inventory of wastewater collection systems, the actual age of sewer pipe is not known; however, it is safe to say that installation of pipe has followed demographic increases in population and growth in metropolitan areas associated with suburbanization.

The vast majority of the nation's pipe network was installed after the World War II, and the first part of this wave of pipe installation is now reaching the end of its useful life. For this reason, even if the pipe system is extended to serve growth and the country invests in the replacement of all pipe as it comes to the end of its useful life, the average age of pipe in the system will still increase until at least 2050.

Although there will be differences based on pipe material and condition, the need to replace pipe will generally echo the original installation wave. Based on the deterioration projections over the next 20 years, if the pipe system is extended to serve growth but there is no renewal or replacement of the existing systems, the amount of pipe classified as "poor," "very poor," or "life elapsed" will increase from 10% of the total network to 44% of the total network.

Many of the wastewater treatment plants in the United States were completely renovated with major plant expansion and upgrade work beginning in the 1970s, responding to new treatment requirements of the 1972 Clean Water Act (CWA) and financed to a great extent by the USEPA's Construction Grants Program. Although plants have shorter useful lives than sewer pipe, plant replacement needs are not projected to be a major part of the renewal and replacement requirements until after 2020.

Because plant equipment (e.g., mechanical and electrical components) is not buried underground and is thus easier to observe, it is subject to more frequent inspection. Some of these visible components will have to be replaced within a 20-year time frame, but relative to the collection systems they are much less significant. However, there are implications to the costs associated with the treatment plants. As the treatment plants continue to age, their operation and maintenance costs will increase at a more rapid rate, having a major impact on future operating budgets.

Furthermore, because so many treatment plants were constructed near the same point in time (i.e., beginning in the 1970s), replacement needs will hit at relatively the same time. The initial treatment plant replacement needs will occur at the same time that many pipes installed post-World War II will begin requiring replacement. Deferral of timely renewal and replacement associated with the oldest pipe over the next 20 years will likely put a system in a difficult financial condition. The typical system could experience a very significant bump in expenditures over a very short period of time to accommodate replacement of old pipes, new pipes, and plant structures in the same time frame.

2.6 Drinking Water Capital Stock

The capital stock of an individual drinking water system can be broken down into four principal components: source, treatment, storage, and transmission and distribution mains. Each of these components fulfills an important function in delivering safe drinking water to the public.

Although there is no study available that directly addresses the capital make-up of our nation's drinking water systems, a general picture can be obtained from the *2007 Drinking Water Infrastructure Needs Survey and Assessment* (USEPA, 2007). The survey found that the total nationwide infrastructure requirement will be $334.8 billion for the 20-year period from January 2007 through December 2026. Although it is the least visible component of a public water system, the buried pipes of a transmission and distribution network generally comprise most of the capital value of a system. Transmission and distribution needs accounted for 60% of the total need reported in the 2007 survey. Treatment facilities necessary to address contaminants with acute and chronic health effects represented the second largest category (22% of the total need). Storage projects required to construct or rehabilitate finished water storage tanks represented 11% of the total need. Projects necessary to address sources of water accounted for 6%. The source category included needs for constructing or rehabilitating surface water intakes, raw water pumping facilities, drilled wells, and spring collectors. Neither the storage nor source categories considered needs associated with the construction or rehabilitation of raw water reservoirs or dams (USEPA, 2007).

The need to replace aging transmission and distribution components is a critical aspect of any drinking water system's capital improvement plan. The AWWA surveyed the inventory of pipe and the year in which the pipe was installed for 20 cities in an effort to predict when the replacement of the pipe would be needed (AWWA, 2001). Although the 20 cities in the sample were not selected at random, the cities likely represent a broad range of systems of various ages and sizes from across the country. More importantly, the study provides the only available data on the age of pipe from a reasonably large number of systems.

Age is one factor that affects the life expectancy of pipe. A simple aging model, therefore, was developed to predict when pipes for these 20 cities would need to be replaced. It was assumed that pipes installed before 1910 last an average of 120 years. Pipe installed from 1911 to 1945 is assumed to last an average of 100 years. Pipe installed after 1945 is assumed to last an average of 75 years. In estimating when the current inventory of pipe will be replaced, the model assumes that the actual life span of the pipe will be distributed normally around its expected average life; that is, pipe expected to last 75 years will last 50 to 100 years, pipe expected to last 100 years will last from 66 to 133 years, and pipe expected to last 120 years will last 80 to 160 years (AWWA, 2001).

DID YOU KNOW?

On average, the replacement cost value of water mains was about $6300 per household in the relatively large utilities studied by the AWWA (2001). If water treatment plants, pumps, etc., are included, the replacement cost value rises to just under $10,000 per household, on average.

This assumption greatly simplifies reality, as the deterioration rates of pipe will vary considerably as a function not only of age but also of climatic conditions, pipe material, and soil properties. Pipe of the same material, for example, can last from 15 years to over 200 years depending on the soil characteristics alone. In the absence of data that would allow for the development of a model to estimate pipe life (i.e., accounting for local variability of pipe deterioration), the application of a normal distribution to an average life expectancy may provide a reasonable approximation of replacement rates. This model also does not account for other factors, most notably inadequate capacity, that may have equal or greater importance than deterioration in determining pipe replacement rates.

Applying this simple aging model to the historical inventory of pipe for the 20 cities reveals that most of the projected replacement needs for those cities will occur beyond the 20-year period of the analysis, with peak annual replacement occurring in 2040. This conclusion makes sense considering that most of the nation's drinking water lines were installed after the 1940s. Moreover, we need to remember that pipes are hearty but ultimately mortal (AWWA, 2001).

2.7 Costs of Providing Service

Although many purveyors of water and wastewater services obtain funds from the federal government to finance the costs of capital improvements, most of the funds that systems use for both capital and operations and maintenance come from revenues derived from user fees. As utilities look to address future capital needs and increasing O&M costs, they need to increase fees to obtain the funding needed for these activities.

Although there is no complete source of national data on how rates have changed through time, the State of Ohio has information that can serve as an example for the purposes of a simple discussion. For more than 15 years, the state has conducted an annual survey of water and sewer rates for communities in the state. Data from communities that reported rates for both 1989 and 1999 revealed that there had been an upward shift in the number of communities paying higher annual fees with time.

User rates that are necessary to meet the cost of providing service have the potential to negatively impact those segments of the population with low incomes. Data from the U.S. Census Bureau (2000) show that between 1980 and 1998 incomes at the lower range (as a percentage share of aggregate income for households) declined or stagnated. If rates increase to fund increasing needs, utilities may be challenged to develop rate structures that will minimize impacts on the less affluent segments of society.

References and Recommended Reading

ASCE. (1999). *Optimization of Collection System Maintenance Frequencies and System Performance*, American Society of Civil Engineers, Reston, VA.

AWWA. (2001). *Dawn of the Replacement Era: Reinvesting in Drinking Water Infrastructure*, American Water Works Association, Denver, CO.

CBO. (1999). *Trends in Public Infrastructure*, Congressional Budget Office, Washington, DC.

U.S. Census Bureau. (2000). *The Changing Shape of the Nation's Income Distribution, 1947–1998*, P60-204, Current Population Reports Series, U.S. Census Bureau, Washington, DC.

USEPA. (2002). *The Clean Water and Drinking Water Infrastructure Gap Analysis*, EPA-816-R-02-020, U.S. Environmental Protection Agency, Washington, DC.

USEPA. (2007). *2007 Drinking Water Infrastructure Needs Survey and Assessment*, U.S. Environmental Protection Agency, Washington, DC.

USEPA. (2008). *Clean Watersheds Needs Survey*, U.S. Environmental Protection Agency, Washington, DC.

3

Water, Wastewater, and Energy*

> What we have here, my young, wide-eyed, and very impressionable students, is not a failure to communicate, no...no...no...well, sort of, maybe...but actually...instead what we have here is full-blown energy-deficit disorder...driven in part by the radical greenies (i.e., the space cadets) who would have us wax and wane in the dim light of those alternative candles (I prefer the scented variety, thank you very much) while we either freeze to death or roast on a spit of ignorance, stupidity, single-mindedness, madness, and/or radical misconception.
>
> —F.R. Spellman (2005)

3.1 Introduction

Providing drinking water and wastewater treatment service to citizens requires energy—a lot of it. The twin problems of steadily rising energy costs and climate change have made the issues of energy management, energy efficiency, and energy sustainability the most salient issues facing drinking water and wastewater utilities today. Energy management, efficiency, and sustainability are at the heart of efforts across the entire sector to ensure that utility operations are ultimately sustainable in the future. More and more utilities are realizing that a systematic approach for managing the full range of energy challenges they face is the best way to ensure that those issues are addressed on an ongoing basis in order to reduce climate impacts, save money, and remain sustainable. In this chapter, we provide a basic primer on and review of energy. In the following chapter, we discuss in detail the efforts of the U.S. Environmental Protection Agency (USEPA) and numerous utilities to provide a systematic approach to reducing energy consumption and energy costs and developing alternative and renewable energy options.

3.2 Energy Basics

Energy (often defined as the ability to do work) is one of the most discussed topics today because of current high prices for hydrocarbon products (gasoline and diesel fuel) and natural gas. These are all forms of energy that we are quite familiar with, but energy also comes in other forms:

---

* Material in this chapter is adapted from USEPA, *Ensuring a Sustainable Future: An Energy Management Guidebook for Wastewater and Water Utilities*, U.S. Environmental Protection Agency, Washington, DC, 2008; Spellman, F.R., *Physics for Non-physicists*, Government Institutes Press, Lanham, MD, 2009.

- *Heat (thermal)* energy is produced by the vibration and movement of atoms and molecules within substances. As an object is heated up, its atoms and molecules move and collide faster. Geothermal is the energy found within the Earth.

- *Light (radiant)* energy is electromagnetic energy that travels in transverse waves. Radiation energy includes visible light, x-rays, gamma rays, and radio waves. Light is one type of radiant energy. Sunshine is radiant energy, which provides the fuel and warmth that make life on Earth possible.

- *Motion (kinetic)* energy is energy stored in the movement of objects. The faster objects move, the more energy is stored. It takes energy to get an object moving, and energy is released when an object slows down. Wind is an example of motion energy. A dramatic example of motion energy is a car crash, when the car comes to a total stop and releases all of its motion energy at once in an uncontrolled instant.

- *Electrical* energy is delivered by electrons, tiny charged particles that typically move through a wire (a conductor). Lightning is an example of electrical energy in nature, so powerful that it is not confined to a wire.

- *Chemical* energy is stored in the bonds of atoms and molecules. Batteries, biomass, petroleum, natural gas, and coal are examples of stored chemical energy. Chemical energy is converted to thermal energy when we burn wood in a fireplace or burn gasoline in the engine of a car.

- *Nuclear* energy is stored in the nucleus of an atom; it is the energy that holds the nucleus together. Very large amounts of energy can be released when the nuclei are combined or split apart. Nuclear power plants split the nuclei of uranium atoms in a process called *fission*. The sun combines the nuclei of hydrogen atoms in a process called *fusion*.

- *Gravitational* energy is stored in the height of an object. The higher and heavier the object, the more gravitational energy is stored. When you ride a bicycle down a steep hill and pick up speed, the gravitational energy is being converted to motion energy. Hydropower is another example of gravitational energy, where a dam gathers and holds water from a river in a reservoir.

Energy is everywhere. All things we do in life and death (decomposition and biodegradation require energy, too) are a result of energy. The two types of energy are stored (*potential*) energy and working or moving (*kinetic*) energy.

3.2.1 Potential Energy

An object can have the ability to do work (have energy) because of position; for example, a weight suspended high from a scaffold can be made to exert a force when it falls. Because gravity is the ultimate source of this energy, it is correctly called *gravitational potential energy*, or GPE (GPE = weight × height), but we usually refer to this as *potential energy*, or PE. Another type of potential energy is *chemical potential energy*, which is the energy stored in a battery or the gas in the gas tank of a vehicle.

Consider Figure 3.1. When the suspended object is released, it will fall on top of the box and crush or squash it, exerting a force on the box over a distance. By multiplying the force exerted on the box by the distance the object falls, we could calculate the amount of work that is done.

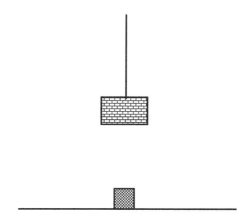

FIGURE 3.1
A box of bricks (gravitational potential energy, GPE) suspended above an empty cardboard box.

3.2.2 Kinetic Energy

Moving objects have energy—the ability to do work. Kinetic energy of an object is related to its motion. Figure 3.2 shows the suspended box of bricks we used earlier to demonstrate potential energy, but now the box of bricks is free falling. The potential energy is converted to kinetic energy because of movement. Specifically, the *kinetic energy* (KE) of an object is defined as half its mass (m) times its velocity (v) squared, or

$$KE = 1/2mv^2$$

From this equation, it is apparent that the more massive an object and the faster it is moving, the more kinetic energy it possesses. The units of kinetic energy are determined by taking the product of the units for mass (kg) and velocity squared (kg·m²/s²); the units of KE, like PE, are joules. Kinetic energy can never be negative and only tells us about speed, not velocity.

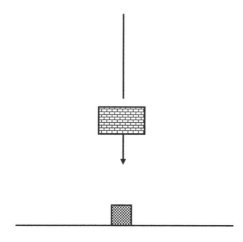

FIGURE 3.2
Free-falling box of bricks (kinetic energy, KE) suspended above an empty cardboard box.

DID YOU KNOW?

To scientists, "conservation of energy" does not mean saving energy. Instead, the law of conservation of energy says that energy is neither created nor destroyed. When we use energy, it does not disappear. We change it from one form of energy into another.

3.3 Renewable and Nonrenewable Energy

When we use electricity in our home, the electrical power was probably generated by burning coal, by nuclear reaction, or by a hydroelectric plant at a dam. Coal, nuclear, and hydro are called *energy sources*. When we fill up a gas tank, the source might be petroleum or ethanol made by growing and processing corn. Energy sources are divided into two groups—*renewable*, an energy source that can be easily replenished, and *nonrenewable*, an energy source that we are using up and cannot recreate. Most of our energy is nonrenewable (see Figure 3.3).

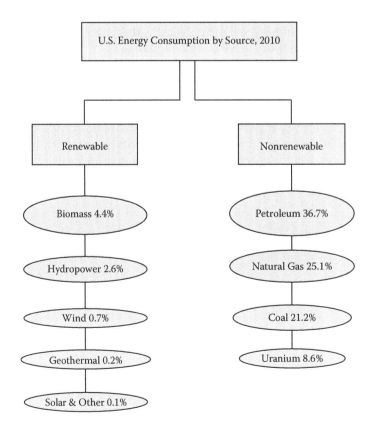

FIGURE 3.3
U.S. energy consumption in 2010. (Adapted from EIA, *Monthly Energy Review, June 2011,* DOE/EIA-0035(2011/06), U.S. Energy Information Administration, Washington, DC, 2011.)

3.3.1 Mix of Energy Production Changes

The nation's overall energy history is one of significant change as new forms of energy have been developed. The three major fossil fuels—petroleum, natural gas, and coal—have dominated the U.S. energy mix for over 100 years. Recent changes in U.S. energy production include the following:

- The share of coal produced from surface mines increased significantly: from 25% in 1949 to 51% in 1971 to 69% in 2010. The remaining share was produced from underground mines.

- In 2010, natural gas production exceeded coal production for the first time since 1981. More efficient, cost-effective drilling techniques, notably in the production of natural gas from shale formations, led to increased natural gas production in recent years.

- Although total U.S. crude oil production has generally decreased each year since it peaked in 1970, it increased by 3%, from 2009 to 2010. The increase in 2010 was led by escalating horizontal drilling programs (hydraulic fracturing or fracking) in U.S. shale plays, notably in the North Dakota section of the Bakken Formation.

- Natural gas plant liquids (NGPLs) are hydrocarbons that are separated as liquids from natural gas at processing plants and used in petroleum refineries. Production of NGPL fluctuates with natural gas produced, but their share of total U.S. petroleum field production increased from 8% in 1950 to 27% in 2010.

- In 2010, total renewable energy consumption and production reached all-time heights of 8 quadrillion Btu each. From 2000 through 2010, biofuels and wind grew faster than other renewable energy sources. In 2010, biofuels production was 8 times greater than in 2000, and wind generation was 16 times greater than in 2000.

3.4 Units for Comparing Energy

Physical units reflect measures of distances, areas, volumes, heights, weights, mass, force, and energy. Different types of energy are measured by different physical units:

- Barrels or gallons for petroleum
- Cubic feet for natural gas
- Tons for coal
- Kilowatt-hours for electricity

To compare different fuels, we need to convert the measurements to the same units. Some popular units for comparing energy include the British thermal unit (Btu), barrels of oil equivalent, metric tons of oil equivalent, metric tons of coal equivalent, and terajoules.

In the United States, the British thermal unit (Btu), a measure of heat energy, is the most commonly used unit for comparing fuels. A Btu is the heat required to raise the temperature of 1 pound of liquid water by 1°F at the temperature at which water has its greatest density (~39°F). Because energy used in different countries comes from different places, the Btu content of fuels varies slightly from country to country.

DID YOU KNOW?

A barrel is a unit of volume or weight that is different depending on who uses the term and what it contains. For example,

- 1 barrel (bbl) of petroleum or related products = 42 gallons
- 1 barrel of Portland cement = 376 pounds
- 1 barrel of flour = 196 pounds
- 1 barrel of pork or fish = 2000 pounds
- 1 barrel of (U.S.) dry measure = 329,122 bushels or 4.2104 cubic feet

A barrel may be called a "drum," but a drum usually holds 55 gallons.

References and Recommended Reading

EIA. (2012). *Energy Explained: Your Guide to Understanding Energy*, Energy Information Administration, Washington, DC (http://www.eia.gov/energyexplained/index.cfm).

Spellman, F.R. (2005). The Science of Renewable Energy, lecture presented to environmental health students, Old Dominion University, Norfolk, VA.

4

Planning for a Sustainable Energy Future

[S]ustainability means running the global environment—Earth Inc.—like a corporation: with depreciation, amortization and maintenance accounts. In other words, keeping the asset whole, rather than undermining your natural capital.

—Maurice Strong, former Under-Secretary-General of the United Nations

The sustainability revolution is nothing less than a rethinking and remaking of our role in the natural world.

—David W. Orr (in Foreword to *The Sustainability Revolution*, by A.R. Edwards)

4.1 Wastewater and Drinking Water Treatment Energy Usage

Energy represents the largest controllable cost of providing wastewater or water services to the public. Most facilities were designed and built when energy costs were not a major concern. With large pumps, drives, motors, and other equipment operating 24 hours a day, water and wastewater utilities can be among the largest individual energy users in the community. In a National Association of Clean Water Agencies (NACWA) survey of energy use in a typical wastewater treatment plant, 38% of energy use was for in-plant pumping, 26% for aeration, 25% for effluent reuse pumping, and 11% for other purposes (Jones, 2006; USEPA, 2008).

4.1.1 Current and Future Challenges

Wastewater or water treatment plant managers face unprecedented challenges that include ever-increasing:

- Public expectations for holding rates/taxes while maintaining service standards
- Population shifts/increases
- Number and complexity of regulatory requirements
- Maintenance and replacement of aging systems/infrastructure
- Concerns about security and emergency preparedness
- Changing work force demographics
- Challenges in managing personnel, operations, and budgets

Overlaying all these issues are steadily rising energy costs for utilities. Dealing with these rising costs requires utilities to better manage their energy consumption and identify areas for improvement. Water and wastewater utility energy consumption is generally the largest single sector of a city's energy bill—on the order of 30 to 60% (EIA, 2008).

When reviewing the energy performance of a facility (i.e., the management of energy consumption and identification of areas for improvement), utility managers may also identify other areas for operational improvements and cost savings, such as labor, chemicals, maintenance, and disposal costs. Additionally, a thorough assessment of the energy performance of a facility may alert managers to other issues. An unexplained increase in energy consumption may be indicative of equipment failure, an obstruction, or some other problem within facility operations.

Given these challenges, it is imperative for water and wastewater treatment facilities to investigate implementing systematic programs to minimize energy usage and cost, without sacrificing performance.

4.2 Fast Facts[*]

Drinking water and wastewater utility energy:

- Water and wastewater industries account for an estimated 75 billion kWh of overall U.S. electricity demand.
- Drinking water and wastewater systems in the United States spend about $4 billion a year on energy to pump, treat, deliver, collect, and clean water.
- Energy efficiency investments often have outstanding rates of return and can reduce costs at a facility by 5, 10, 25% or more.
- Loads are expected to increase by 20% in the next 15 years due to increased populations and more stringent regulations.
- Energy costs for water and wastewater can represent a third of a municipality's total energy bill.
- If drinking water and wastewater systems reduce use by just 10% through cost-effective investments, collectively they could save approximately $400 million and 5 billion kWh annually.

Wastewater utilities:

- There are 15,000 wastewater systems, including 6000 publicly owned treatment works (POTWs), in the United States.
- The majority of energy use occurs in the treatment process (aeration) and pumping.
- Energy use is affected by population, influent loading, effluent quality, process type, size, and age.
- Major processes are collection systems (sewers and pumping stations); wastewater treatment (primary, secondary, and/or tertiary/advanced); and biosolids processing, disposal, or reuse.

---

[*] Based on information from USEPA, *Ensuring a Sustainable Future: An Energy Management Guidebook for Wastewater and Water Utilities*, U.S. Environmental Protection Agency, Washington, DC, 2008.

Drinking water utilities:

- There are 60,000 community drinking water systems in the United States.
- The majority of energy use occurs in pumping.
- Energy use is affected by water source, quality, storage, elevation, distance, age, and process.
- Major processes are production, treatment (disinfection), and distribution.

4.3 Benchmark It!

Cobblers are credited with coining the term *benchmarking*. They used the term to measure people's feet for new shoes. They would place the person's foot on a bench and trace it to make the pattern for shoes. Benchmarking is still used to measure but now it specifically gauges performance based on specific indicators such as cost per unit of measure, productivity per unit of measure, cycle time of some value per unit of measure, or defects per unit of measure.

It is interesting to note that there is no specific benchmarking process that has been universally adopted, primarily because of its wide appeal and acceptance. Accordingly, benchmarking manifests itself via various methodologies. Robert Camp (1989) wrote one of the earliest books on benchmarking and developed a 12-stage approach to benchmarking:

1. Select subject.
2. Define the process.
3. Identify potential partners.
4. Identify data sources.
5. Collect data and select partners.
6. Determine the gap.
7. Establish process differences.
8. Target future performance.
9. Communicate.
10. Adjust goal.
11. Implement.
12. Review and recalibrate.

With regard to improving energy efficiency and sustainability in drinking water and wastewater treatment operations, benchmarking is simply defined (in this text) as the process of comparing the energy usage of a particular drinking water or wastewater treatment operation to similar operations. Local utilities of similar size and design are excellent points of comparison. Broadening the search, one can find several resources discussing the typical energy consumption across the United States for a water or wastewater utility of a particular size and design.

Keep in mind that in drinking water and wastewater treatment utilities (and other utilities and industries), benchmarking is often used by management personnel to increase efficiency and ensure sustainability of energy resources, but it is also used to ensure a utility's self-preservation (i.e., to remain lucrative). With self-preservation as the motive, benchmarking is used as a tool to compare operations with best-in-class similar facilities or operations to improve performance to avoid the current (and ongoing) trend to privatize water, wastewater, and other public operations. Based on personal observation, usually the real work to prevent privatization is delegated to the individual managers in charge of each specific operation because they also have a stake in making sure that their relatively secure careers are not affected by privatization. Moreover, these front-line managers are best positioned to make nut-and-bolt decisions that can increase efficiency in all operations and in managing to conserve and sustain energy supplies. It can be easily seen that working against privatization by these local managers is in their own self-interest and in the interest of their workers because their jobs may be at stake.

The question is, of course, how do these managers go about preventing their water and wastewater operation from being privatized? The answer is rather straightforward and clear: Efficiency must be improved at reduced cost of operations. In the real world, this is easier said than done but is not impossible; for example, for those facilities under properly implemented and managed Total Quality Management (TQM), the process can be much easier. The advantage TQM offers the plant manager lies in the variety of tools provided to help plan, develop, and implement water and wastewater energy efficiency measures. These tools include self-assessments, statistical process control, International Organization for Standards (ISO) 9000 and 14000 certification, process analysis, quality circles, team forming, and benchmarking.

In the pursuit of energy efficiency and sustainability in drinking water and wastewater treatment operations, Camp's 12 stages listed earlier can be simplified into the following six-step process:

$$\text{Start} \rightarrow \text{Plan} \rightarrow \text{Research} \rightarrow \text{Observe} \rightarrow \text{Analyze} \rightarrow \text{Adapt}$$

In this text, of course, the focus is on use of the benchmarking tool to improve water and wastewater operation efficiency and ensuring a sustainable future for wastewater and water treatment utilities. Before applying the benchmarking process, an Energy Team should be formed and assigned the task of studying how to implement energy-saving strategies and how to ensure sustainability in the long run. Keep in mind that forming such a team is not the same as fashioning a silver bullet; the team is only as good as its leadership and its members. Again, benchmarking is a process for rigorously measuring performance vs. best-in-class operations and then using the analysis to meet and exceed the best in class; thus, those involved in the benchmarking process should be the best of the best (Spellman, 2009).

4.3.1 What Benchmarking Is

1. Benchmarking vs. best practices gives water and wastewater operations a way to evaluate their operations overall with regard to
   a. How effective they are
   b. How cost effective they are

2. Benchmarking shows plants both how well their operations stack up and how well those operations are implemented.

3. Benchmarking is an objective-setting process.

4. Benchmarking is a new way of doing business.

5. Benchmarking forces an external view to ensure correctness of objective-setting.

6. Benchmarking forces internal alignment to achieve plant goals.

7. Benchmarking promotes teamwork by directing attention to those practices necessary to remain competitive.

4.3.2 Potential Results of Benchmarking

Benchmarking may indicate a direction of required change rather than specific metrics; for example, perhaps costs must be reduced, customer satisfaction increased, return on assets increased, maintenance improved, or operational practices improved. Best practices translate into operational units of measure.

4.3.3 Targets

Consideration of available resources converts benchmark findings to targets. A target represents what can realistically be accomplished in a given time frame, and it can show progress toward benchmark practices and metrics. Quantification of precise targets should be based on achieving the benchmark.

Note: Benchmarking can be performance based, process based, or strategic based and can compare financial or operational performance measures, methods, or practices, or strategic choices.

4.3.4 Process of Benchmarking

When a benchmarking team is being assembled, the goal should be to utilize a benchmark that evaluates and compares privatized and reengineered water and wastewater treatment operations to operations within the team's utility in order to be more efficient and remain competitive and make continual improvements. It is important to point out that benchmarking is more than simply setting a performance reference or comparison; it is a way to facilitate learning for continual improvements. The key to the learning process is looking outside one's own plant to other plants that have discovered better ways of achieving improved performance.

4.3.5 Benchmarking Steps

As shown earlier, the benchmarking process consists of five major steps:

1. *Planning*—Managers must select a process (or processes) to be benchmarked and form a benchmarking team. The process of benchmarking must be thoroughly understood and documented. A performance measure for the chosen process should be established (e.g., cost, time, quality).

2. *Research*—Information on the best-in-class performer must be acquired through research. The information can be derived from the industry's network, industry experts, industry and trade associations, publications, public information, and other award-winning operations.

3. *Observation*—The observation step is a study of the benchmarking subject's performance level, processes, and practices that have achieved those levels, as well as other enabling factors.

4. *Analysis*—In this phase, comparisons in performance levels among facilities are determined. The root causes for the performance gaps are studied. To make accurate and appropriate comparisons, the comparison data must be sorted, controlled for quality, and normalized.

5. *Adaptation*—This phase is putting what is learned throughout the benchmarking process into action. The findings of the benchmarking study must be communicated to gain acceptance, functional goals must be established, and a plan must be developed. Progress should be monitored and, as required, corrections in the process made.

Note: Benchmarking should be interactive. It should also recalibrate performance measures and improve the process itself.

4.3.6 Collection of Baseline Data and Tracking Energy Use

Using the five-stage benchmarking procedure detailed above, the benchmarking team identifies, locates, and assembles baseline data information that can help determine what is needed to improve the utility's energy performance. Keep in mind that the data collected will be compared to like operations in the benchmarking process. The point is that it is important to collect data that are comparable, like oranges to oranges, apples to apples, grapes to grapes, and so forth. It does little good, makes no sense, and wastes time and money to collect data for equipment, machinery, and operations that are not comparable to those of the utility or utilities to which the data will be compared.

The first step is to determine what data are already available. At a minimum, it is necessary to have one full year of monthly data for consumption of electricity, natural gas, and other fuels—three years of data are even better. However, if data going that far back are not available, use what you have or can easily collect. In addition, gathering data at daily or hourly intervals may be helpful in identifying a wider range of energy opportunities (USEPA, 2008).

Here are several data elements to document and track for your utility in order to review energy improvement opportunities:

- Water and/or wastewater flows are key to determining energy performance per gallon treated. For drinking water, the distance of travel and number of pumps are also key factors.
- Electricity data include overall electricity consumption (kWh) as well as peak demand (kW) and load profiles, if available.
- Other types of energy data include purchases of diesel fuel, natural gas, or other energy sources, including renewables.

DID YOU KNOW?

Benchmarking can be useful, but no two utilities are ever exactly the same. Each utility will have particular characteristics affecting its relative performance that are beyond its control.

- Design specifications can help to determine how much energy a given process or piece of equipment should be using.
- Operating schedules for intermittent process will help you make sense of your load profile and possibly plan an energy-saving or cost-saving alternative.

Along with making sure that the data you collect are comparable (e.g., apples to apples), keep in mind that energy units may vary. If you are comparing apples to apples, are you comparing bushel to bushel, pound to pound, or quantity to quantity? In an energy efficiency and sustainability benchmarking study comparison, for example, captured methane or purchased natural gas may be measured in 100 cubic feet (Ccf) or millions of British thermal units (MMBtu). Develop a table like Table 4.1 to document and track your data needs (USEPA, 2008). Remember—keep units consistent!

Consider any other quantities that should be measured. Should anything else be added to Table 4.1? Chances are good that you will want to add some other quantities. Let's get back to unit selection for your tables. Be sure to select units that the Energy Team is comfortable with and that the data are typically available in. If the data are reported using different units, the team may obtain some conflicting or confusing results.

Keep in mind that units by themselves are not that informative; to be placed in proper context, they need to be associated with an interval of time. The next step, then, is simply to grow Table 4.1 by adding a "Desired Frequency of Data" column (Table 4.2).

Remember, although it is useful to know the utility's energy consumption per month, knowing it in kWh per day is better. Hourly consumption data can be used to develop a "load profile," or a breakdown of your energy demand during the day. If the load profile is

TABLE 4.1

Data Needs and Units

Data Need	Units
Wastewater flow	MGD
Electricity consumption	kWh
Peak demand	kW
Methane capture (applies to plants that digest biosolids)	MMBtu
Microturbine generation	kWh
Natural gas consumed	MMBtu
Fuel oil consumed	Gallons
Diesel fuel consumed	Gallons
Design specifications	N/A
Operating schedules	N/A
Grease trap waste collected (future renewable fuel source)	Gallons
Other (based on your operation)	TBD

TABLE 4.2

Data Needs, Units, and Desired Frequency of Data

Data Need	Units	Desired Frequency of Data
Wastewater flow	MGD	Daily
Electricity consumption	kWh	Hourly if possible or daily if not
Peak demand	kW	Monthly
Methane capture	MMBtu	Monthly
Microturbine generation	kWh	Monthly
Natural gas consumed	MMBtu	Monthly
Fuel oil consumed	Gallons	Monthly
Diesel fuel consumed	Gallons	Monthly
Design specifications	N/A	N/A
Operating schedules	N/A	N/A

relatively flat or if the energy demand is greater in the off-peak hours (overnight and early morning) than in the peak hours (daytime and early evening), the utility may qualify for special pricing plants from its energy provider.

Typically, water and wastewater treatment operations have a predictable diurnal variation (i.e., fluctuations that occur during each day). Usage is least heavy during the early overnight hours. It is heaviest during the early morning, lags during the afternoon, has another less intensive peak in the early evening, and then hits the lowest point overnight. Normally, energy use for water and wastewater treatment operations could be expected to follow a pattern of water flows, but this effect can be delayed by the travel time from the source through the collection system to the plant, or by storage tanks within the distribution system. A larger system will have varying travel times, whereas a smaller system will have lower variability. Moreover, this effect can be totally eliminated if the plant has equalization tanks (USEPA, 2008). A utility that is paying a great deal of money for peak demand charges might consider the capital investment of an equalization tank. Demand charges can be significant for wastewater utilities, as they are generally about 25% of the utility's electricity bill (WEF, 1997).

The next step is to determine how to collect the baseline data. Energy data are recorded by energy providers (e.g., electric utility, natural gas utility, heating oil and diesel oil companies). A monthly energy bill contains the total consumption for that month, as well as the peak demand. In some cases, local utilities will record the demand on every meter for every 15-minute interval of the year. Similar data may be available to utilities that have a system that monitors energy performance. Sources of energy data include the following:

- *Monthly energy bills* vary in detail but all contain the most essential elements.
- *The energy provider* may be able to provide more detailed information.
- *An energy management program* (e.g., supervisory control and data acquisition, or SCADA) automatically tracks energy data, often with submeters to identify the load on individual components. A utility that has such a system in place will have a large and detailed dataset on hand.

Other data needs may also have a range of sources. Design specifications for equipment may be found in manuals at the utility, but it may still be necessary to contact manufacturers for specific items. In addition to providing raw data, energy providers can offer extensive expertise on energy-saving technologies, practices, and programs, and contractors can help implement certain types of improvements (USEPA, 2008).

What Is SCADA?

Simply, SCADA is a computer-based system that remotely controls processes previously controlled manually. SCADA allows an operator using a central computer to supervise (control and monitor) multiple networked computers at remote locations. Each remote computer can control mechanical processes (e.g., pumps, valves) and collect data from sensors at its remote location, thus the phrase *supervisory control and data acquisition*, or SCADA.

The central computer is the *master terminal unit*, or MTU. The operator interfaces with the MTU using software referred to as the *human–machine interface*, or HMI. The remote computer is the *programmable logic controller* (PLC) or *remote terminal unit* (RTU). The RTU activates a relay (or switch) that turns mechanical equipment on and off. The RTU also collects data from sensors.

In the initial stages, utilities ran wires, also known as hardwire or land lines, from the central computer (MTU) to the remote computers (RTUs). Because remote locations can be located hundreds of miles from the central location, utilities began to use public phone lines and modems, leased telephone company lines, and radio and microwave communication. More recently, they have also begun to use satellite links, the Internet, and newly developed wireless technologies.

Because SCADA system sensors provided valuable information, many utilities established connections between their SCADA systems and their business systems. This allowed utility management and other staff access to valuable statistics, such as water usage. When utilities later connected their systems to the Internet, they were able to provide stakeholders with water/wastewater statistics on the utilities' web pages.

SCADA APPLICATIONS IN WATER/WASTEWATER SYSTEMS

SCADA systems can be designed to measure a variety of equipment operating conditions and parameters, volumes and flow rates, or water quality parameters and to respond to changes in those parameters either by alerting operators or by modifying system operation through a feedback loop system without having personnel physically visit each process or piece of equipment on a daily basis to check it and ensure that it is functioning properly. SCADA systems can also be used to automate certain functions so they can be performed without being initiated by an operator (e.g., injecting chlorine in response to periodic low chlorine levels in a distribution system, turning on a pump in response to low water levels in a storage tank). In addition to process equipment, SCADA systems can also integrate specific security alarms and equipment, such as cameras, motion sensors, lights, data from card reading systems, etc., thereby providing a clear picture of what is happening at areas throughout a facility. Finally, SCADA systems also provide constant, real-time data on processes, equipment, location access, etc., so the necessary response can be made quickly. This can be extremely useful during emergency conditions, such as when distribution mains break or when potentially disruptive BOD spikes appear in wastewater influent.

Because these systems can monitor multiple processes, equipment, and infrastructure and then provide quick notification of, or response to, problems or upsets. SCADA systems typically provide the first line of detection for atypical or abnormal conditions. For example, a SCADA system connected to sensors that measure specific water quality parameters would indicate when the parameters are outside of a specific range. A real-time customized operator interface screen could display and control critical systems monitoring parameters.

The system could transmit warning signals back to the operators, such as by initiating a call to a personal pager. This might allow the operators to initiate actions to prevent contamination and disruption of the water supply. Further automation of the system could ensure that the system initiated measures to rectify the problem. Preprogrammed control functions (e.g., shutting a valve, controlling flow, increasing chlorination, or adding other chemicals) can be triggered and operated based on SCADA utility.

4.4 Baseline Audit

The energy audit is an essential step in energy conservation and energy management efforts. Drinking water or wastewater operations may have had energy audits or energy program reviews conducted at some point. If so, find the final report and have the Energy Team review it. How long did the process take? Who participated in it—your team, the electric utility, independent contractors? What measures were suggested to improve energy efficiency? What measures were actually implemented? Did they meet expectations? Were there lessons learned from the process that should be applied to future audits? In addition, if your facility's previous energy audit had recommended measures, determine if they are still viable.

DID YOU KNOW?

On April 23, 2000, police in Queensland, Australia stopped a car on the road and found a stolen computer and radio inside. Using commercially available technology, a disgruntled former employee had turned his vehicle into a pirate command center for sewage treatment along Australia's Sunshine Coast. The former employee's arrest solved a mystery that had troubled the Maroochy Shire wastewater system for two months. Somehow the system was leaking hundreds of thousands of gallons of putrid sewage into parks, rivers, and the manicured grounds of a Hyatt Regency hotel. Marine life died, creek water turned black, and the stench was unbearable for residents. Until the former employee's capture—during his 46th successful intrusion—the utility's managers did not know why this was happening.

Specialists study this case of cyberterrorism because it is the only one known in which someone used a digital control system deliberately to cause harm. The former employee's intrusion shows how easy it is to break in—and how unrestrained he was with his power.

To sabotage the system, the former employee set the software on his laptop to identify it as a pumping station and then suppressed all alarms. The former employee became the central control station during his intrusions, with unlimited command of 300 SCADA nodes governing sewage and drinking water alike.

The bottom line: As serious as the former employee's intrusions were they pale in comparison with what he could have done to the freshwater system—he could have done anything he liked (Gellman, 2002).

In many cases, electrical utilities offer audits as part of their energy conservation programs. Independent energy service companies also provide these services. An outside review from an electric utility or an engineering company can provide useful input, but it is important to ensure that any third party is familiar with the water and wastewater systems.

Some energy audits focus on specific types of equipment such as lighting, HVAC, or pumps. Others look at the processes used and take a more systematic approach. Audits focused on individual components, as well as in-depth process audits, will include testing equipment. For example, in conducting the baseline energy audit, the Energy Team may compare the nameplate efficiency of a motor or pump to its actual efficiency.

In a process approach, a preliminary walk-through or walk-around audit is often used as a first step to determine if there are likely to be opportunities to save energy. If such opportunities exist, then a detailed process audit is conducted. This may include auditing the performance of the individual components as well as considering how they work together as a whole. Much like an environmental management system's initial assessment that reviews current status of regulatory requirements, training, communication, operating conditions, and current practices and processes, a preliminary energy audit or energy program review will provide a baseline of what the utility's energy consumption is at that point in time.

Once the utility's baseline data have been collected and monthly and annual energy use has been tracked, two additional steps remain to complete the energy assessment or baseline energy audit: (1) conduct a field investigation, and (2) create an equipment inventory and distribution of demand and energy (USEPA, 2008).

4.4.1 Field Investigation

The field investigation is the heart of an energy audit. It includes obtaining information for an equipment inventory, discussing process operations with the individuals responsible for each operation, discussing the impact of specific energy conservation ideas, soliciting ideas from the Energy Team, and identifying the energy profiles of individual system components. The Electric Power Research Institute (EPRI) recommends evaluating how each process or piece of equipment could otherwise be used. It might be possible, for example, for a given system to be complemented by one of lower capacity during normal operation, to be run fewer hours, to be run during off-peak hours, to employ a variable speed drive, or to be replaced by a newer or more efficient system. Depending on the situation, one or more of these changes might be appropriate.

4.4.2 Create Equipment Inventory and Distribution of Demand and Energy

This is a record of your operation's equipment, equipment names, nameplate horsepower (if applicable), hours of operation per year, measured power consumption, and total kilowatt-hours (kWh) of electrical consumption per year. Other criteria such as age may also be included. In addition, different data may be appropriate for other types of systems, such as methane-fired heat and power systems.

You may find that much of this information is already available in the utility's maintenance management system (if applicable). A detailed approach for developing an equipment inventory and identifying the energy demand of each piece of equipment has been provided by the Water Environment Federation (WEF, 1997). The basics are presented here, but readers are encouraged to review the WEF manual for a more thorough explanation.

Example drinking water and/or wastewater treatment operations equipment inventories and the relevant energy data to collect could include the following:

- Motors and related equipment:
  - Start at each motor control center (MCC) and itemize each piece of equipment in order as listed on the MCC.
  - Itemize all electric meters on MCCs and local control panels.
  - Have a qualified electrician check the power draw of each major piece of equipment.
- Pumps:
  - From the equipment manufacturer's literature, determine the pump's power ratio (this may be expressed in kW/MGD).
  - Multiply horsepower by 0.746 to obtain kilowatts.
  - Compare the manufacturer's data with field-obtained data.
- Aeration equipment:
  - Because the power draw of aeration equipment is difficult to estimate, it should be measured.
  - Measure aspects related to biochemical oxygen demand (BOD) loading, food-to-microorganism ratio, and oxygen transfer efficiency (OTE). Note that OTE levels depend on the type and condition of aeration equipment. Actual OTE levels are often considerably lower than described in the literature or in manufacturers' materials.

CASE STUDY 4.1. BENCHMARKING: AN EXAMPLE

To gain a better understanding of benchmarking, the following example is provided. It is a summary only, as discussion of a full-blown study is beyond the scope of this text.

Rachel's Creek Sanitation District

Introduction

In January 1997, Rachel's Creek Sanitation District formed a benchmarking team with the goal of providing a benchmark that evaluates and compares privatized and reengineered wastewater treatment operations to Rachel's Creek operations in order to be more efficient and remain competitive. After 3 months of evaluating wastewater facilities using the benchmarking tool, our benchmarking is complete. This report summarizes our findings and should serve as a benchmark by which to compare and evaluate Rachel's Creek Sanitation District operations.

Facilities

41 wastewater treatment plants throughout the United States

Target Areas

The benchmarking team focused on the following target areas for comparison:

1. Reengineering
2. Organization
3. Operations and maintenance

  a. Contractual services
  b. Materials and supplies
  c. Sampling and data collection
  d. Maintenance
 4. Operational directives
 5. Utilities
 6. Chemicals
 7. Technology
 8. Permits
  a. Water quality
  b. Solids quality
  c. Air quality
  d. Odor quality
 9. Safety
10. Training and development
11. Process
12. Communication
13. Public relations
14. Reuse
15. Support services
  a. Pretreatment
  b. Collection systems
  c. Procurement
  d. Finance and administration
  e. Laboratory
  f. Human resources

Summary of Findings

Our overall evaluation of Rachel's Creek Sanitation District as compared to our benchmarking targets is a good one; that is, we are in good standing as compared to the 41 target facilities we benchmarked against. In the area of safety, we compare quite favorably. Only plant 34, with its own full-time safety manager, appeared to be better than we are. We were very competitive with the privatized plants in our usage of chemicals and far ahead of many public plants. We were also competitive in the use of power. Our survey of what other plants are doing to cut power costs showed that we had clearly identified those areas of improvement and that our current effort to further reduce power costs is on track. We were far ahead in the optimization of our unit processes, and we were leaders in the area of odor control.

 We also found areas where we need to improve. To the Rachel's Creek employee, reengineering applies only to the treatment department and has been limited to cutting staff while plant practices and organizational practices remain outdated and inefficient. Under the reengineering section of this report, we have provided a summary of reengineering efforts at the reengineered plants visited. The experiences of these plants can be used to improve our own reengineering effort. The next area we examined is our organization and staffing levels. A private company could reduce the entire treatment department staff by about 18 to 24%, based on the number of employees and not costs. In the organization section of this report, organizational models and their staffing levels are provided as guidelines to improving our organization and determining optimum staffing levels. The last big area where we need to improve is in the way we accomplish the work we perform. Our people are not used efficiently because of outdated and inefficient policies and work practices. Methods to improve the way we do work are found throughout this report. We noted that efficient work practices used by private companies allow plants to operate with small staffs.

DID YOU KNOW?

Some utilities will have an inherently higher or lower energy demand due to factors beyond their control. For example, larger plants will, in general, have a lower energy demand per million gallons treated due to economies of scale. A plant that is large relative to its typical load is going to have a higher energy demand per million gallons treated. Some secondary treatment processes require greater energy consumption than others. Still, benchmarking allows a rough estimate of the utility's relative energy performance. Benchmarking of individual components is also useful. A survey of one's peers may identify what level of performance can realistically be expected from, say, a combined heat and power system or a specific model of methane-fueled microturbine (USEPA, 2008).

Overall, Rachel's Creek Sanitation District's treatment plants are performing much better than other public service plants. Although some public plants may have better equipment, better technology, and cleaner effluents, their labor and materials costs are much higher than ours. Several of the public plants were in bad condition. Contrary to popular belief, the privately operated plants had good to excellent operations. These plants met permit, complied with safety regulations, maintained plant equipment, and kept the plant clean. Due to their efficiency and low staff, we feel that most of the privately operated plants are performing better than we are. We agree that this needs to be changed. Using what we learned during our benchmarking effort, we can be just as efficient as a privately operated plant and still maintain our standards of quality (Spellman, 2009).

References and Recommended Reading

Camp, R.C. (1989). *The Search for Industry Best Practices That Lead to Superior Performance*, ASQC Quality Press, Milwaukee, WI.

EIA. (2008). *Current and Historical Monthly Retail Sales, Revenues and Average Revenue per Kilowatt Hour by State and by Sector*, EIA-826, Energy Information Administration, Washington, DC (http://www.eia.gov/cneaf/electricity/page/sales_revenue.xls).

Gellman, B. (2002). Cyber-attacks by Al Qaeda feared: terrorists at threshold of using Internet as tool of bloodshed, experts say, *Washington Post*, June 27, p. A01.

Jones, T. (2006). Water–Wastewater Committee: Program Opportunities in the Municipal Sector: Priorities for 2006, presentation to CEE June Program Meeting, Boston, MA.

Spellman, F.R. (2009). *Handbook of Water and Wastewater Treatment Plant Operations*, 2nd ed., CRC Press, Boca Raton, FL.

USEPA. (2008). *Ensuring a Sustainable Future: An Energy Management Guidebook for Wastewater and Water Utilities*, U.S. Environmental Protection Agency, Washington, DC.

WEF. (1997). *Energy Conservation in Wastewater Treatment Facilities*, Manual of Practice No. MFD-2, Water Environment Federation, Alexandria, VA.

Section II

Energy-Efficient Equipment, Technology, and Operating Strategies

5

Energy-Efficient Equipment

"Energy efficiency" is the amount of useful energy you get from any type of system. A perfect energy-efficient machine would change all the energy put in it into useful work. In reality, converting one form of energy into another form always involves a loss of useable energy.

—**EIA (2011)**

The constant-speed approach to AC motors made a lot of sense when energy was cheap and drives technology was in its infancy. With their latest advances, adjustable speed drives can provide one of the best energy efficiency options for a plant.

—**Spear (2005)**

5.1 Introduction

In Section I, we discussed benchmarking and developing audit results that can identify a number of changes, large and small, that can be made to save on energy costs. Drinking water and wastewater facilities have many options to conserve energy, ranging from changing light bulbs and upgrading pumps and motors to installing co-generation systems and renewable energy technologies. We discuss many of these energy-saving options in this and subsequent chapters and provide case studies to demonstrate how some facilities have used their equipment, technology, and operating strategies to save money and reduce their impact.

5.2 Motors

Before explaining how high-efficiency motors and variable-frequency drives function and how they are used in water and wastewater treatment processes, including the benefits of their use, their performance history, cost considerations, and demonstrated success of their application, we provide a basic discussion of alternating current (AC) motors.

5.2.1 AC Motors

At least 60% of the electrical power fed to a typical waterworks or wastewater treatment plant is consumed by electric motors. One thing is certain: Electric motors perform an almost endless variety of tasks in water and wastewater treatment. An *electric motor* is a machine used to change electrical energy to mechanical energy to do the work.

Note: Do not confuse an electric motor with an electric generator. A generator does just the opposite of a motor; that is, a generator changes mechanical energy to electrical energy.

Electric motor action—changing electrical energy into mechanical energy—occurs when a current passes through a wire, and a magnetic field is produced around the wire. If this magnetic field passes through a stationary magnetic field, the fields either repel or attract, depending on their relative polarity. If both are positive or negative, they repel. If they are opposite polarity, they attract. Applying this basic information to motor design, an electromagnetic coil, the *armature*, rotates on a shaft. The armature and shaft assembly is called the *rotor*. The rotor is assembled between the poles of a permanent magnet, and each end of the rotor coil (armature) is connected to a commutator also mounted on the shaft. A commutator is composed of copper segments insulated from the shaft and from each other by an insulating material. As like poles of the electromagnet in the rotating armature pass the stationary permanent magnet poles, they are repelled, continuing the motion. As the opposite poles near each other, they attract, continuing the motion.

In comparison to direct current (DC), alternating voltage can be easily transformed from low voltages to high voltages or vice versa, and AC current can be forced to move (voltage does not move; think of voltage as the pump and current as the resulting flow) over a much greater distance without too much loss in efficiency. Most of the power generating systems today, therefore, produce alternating current. Thus, it logically follows that a great majority of the electrical motors utilized today are designed to operate on alternating current; however, there are other advantages to using AC motors besides the wide availability of AC power.

In general, AC motors are less expensive than DC motors. Also, most types of AC motors do not employ brushes and commutators, which eliminates many problems of maintenance and wear and eliminates dangerous sparking. AC motors are manufactured in many different sizes, shapes, and ratings for use on a greater number of jobs. They are designed for use with either polyphase or single-phase power systems. This section cannot possibly cover all aspects of the subject of AC motors; consequently, it deals mainly with the operating principles of the two most common types—induction motors and synchronous motors.

5.2.1.1 Induction Motors

The induction motor is the most commonly used type of AC motor because of its simple, rugged construction and good operating characteristics. It consists of two parts: the *stator* (stationary part) and the *rotor* (rotating part). The most important type of polyphase induction motor is the three-phase motor.

The driving torque of both DC and AC motors is derived from the reaction of current-carrying conductors in a magnetic field. In the DC motor, the magnetic field is stationary, and the armature, with its current-carrying conductors, rotates. The current is supplied to the armature through a commutator and brushes. In induction motors, the rotor currents are supplied by electromagnet induction. The stator windings, connected to the AC supply, contain two or more out-of-time-phase currents, which produce corresponding magneto-motive force (mmf). The mmf establishes a rotating magnetic field across the air gap. This magnetic field rotates continuously at constant speed regardless of the load on the motor. The stator winding corresponds to the armature winding of a DC motor or to the primary winding of a transformer. The rotor is not connected electrically to the power supply.

DID YOU KNOW?

A three-phase (3-θ) system is a combination of three single-phase (1-θ) systems. In a 3-θ balanced system, the power comes from an AC generator that produces three separate but equal voltages, each of which is out of phase with the other voltages by 120°. Although 1-θ circuits are widely used in electrical systems, most generation and distribution of AC current is 3-θ.

The induction motor derives its name from the fact that mutual induction (or transformer action) takes place between the stator and the rotor under operating conditions. The magnetic revolving field produced by the stator cuts across the rotor conductors, inducing a voltage in the conductors. This induced voltage causes rotor current to flow. Hence, motor torque (i.e., twisting force) is developed by the interaction of the rotor current and the magnetic revolving field.

5.2.1.2 Motor Power Distribution System

Figure 5.1 shows a typical motor power distribution system with a variable-frequency drive controller. The figure also shows various protective devices that can be installed in motor distribution systems to protect against circuit faults or interruptions. Interruptions are very rare in drinking water and wastewater treatment plant power distribution systems that have been properly designed. Still, protective devices are necessary because of the load diversity. Most installations are quite complex. In addition, externally caused variations might overload them or endanger personnel. Figure 5.1 shows the general relationship

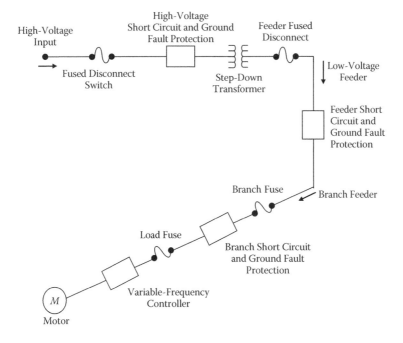

FIGURE 5.1
Motor power distribution system.

DID YOU KNOW?

A fuse is a thin strip of easily melted material. It protects a circuit from large currents by melting quickly, thereby breaking the circuit.

between protective devices and different components of a complete distribution system. Each part of the circuit has its own protective device or devices that protect not only the load (the motor) but also the wiring and control devices themselves. These disconnect and protective devices are described in the following sections.

5.2.1.2.1 Fuses

The passage of an electric current produces heat. The larger the current, the more heat is produced. In order to prevent large currents from accidentally flowing through expensive apparatus and burning it up, a fuse is placed directly into the circuit, as shown in Figure 5.1, and all the current must flow through the fuse. The fuse will permit currents smaller than the fuse value to flow but will melt and therefore break the circuit if a larger, danger-ous current ever appears; for example, a dangerously large current will flow when a *short circuit* occurs. A short circuit is usually caused by an accidental connection between two points in a circuit offering very little resistance to the flow of electrons. If the resistance is small, there will be nothing to stop the flow of the current, and the current will increase enormously. The resulting heat generated might cause a fire; however, if the circuit is pro-tected by a fuse, the heat caused by the short-circuit current will melt the fuse wire, thus breaking the circuit and reducing the current to zero.

The number of amperes of current that can flow through them before they melt and break the circuit determines the rating of a fuse; for example, we have 10-, 15-, 20-, and 30-amp fuses. We must be careful that any fuse inserted in a circuit be rated low enough to melt, or blow, before the apparatus is damaged. In a plant building wired to carry a cur-rent of 10 amps, for example, it is best to use a fuse no larger than 10 amps so that a current larger than 10 amps could never flow.

Some equipment, such as the electric motor shown in Figure 5.1, requires more current during start-up than for normal running; thus, a fast-time or medium-time fuse rating that will give running protection might blow during the initial period when high starting cur-rent is required. *Delayed action* fuses are used to handle these situations.

5.2.1.2.2 Circuit Breakers

Circuit breakers are protective devices that open automatically at a preset ampere rating to interrupt an overload or short circuit. Unlike fuses, they do not require replacement when they are activated. They are simply reset to restore power after the overload has been cleared.

DID YOU KNOW?

A circuit breaker is designed to break the circuit and stop the current flow when the current exceeds a predetermined value.

Circuit breakers are made in both plug-in and bolt-on designs. Plug-in breakers are used in load centers. Bolt-on breakers are used in panelboards and exclusively for high interrupting current applications.

Circuit breakers are rated according to current and voltage, as well as short-circuit interrupting current. A single handle opens or closes contacts between two or more conductors. Breakers are single pole but single-pole units can be ganged to form double- or triple-pole devices opened with a single handle.

Several types of circuit breakers are commonly used. They may be thermal, magnetic, or a combination of the two. Thermal breakers are tripped when the temperature rises because of heat created by the overcurrent condition. Bimetallic strips provide the time delay for overload protection. Magnetic breakers operate on the principle that a sudden current rise creates enough magnetic field to turn an armature, tripping the breaker and opening the circuit. Magnetic breakers provide the instantaneous action needed for short-circuit protection. They are also used in circumstances where ambient temperature might adversely affect the action of a thermal breaker. Thermal–magnetic breakers combine features of both types of breakers.

An important feature of the circuit breaker is its *arc chutes*, which enable the breaker to extinguish very hot arcs harmlessly. Some circuit breakers must be reset by hand, but others reset themselves automatically. If the overload condition still exists when the circuit breaker is reset, the circuit breaker will trip again to prevent damage to the circuit.

5.2.1.2.3 Control Devices

Control devices are those electrical accessories (switches and relays) that govern the power delivered to any electrical load. In its simplest form, the control device applies voltage to or removes it from a single load. In more complex control systems, the initial switch may set into action other control devices (relays) that govern motor speeds, servomechanisms, temperatures, and numerous other equipment. In fact, all electrical systems and equipment are controlled in some manner by one or more controls. A controller is a device or group of devices that serves to govern, in some predetermined manner, the device to which it is connected. In large electrical systems, it is necessary to have a variety of controls for operation of the equipment. These controls range from simple pushbuttons to heavy-duty contactors that are designed to control the operation of large motors. The pushbutton is manually operated and a contactor is electrically operated.

5.2.2 Electric Motor Load and Efficiency[*]

Most likely your drinking water or wastewater treatment plant operations account for a large part of your monthly electric bill. Far too often motors are mismatched (over- or undersized) for the load they are intended to serve, or they have been rewound multiple times. To compare the operating costs of an existing standard motor with an appropriately sized, energy-efficient replacement, it is necessary to determine operating hours, efficiency improvement values, and load. *Part-load* is a term used to describe the actual load served by the motor as compared to the rated full-load capability of the motor. Motor part-loads may be estimated by using input power, amperage, or speed measurements. This section briefly discusses several load estimation techniques.

---

[*] Adapted from USDOE, *Determining Electric Motor Load and Efficiency*, Fact Sheet, U.S. Department of Energy, Washington, DC, 2001.

DID YOU KNOW?

Instantaneous power is proportional to instantaneous voltage times instantaneous current. AC voltage causes the current to flow in a sine wave replicating the voltage wave; however, inductance in the motor windings somewhat delays current flow, resulting in a phase shift. This transmits less net power than perfectly time-matched voltage and current of the same root mean square (RMS) values. *Power factor* is that fraction of power actually delivered in relation to the power that would be delivered by the same voltage and current without the phase shift. Low power factor does not imply lost or wasted power, just excess current. The energy associated with the excess current is alternately stored in the windings' magnetic field and regenerated back to the line with each AC cycle. This exchange is called *reactive power*.

5.2.2.1 Reasons to Determine Motor Loading

Most electric motors are designed to run at 50 to 100% of rated load. Maximum efficiency is usually near 75% of rated load; thus, a 10-horsepower (hp) motor has an acceptable load range of 5 to 10 hp, and peak efficiency is at 7.5 hp. The efficiency of a motor tends to decrease dramatically below about 50% load; however, the range of good efficiency varies with individual motors and tends to extend over a broader range for larger motors. A motor is considered underloaded when it is within the range where efficiency drops significantly with decreasing load. Power factor tends to drop off sooner, but less steeply than efficiency, as load decreases.

Overloaded motors can overheat and lose efficiency. Many motors are designed with a *service factor* that allows occasional overloading. Service factor is a multiplier that indicates how much a motor can be overloaded under ideal ambient conditions. For example, a 10-hp motor with a 1.15 service factor can handle an 11.5-hp load for short periods of time without incurring significant damage. Although many motors have service factors of 1.15, running the motor continuously above rated load reduces efficiency and motor life. Never operate overloaded when voltage is below nominal or when cooling is impaired by altitude, high ambient temperature, or dirty motor surfaces.

If your operation uses equipment with motors that operate for extended periods under 50% load, consider making modifications. Sometimes motors are oversized because they must accommodate peak conditions, such as when a pumping system must satisfy occasionally high demands. Operations available to meet variable loads include two-speed motors, adjustable speed drives, and load management strategies that maintain loads within an acceptable range.

DID YOU KNOW?

Even though reactive power is theoretically not lost, the distribution system must be sized to accommodate it, which is a cost factor. To reduce these costs, capacitors are used to "correct" low power factor. Capacitors can be thought of as electrical reservoirs to capture and reflect reactive power back to the motor.

DID YOU KNOW?

AC voltage rises positive and falls negative 60 times per second, so how do you state its value? Industry practice is to quote the root mean square (RMS) voltage. RMS is a value 70.7% of the peak positive voltage. An RMS voltage will produce exactly the same heat rate in a resistive load as a DC voltage of the same value. RMS is an acronym for the mathematical steps used in its derivation. Square the voltage at all moments in an AC cycle, take the mean of these, and then take the square root of the mean. For reasons lost in obscurity, the steps are stated in reverse sequence, root mean square.

Determining if your motors are properly loaded allows making informed decisions about when to replace motors and which replacements to choose. Measuring motor loads is relatively quick and easy when you use the techniques discussed here; also refer to USDOE (2001). You should perform a motor load and efficiency analysis of all major working motors as part of the utiilty's preventive maintenance and energy conservation program.

To maintain good engineering principles and efficient operations, this text recommends that drinking water and wastewater treatment managers require surveying and testing all motors operating over 1000 hours per year. Using the analysis results, divide motors into the following categories:

- *Motors that are significantly oversized and underloaded*—Replace with more efficient, properly sized models at the next opportunity, such as scheduled plant downtime.
- *Motors that are moderately oversized and underloaded*—Replace with more efficient, properly sized models when they fail.
- *Motors that are properly sized but standard efficiency*—Replace most of these with energy-efficient models when they fail. The cost effectiveness of an energy-efficient motor purchase depends on the number of hours the motor is used, the price of electricity, and the price premium of buying an energy-efficient motor.

5.2.3 Determining Motor Loads

5.2.3.1 Input Power Measurements

When direct-read power measurements are available, use them to estimate motor part-load. With measured parameters taken from hand-held instruments, use Equation 5.1 to calculate the three-phase input power to the loaded motor. You can then quantify the motor's part-load by comparing the measured input power under load to the power required when the motor operates at rated capacity (Equation 5.2). That relationship is shown in Equation 5.3.

$$P_i = \frac{V \times I \times PF \times \sqrt{3}}{1000} \tag{5.1}$$

where

P_i = Three-phase power (kW)

V = RMS voltage, mean line-to-lie of three phases

I = RMS current, mean of three phases

PF = Power factor as a decimal

$$P_{ir} = \text{hp} \times \frac{0.7457}{\eta_{fl}} \qquad (5.2)$$

where

    P_{ir} = Input power at full-rated load (kW)

    hp = Nameplate rated horsepower

    η_{fl} = Efficiency at full-rated load

$$\text{Load} = \frac{P_i}{P_{ir}} \times 100\% \qquad (5.3)$$

where

    Load = Output power as a percentage of rated power

    P_i = Measured three-phase power (kW)

    P_{ir} = Input power at full-rated load (kW)

■ *Example 5.1*

Problem: An existing motor is identified as a 40-hp, 1800-rpm unit with an open drip-proof enclosure. The motor is 12 years old and has not been rewound. The electrician makes the following measurements:

V_{ab} = 467 V	I_a = 36 amps	PF_a = 0.75
V_{bc} = 473 V	I_b = 38 amps	PF_b = 0.78
V_{ca} = 469 V	I_c = 37 amps	PF_c = 0.76

What is the input power?

Solution:

$$V = (467 + 473 + 469)/3 = 469.7 \text{ volts}$$
$$I = (36 + 38 + 37)/3 = 37 \text{ amps}$$
$$PF = (0.75 + 0.78 + 0.76)/3 = 0.763$$

Equation 5.1 reveals:

$$P_i = \frac{469.7 \times 37 \times 0.763 \times \sqrt{3}}{1000} = 22.9 \text{ kWh}$$

5.2.3.2 Line Current Measurements

The current load estimation method is recommended when only amperage measurements are available. The amperage draw of a motor varies approximately linearly with respect to load, down to about 50% of full load. Below the 50% load point, due to reactive magnetizing current requirements, power factor degrades and the amperage curve becomes increasingly nonlinear. In the low load region, current measurements are not a useful indicator of load.

Nameplate full-load current value applies only at the rated motor voltage. Thus, root mean square (RMS) current measures should always be corrected for voltage. If the supply voltage is below that indicated on the motor nameplate, the measured amperage value is correspondingly higher than expected under rated conditions and must be adjusted downward. The converse holds true if the supply voltage at the motor terminals is above the motor rating. The equation that relates motor load to measured current values is shown in Equation 5.4.

$$\text{Load} = \frac{I}{I_r} \times \frac{V}{V_r} \times 100\% \tag{5.4}$$

where

Load = Output power as a percentage of rated power
I = RMS current, mean of three phases
I_r = Nameplate rated current
V = RMS voltage, mean line-to-line of three phases
V_r = Nameplate rated voltage

5.2.3.3 Slip Method

The slip method for estimating motor load is recommended when only operating speed measurements are available. The synchronous speed of an induction motor depends on the frequency of the power supply and on the number of poles for which the motor is wound. The higher the frequency, the faster a motor runs. The more poles the motor has, the slower it runs. Typical induction motor synchronous speeds are provided below:

Poles	60 Hertz
2	3600
4	1800
6	1200
8	900
10	720
12	600

The actual speed of the motor is less than its synchronous speed; the difference between the synchronous and actual speed referred to as *slip*. The amount of slip present is proportional to the load imposed upon the motor by the drive equipment. For example, a motor running with a 50% load has a slip halfway between the full load and synchronous speeds.

By using a tachometer to measure actual motor speed, it is possible to calculate motor loads. The safest, most convenient, and usually most accurate tachometer is a battery-powered stroboscopic tachometer. Mechanical tachometers, plug-in tachometers, and tachometers that require stopping the motor to apply a paint or reflective tape should be avoided. The motor load can be estimated with slip measurements as shown in Equation 5.5 and in Example 5.2.

$$\text{Load} = \frac{\text{Slip}}{S_s - S_r} \times 100\% \tag{5.5}$$

where

Load = Output power as a percentage of rated power

Slip = Synchronous speed (measured speed) (rpm)

S_s = Synchronous speed (rpm)

S_r = Nameplate full-load speed

■ *Example 5.2*

Problem: Determine the actual output horsepower, given the following:

Synchronous speed (rpm) = 1800

Nameplate full load speed = 1750

Measured speed (rpm) = 1770

Nameplate rated horsepower = 25 hp

Solution: From Equation 5.5,

$$\text{Load} = \frac{1800 - 1770}{1800 - 1750} \times 100\% = 60\%$$

Actual output horsepower would be 60% × 25 hp = 15 hp.

5.2.3.3.1 *Slip Method Precision*

The speed/slip method of determining motor part-load is often favored due to its simplicity and safety advantages. Most motors are constructed such that the shaft is accessible to a tachometer or a strobe light. *The accuracy of the slip method, however, is limited.* The largest uncertainty relates to the 20% tolerance that the National Electrical Manufacturers Association (NEMA) allows manufacturers in their reporting of nameplate full-load speed.

Given this broad tolerance, manufacturers generally round their reported full-load speed values to some multiple of 5 rpm. Although 5 rpm is but a small percent of the full-load speed and may be thought of as insignificant, the slip method relies on the difference between the full-load nameplate and synchronous speeds. Given a 40-rpm "correct" slip, a seemingly minor 5-rpm disparity causes a 12% change in calculated load.

Slip also varies inversely with respect to the motor terminal voltage squared—and voltage is subject to a separate NEMA tolerance of ±10% at the motor terminals. A voltage correction factor can, of course, be inserted into the slip load equation. The voltage compensated load can be calculated as shown in Equation 5.6.

$$\text{Load} = \frac{\text{Slip}}{\left(S_s - S_r\right) \times \left(V_r / V\right)^2} \times 100\%$$ (5.6)

where

Load = Output power as a percentage of rated power

Slip = Synchronous speed (measured speed) (rpm)

S_s = Synchronous speed in rpm

S_r = Nameplate full-load speed

V = RMS voltage mean line-to-line of three phases

V_r = Nameplate rated voltage

An advantage of using the current-based load estimation technique is that NEMA MG1-12.47 allows a tolerance of only 10% when reporting nameplate full-load current. In addition motor terminal voltages only affect current to the first power, whereas slip varies with the square of the voltage. Although the voltage-compensated slip method is attractive for its simplicity, its precision should not be overestimated. The slip method is generally not recommended for determining motor loads in the field.

5.2.4 Determining Motor Efficiency

The NEMA definition of energy efficiency is the ratio of its useful power output to its total power input and is usually expressed in percentage as shown in Equation 5.7.

$$\eta = \frac{0.7457 \times \text{hp} \times \text{Load}}{P_i}$$ (5.7)

where

η = Efficiency as operated (%)

hp = Nameplate rated horsepower

Load = Output power as a percentage of rated power

P_i = Three-phase power (kW)

By definition, a motor of a given rated horsepower is expected to deliver that quantity of power in a mechanical form at the motor shaft.

Figure 5.2 provides a graphical depiction of the process of converting electrical energy to mechanical energy. Motor losses are the difference between the input and output power. Output power can be calculated when the motor efficiency has been determined and the input power is known. NEMA design A and B motors up to 500 hp in size are required to have a full-load efficiency value (selected from a table of nominal efficiencies) stamped on the nameplate. Most analyses of motor energy conservation savings assume that the existing motor is operating at its nameplate efficiency. This assumption is reasonable above the 50% load point as motor efficiencies generally peak at around 3/4 load, with performance at half load almost identical to that a full load. Larger horsepower motors exhibit a relatively flat efficiency curve down to 25% of full load.

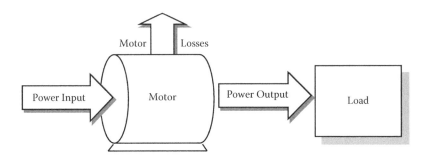

FIGURE 5.2
Depiction of motor losses.

It is more difficult to determine the efficiency of a motor that has been in service a long time. It is not uncommon for the nameplate on the motor to be lost or painted over. In that case, it is almost impossible to locate efficiency information. Also, if the motor has been rewound, the motor efficiency may have been reduced.

When nameplate efficiency is missing or unreadable, it is necessary to determine the efficiency value at the operating load point for the motor. If available, record significant nameplate data and contact the motor manufacturer. With the style, type, and serial number the manufacturer can identify approximately when the motor was manufactured. Often the manufacturer will have historical records and can supply nominal efficiency values as a function of load for a family of motors.

Three steps are used to estimate efficiency and load. First, use power amperage or slip measurements to identify the load imposed on the operating motor. Second, obtain a motor part-load efficiency value consistent with the approximated load from the manufacturer. Finally, if direct-read power measurements are available, derive a revised load estimate using both the power measurement at the motor terminals and the part-load efficiency value as shown in Equation 5.8:

$$\text{Load} = \frac{P_i \times \eta}{\text{hp} \times 0.7457} \tag{5.8}$$

DID YOU KNOW?

Inside a three-phase motor there are three windings, one for each phase. The easiest three-phase motor connection to visualize is each of the three windings being connected line to neutral. This is called *wye* because schematically it looks like the letter Y. A more common connection eliminates the neutral tie and connects the three windings from line to line. This is called *delta* because, schematically, this looks like a triangle or the Greek letter Delta (Δ). The winding experiences 73% higher voltage when connected line to line, so it must be designed for the type of connection it will have. Even if a motor's windings are internally wye connected, its nameplate voltage rating is the line-to-line value.

where

Load = Output power as a percentage of rated power

P_i = Three-phase power (kW)

η = Efficiency as operated (%)

hp = Nameplate rated horsepower

5.2.4.1 Computerized Load and Efficiency Estimation Techniques

There are several sophisticated methods for determining motor efficiency. These fall into three categories: special devices, software methods, and analytical methods. The special devices package all or most of the required instrumentation in a portable box. Software and analytical methods require generic portable instruments for measuring watts, vars (unit of reactive power of an alternating current), resistance, volts, amps, and speed. These must be instruments of premium accuracy, especially the wattmeter, which must have a broad range and good accuracy at low power and low power factor.

5.3 Variable-Frequency Drives*

A variable-frequency drive (VFD) is an electronic controller that adjusts the speed of an electric motor by modulating (i.e., regulating) the power being delivered. These drives provide continuous control, matching motor speed to the specific energy demands. In other words, the VFD controls the rotational speed of an AC electric motor by controlling the frequency of the electrical power supplied to the motor (Campbell, 1987; Jaeschke, 1978; Siskind, 1963). A variable-frequency drive has the potential to be a significant cost-saving device for water and wastewater facilities. Variable-frequency drives are an excellent choice for adjustable-speed drive users because they allow operators to fine-tune processes while reducing costs for energy and equipment maintenance.

Variable-frequency drives are used in a wide number of applications to control pumps, fans, hoists, conveyors, and other machinery. They are enjoying increasing popularity at water and wastewater facilities, where the greatest energy draw comes from pumping and aeration—two applications particularly suited to variable-frequency drives.

---

* Adapted from USEPA, *Water & Energy Efficiency in Water and Wastewater Facilities*, U.S. Environmental Protection Agency, Washington, DC, 2012 (http://www.epa.gov/region9/waterinfrastructure/howto.html); CEC, *Variable-Frequency Drive*, California Energy Commission, Sacramento, 2000 (http://www.energy.ca.gov/process/pubs/vfds.pdf).

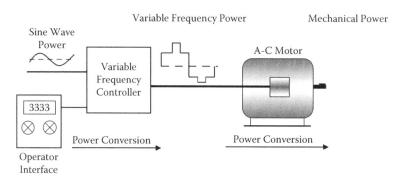

FIGURE 5.3
Variable-frequency drive (VFD) system.

For applications where flow requirements vary, mechanical devices such as flow-restricting valves or moveable air vanes are often used to control flow, which is akin to driving a car at full throttle while using the brake to control speed. This process uses excessive energy and may create punishing conditions for the mechanical equipment involved. Variable-frequency drives enable pumps to accommodate fluctuating demand by running pumps at lower speeds and drawing less energy while still meeting pumping needs.

As shown in Figure 5.3, a variable-frequency drive system generally consists of an AC motor, a controller, and an operator interface. Variable-frequency drives work with most three-phase electric motors, so existing pumps and blowers that use throttling devices can be retrofit with these controls. VFDs can also be specified for new equipment.

A huge advantage of variable-frequency drives is that they give a single-speed drive motor a "soft-start" capability, gradually increasing the motor to operating speed. In contrast, single-speed drives start motors abruptly, subjecting the motor to high torque and current surges up to 10 times the full-load current. If normal operating load current is 10 amps, for example, that motor could draw 100 amps at start. The VFD soft-start capability lessens mechanical and electrical stress on the motor system, can reduce maintenance and repair costs, and can extend motor life.

In addition to their soft-start capability, variable-frequency drives allow more precise control of processes such as water distribution, aeration, and chemical feed. Pressure in water distribution systems can be maintained to closer tolerances. Wastewater treatment plants can consistently maintain desired dissolved oxygen concentrations over a wide range of flow and biological loading conditions by using automated controls to link dissolved oxygen sensors to variable-frequency drives on the aeration blowers.

Energy savings from variable-frequency drives can be significant. Affinity laws for centrifugal pumps suggest that even a small reduction in motor speed will highly leverage energy savings. Variable-frequency drives can reduce a pump's energy use by as much as 50%. A VFD controlling a pump motor that usually runs less than full speed can substantially reduce energy consumption compared to a motor running at constant speed for the same period. For a 25-hp motor running 23 hours per day (2 hours at 100% speed, 8 hours at 75%, 8 hours at 67%, and 5 hours at 50%), a variable-frequency drive can reduce energy use by 45%. At $0.10 per kilowatt hour, this saves $5374 annually. Because this benefit varies depending on system variables such as pump size, load profile, amount of static head, and friction, it is important to calculate benefits for each application before specifying a variable-frequency drive.

DID YOU KNOW?

Any machine that imparts velocity and converts a velocity to pressure can be categorized by a set of relationships that apply to any dynamic conditions. These relationships are referred to as the *affinity laws*. They can be described as similarity processes, which follow these rules:

1. Capacity varies with rotating speed (i.e., peripheral velocity of the impeller).
2. Head varies as the square of the rotating speed.
3. Brake horsepower (BHP) varies as the cube of the rotating speed.

Experience has demonstrated that variable-frequency drives are reliable and easy to operate, increase the degree of flow control, and reduce pump noise. On the other hand, because of the nature of this technology, variable-frequency drives can produce harmonic distortion (harmonic distortion is a measurement and exists only with pure sine waves)—adversely affecting power quality and, subsequently, other electrical machinery. However, manufacturers have developed many solutions to correct this problem; for example, installing an isolation transformer in conjunction with the variable-frequency drive can reduce distortion to an inconsequential level.

With regard to cost, initially variable-frequency drives are relatively expensive. Installed drives range from about $3000 for a 5-hp motor to almost $45,000 for a custom-engineered 300-hp motor and more for larger versions. Variable-frequency drive installation can take from 10 to over 70 labor-hours, depending on system size and complexity. The payback period for these drives can range from just a few months to less than 3 years for 25- to 250-hp models. Because each VFD can drive more than one motor, some costs can be consolidated. In addition, savings from reduced maintenance and longer equipment life contribute significantly to achieving a rapid payback and long-term savings. Many electric utilities offer financial incentives that can reduce the installed costs of variable-frequency drives (USEPA, 2005).

5.4 HVAC Enhancements

Drinking water and wastewater treatment facilities can upgrade their heating, cooling, and ventilation systems to improve energy efficiency and save money while keeping the work environment comfortable (CEC, 2000; USEPA, 2005). Many municipal drinking water and wastewater treatment operations can reduce building energy requirements by operating heating, ventilation, and air conditioning (HVAC) equipment more effectively or by replacing older units with new, high-efficiency systems. Such enhancements can provide immediate energy savings while improving indoor air quality and general workplace comfort. These heating, ventilation, and air conditioning strategies, to be described in greater detail, can also extend equipment life and reduce maintenance costs.

In drinking water and wastewater treatment operations, experience has shown that the latest heating, ventilation, and air conditioning systems can dramatically reduce energy used compared to typical 10- to 20-year-old systems. New air conditioners have high efficiencies—as high as 11.5 energy efficiency ratio—and can reduce cooling energy use about 30 to 40%. Air-source heat pumps are also very efficient (10.5 energy efficiency ratio) and can reduce heating energy use by about 20 to 35%. Water-source heat pumps also have superior ratings (15.2 energy efficiency ratio) and can use heat from treated effluent to supply space heating. These systems can be significantly more efficient than air-to-air heat pumps, especially when outside air temperatures drop below 20°F, when air-to-air units are more than 30% less efficient than at 50°F.

Facilities using large, older chillers can optimize energy use by replacing those units with several small package models or by using an oversized cooling tower. Converting to small electric chillers improves efficiency and enables users to sequence units to meet load demand, reducing energy use as much as 0.4 kilowatts/ton, or about 40%. Oversizing the cooling tower can improve chiller efficiency by about 10%.

A cogeneration system, which can burn facility byproduct gases to generate electricity and heat for process use, can also provide space heating in office buildings or plant workspaces, in some cases completely eliminating the traditional heating system. Absorption chillers, which use heat as an energy source, can draw that heat from an onsite cogeneration system.

Heating, ventilation, and air conditioning system operations can be improved through the use of various controls and controllers such as timers and electronic time clocks. These devices can stop equipment operation when the building is unoccupied to reduce energy use during periods of low occupancy. Alternatively, electronic thermostats can automatically reduce temperature set points during unoccupied periods, effectively turning equipment off. A computerized energy management system can manage energy use throughout a building on the basis of weather conditions, building use patterns, and a host of other variables, potentially reducing building energy use by 10 to 20%.

In regard to ventilation, buildings require a certain amount of outside air to remove odor and contaminants, yet excessive flows increase heating and cooling costs most of the time. Installing an outside air economizer that automatically controls air flow will minimize energy consumption and improve indoor air quality. For systems with manually controlled dampers, setting air flow levels to match ventilation needs will minimize heating and cooling energy needs. In laboratories, careful control of exhaust hoods can help prevent air loss. In addition, variable-frequency drives can modulate laboratory exhaust fans to minimize energy use.

For older space conditioning systems, replacing the pilot light with an electronic intermittent ignition device will eliminate unnecessary energy use. To prevent energy losses caused by dirt, maintenance routines should include regular cleaning of the condenser, evaporator coils, and intake louvers. Regular cleaning of air filters alone can lower energy use as much as 20% and extend equipment life. Outside air economizers should be cleaned regularly and checked to ensure that they are functioning properly.

5.5 Energy-Smart Lighting

Recent advances in lamps, luminaries, controls, and lighting design provide numerous advantages over the lighting systems originally installed in many facilities (CEC, 2000). Offering high-efficiency alternatives for nearly every plant or office building, these technologies can enhance light levels, improve comfort and safety in the workplace, and reduce routine maintenance costs.

Before discussing reduction of operating costs at water and wastewater treatment facilities by installing high-efficiency lights and fixtures and by changing how lights are used, first it is important to point out that the tragic events of 9/11 caused many facilities to increase their operating costs for upgrades to security lighting. After 9/11, managers of critical infrastructures, such as drinking water and wastewater treatment plants (and other potential terrorist targets), realized that certain security upgrades were needed. It can be said that security begins and ends (or certainly depends on) lighting up and making clearly visible all potential target areas; that is, lighting is integral to crime and terrorism prevention through environmental design. Simply, use of security lighting is seen by some as a preventive and corrective measure against intrusions, terrorism, or other criminal activity on a physical plant site.

Lighting up potential target areas to make them clearly visible also aids in providing visual surveillance monitoring. Visual surveillance is used to detect threats through continuous observation of important or vulnerable areas of an asset. The observations can also be recorded for later review or use (for example, in court proceedings). Visual surveillance systems can be used to monitor various parts of collection, distribution, or treatment systems, including the perimeter of a facility, outlying pumping stations, or entry or access points into specific buildings. These systems are also useful in recording individuals who enter or leave a facility, thereby helping to identify unauthorized access. Images can be transmitted live to a monitoring station, where they can be monitored in real time, or they can be recorded and reviewed later. Many facilities have found that a combination of electronic surveillance and security guards provides an effective means of facility security.

Visual surveillance is provided through a closed-circuit television (CCTV) system, in which the capture, transmission, and reception of an image are localized within a closed "circuit." This is different than other broadcast images, such as over-the-air television, which is broadcast over the air to any receiver within range. At a minimum, a CCTV system consists of

- One or more cameras
- A monitor for viewing the images
- A system for transmitting the images from the camera to the monitor

Specific attributes and features of camera systems, lenses, and lighting systems are presented in Table 5.1.

Notwithstanding the need to ensure facility security by increasing funding to upgrade site lighting, visual surveillance (aided by enhanced lighting systems), motion detectors, and other security measures, drinking water and wastewater treatment facilities can still reduce overall operating costs by installing high-efficiency lights and fixtures and by changing how lights are used:

- *Advanced fluorescent lighting*—In most interior spaces, facilities can replace or upgrade existing fixtures to include high-efficiency fluorescent lamps, electronic ballasts, custom-designed reflectors, and appropriate lenses or louvers. New systems are particularly beneficial in computer-intensive environments where older lighting systems can cause glare on video screens.

- *High-intensity discharge lighting*—In outdoor applications and in warehouses and indoor areas with ceilings over 15 feet, facilities can replace highly inefficient incandescent or mercury vapor lamps with metal halide and high-pressure

TABLE 5.1

Attributes of Camera, Lenses, and Lighting Systems

Attribute	Discussion
Camera System Attributes	
Camera type	Major factors in choosing the correct camera are the resolution of the image required and lighting of the area to be viewed:
	• *Solid state* (including charge-coupled devices, charge priming devices, charge injection devices, and metal oxide substrate)—These cameras are becoming predominant in the marketplace because of their high resolution and their elimination of problems inherent in tube cameras.
	• *Thermal*—These cameras are designed for night vision. They require no light and use differences in temperature between objects in the field of view to produce a video image. Resolution is low compared to other cameras, and the technology is currently expensive relative to other technologies.
	• *Tube*—These cameras can provide high resolution but the tubes burn out and must be replaced after 1 to 2 years. In addition, tube performance can degrade over time. Finally, tube cameras are prone to burn images in the tube replacement.
Resolution (ability to see fine details)	User must determine the amount of resolution required depending on the level of detail required for threat determination. A high-definition focus with a wide field of vision will give an optimal viewing area.
Field of vision width	Cameras are designed to cover a defined field of vision, which is usually defined in degrees. The wider the field of vision, the more area a camera will be able to monitor.
Type of image produced (color, black and white, thermal)	Color images may allow the identification of distinctive markings, whereas black and white images may provide sharper contrast. Thermal imaging allows the identification of heat sources (such as human beings or other living creatures) from low-light environments; however, thermal images are not effective in identifying specific individuals (i.e., for subsequent legal processes).
Pan/tilt/zoom (PTZ)	Panning (moving the camera in a horizontal plane), tilting (moving the camera in a vertical plane), and zooming (moving the lens to focus on objects that are at different distances from the camera) allow the camera to follow a moving object. Different systems allow these functions to be controlled manually or automatically. Factors to be considered in PTZ cameras are the degree of coverage for pan and tilt functions and the power of the zoom lens.
Lens Attributes	
Format	Lens format determines the maximum image size to be transmitted.
Focal length	This is the distance from the lens to the center of the focus. The greater the focal length, the higher the magnification but the narrower the field of vision.
f number	The f number is the ability to gather light. Smaller f numbers may be required for outdoor applications where light cannot be controlled as easily.
Distance and width approximation	The distance and width approximations are used to determine the geometry of the space that can be monitored at the best resolution.
Lighting System Attributes	
Intensity	Light intensity must be great enough for the camera type to produce sharp images. Light can be generated from natural or artificial sources. Artificial sources can be controlled to produce the amount and distribution of light required for a given camera and lens.
Evenness	Light must be distributed evenly over the field of view so that there are no darker or shadowy areas. If there are lighter vs. darker areas, brighter areas may appear washed out (i.e., details cannot be distinguished) while no specific objects can be viewed from darker areas.
Location	Light sources must be located above the camera so light does not shine directly into the camera.

Source:　USEPA, *Water and Wastewater Security Product Guide*, U.S. Environmental Protection Agency, Washington, DC, 2005.

sodium lamps. High-intensity discharge lamps generate high lighting output, use a fraction of the energy required for incandescent or mercury vapor equivalents, and have substantially longer lamp life than incandescents.

- *Lighting controls*—Simple controls can eliminate unnecessary lighting in the many facility areas that do not require continuous light. Occupancy sensors detect the presence of personnel within an area and turn lights on and off accordingly. Time switches that turn lighting systems on and off are useful for outdoor signs, security lighting, and corridors. Dimming systems take advantage of daylight to further reduce energy use and costs. Photocell controls provide easy, effective on/off switching of outdoor lighting. In addition, photocells can be combined with time switch controls for areas that do not require lighting all night.

- *Maintenance*—A program of regular cleaning, replacement, and maintenance of lamps and luminaries can significantly save energy. A typical lamp, as it reaches 80% of its useful life, produces 15 to 35% less light due to lamp degradation. Dust, dirt, and other materials on lamps, reflectors, and lenses can decrease lighting output by 30% or more. Photocells used to activate outdoor lights should also be cleaned regularly.

- *General operations and maintenance*—For older space conditioning systems, replacing the pilot light with an electronic intermittent ignition device will eliminate unnecessary energy use. To prevent energy losses caused by dirt, maintenance routines should include regular cleaning of the condenser, evaporator coils, and intake louvers. Regular cleaning of air filters alone can lower energy use as much as 20% and extend equipment life. Outside air economizers should be cleaned regularly and checked to ensure that they are functioning properly.

Lighting accounts for 35 to 45% of the energy use of an office building. Retrofitting existing lighting systems with high-efficiency alternatives is a strategic approach to helping improve a facility's profitability. In addition, new lighting technologies that improve light levels, eliminate flicker, or reduce glare can potentially improve worker productivity by decreasing eye strain and fatigue. Systems that enhance the reliability of lighting in industrial facilities can improve worker safety. Some potential benefits for lighting changes follow:

- *Fluorescent systems*—Replacing or upgrading individual fluorescent lighting systems offers high potential for energy savings. The most cost-effective retrofit application is replacing T-12 lamps and older standard magnetic ballasts with T-8 lamps and electronic ballasts. Although the higher efficiency lamps cost slightly more, they provide higher quality light while using 34% less energy.

- *High-intensity discharge lamps*—Compact metal halide or high-pressure sodium lamps are three to five times more efficient than incandescents and produce three times the illumination. Because of longer bulb life, maintenance and replacement costs are lower. Metal halide lamps can approximate incandescent or fluorescent lamps in color quality and useful life (2000 to 20,000 hours). High-pressure sodium lamps produce a golden-white lighting color that is preferable in warehouses and outdoor applications where color rendering is not critical; these have a long useful life (16,000 to 24,000 hours).

- *Controls*—Occupancy sensors can reduce lighting use by 25 to 50% compared to manual switching. Dimmable electronic ballasts, although fairly new, have proven quite successful.

- *Maintenance*—Building owners planning to follow a regular lighting maintenance program can design lighting systems using fewer fixtures. Implementing group lamp replacement and annual cleaning will reduce the amount of lighting needed to achieve minimum light levels, resulting in lower first costs and energy savings of about 15%.

When approaching lighting enhancements, consider lighting as an interrelated system, rather than as individual components, to yield more satisfying and cost-effective results. Retrofits will save enough electricity to provide payback in 2 to 3 years, less if financial incentives are available. Many electric utilities offer rebates and energy services to help facilities identify and implement methods for reducing lighting system energy use (USEPA, 2005).

References and Recommended Reading

Anon. (1995). Commercial cooling update: chiller selection, SU-103433-RI, *EPRI Journal*, 7(Rev. I).

Campbell, S.J. (1987). *Solid-State AC Motor Controls*, Marcel Dekker, New York.

CEC. (2000). *Heating, Ventilation, and Air Conditioning Enhancements*, California Energy Commission, Sacramento (http://www.energy.ca.gov/process/pubs/hvac.pdf).

EIA. (2012). *What Is Energy? Explained*, Energy Information Administration, Washington, DC (http://www.eia.gov/energyexplained/print.cfm?page=about_laws_of_energy).

Jaeschke, R.L. (1978). *Controlling Power Transmission Systems*, Penton/IPC, Cleveland OH.

Nailen R.L. (1994). Finding true power output isn't easy, *Electrical Apparatus*, 47(2), 31–36.

Siskind, C.S. (1963). *Electrical Control Systems in Industry*, McGraw-Hill, New York.

Spear, M. (2005). Drive up energy efficiency, *ChemicalProcessing.com*, 2005, http://www.chemicalprocessing.com/articles/2005/489/.

Spellman, F.R. (2007). *Handbook of Water and Wastewater Treatment Plant Operations*, 2nd ed., CRC Press, Boca Raton, FL.

USDOE. (2001). *Determining Electric Motor Load and Efficiency*, Fact Sheet, U.S. Department of Energy, Washington, DC.

USEPA. (2005). *Water and Wastewater Security Product Guide*, U.S. Environmental Protection Agency, Washington, DC (http://cfpub.epa.gov/safewater/watersecurity/guide/productguide.cfm?page=visualsurveillance).

WSEO. (1993). *Improving the Energy Efficiency of Wastewater Treatment Facilities*, WSEO-192, Washington State Energy Office, Olympia.

6

Energy-Efficient Operating Strategies*

For it matters not how small the beginning may seem to be: what is once well done is well done forever.

—Henry David Thoreau

Man is born to die. His works are short lived. Buildings crumble, monuments decay, wealth vanishes, but Katahdin in all its glory Forever shall remain the Mountain of the People of Maine.

— **Former Governor Percival Proctor Baxter, who donated land for Baxter State Park in Maine**

6.1 Introduction

Electrical energy consumption at water and wastewater treatment plants is increasing because of more stringent regulations and customer concerns about water quality (CEC, 2001). As a result, more facility managers are turning to energy management to reduce operating costs; however, reducing energy consumption by managing the facility's electrical load is only part of the equation. Operating strategies to reduce energy usage also include biosolids management, operation and maintenance practices, and inflow and infiltration control; these operating strategies are discussed in this chapter.

6.2 Electrical Load Management

By choosing when and where to use electricity, drinking water and wastewater facilities can often save as much (or more) money as they could by reducing energy consumption. Note that electricity is typically billed in two ways: by the quantity of *energy* used over a period of time (measured in kilowatt-hours) and by *demand*, the rate of flow of energy (measured in kilowatts).

6.2.1 Rate Schedules

Electric utilities often structure rates to encourage customers to minimize demand during peak periods, because it is costly to provide generating capacity for use during periods of peak demand. That is why drinking water and wastewater treatment plants should

---

* Material in this chapter is adapted from USEPA, *Water & Energy Efficiency in Water and Wastewater Facilities*, U.S. Environmental Protection Agency, Washington, DC, 2012.

investigate the variety of rate schedules offered by electric utilities. They may achieve substantial savings simply by selecting a rate schedule that better fits their pattern of electricity use.

- *Time-of-use rates*—Time-of-use rates, which favor off-peak electrical use, are available in most areas of the country. Under the time-of-use rates, energy and demand charges vary during different block periods of the day. Energy charges in the summer may be only 5¢ per kilowatt-hour with no demand charge between 9:30 p.m. and 8:30 a.m., but they may increase to 9¢ per kilowatt-hour for a demand charge of $10 per kilowatt between noon and 6:00 p.m. The monthly demand charge is often based on the highest 15-minute average demand for the month.

- *Interruptible rates*—Interruptible rates offer users discounts in exchange for a user commitment to reduce demand on request. On the rare occasions when a plant receives such a request, it can run standby power generators.

- *Power factor charges*—As mentioned earlier, power factor, also known as *reactive power* or kVAR, reflects the extent that current and voltage cycle in phase. Low power factor, such as that caused by a partly loaded motor, results in excessive current flow. Many electric utilities charge extra for low power factor because of the cost of providing the extra current.

- *Future pricing options*—As the electrical industry is deregulated, many new pricing options will be offered. *Real-time pricing*, where pricing varies continuously based on regional demand, and *block power*, or electricity priced in low-cost, constant-load increments, are only two of the many rate structures that may be available. Facilities that know how and when they use energy and have identified flexible electric loads can select a rate structure that offers the highest economy, while meeting their energy needs.

6.2.2 Energy Demand Management

Energy demand management, also known as *demand side management* (DSM), is the modification of consumer (utility) demand for energy through various methods. DSM programs consist of the planning, implementing, and monitoring activities of electric utilities that are designed to encourage consumers to modify their level and pattern of electricity usage.

6.2.2.1 Energy Management Strategies

- *Conduct an energy survey.* The first step to an effective energy management program for a facility is to learn how and when each piece of equipment uses energy. Calculate the demand and monthly energy consumption for the largest motors in

DID YOU KNOW?

Electricity prices generally reflect the costs to build, finance, maintain, manage, and operate power plants and the electricity grid (the complex system of power transmission and distribution lines) and to operate and administer the utilities that supply electricity to consumers. Some utilities are for-profit, and their prices include a return for the owners and shareholders.

the plant. Staff may be surprised at the results; for example, a 100-hp motor may cost over $4500 per month if run continuously. The rate at which energy is used will vary throughout the day, depending upon factors such as demand from the distribution system and reservoir and well levels for water systems or influent flows and biological oxygen demand loading for wastewater systems. Plot daily electrical load as a function of time for different plant loading conditions and note which large equipment can be operated off-peak. Examine all available rate schedules to determine which can provide the lowest cost in conjunction with appropriate operational changes.

- *Reduce peak demand.* Look for opportunities to improve the efficiency of equipment that must run during the peak period, such as improving pump efficiency or upgrading the aeration system of a wastewater plant. During on-peak periods, avoid using large equipment simultaneously. Two 25-kilowatt pumps that run only 2 hours each day can contribute 50-kilowatts to demand if run at the same time.

- *Shift load to off-peak.* Many large loads can be scheduled for off-peak operation; for example, plants can use system storage to ride out periods of highest load rather than operating pumps. Avoid running large intermittent pumps when operating the main pumps.

- *Improve power factor.* Low power factor is frequently caused by motors that run less than fully loaded. This also wastes energy because motor efficiency drops off below full load. Examine motor systems to determine if the motor should be resized or if a smaller motor can be added to handle lower loads. Power factor can also be corrected by installing a capacitor in parallel with the offending equipment.

6.2.3 Electrical Load Management Success Stories

6.2.3.1 Encina Wastewater Authority*

Service area: 125 square miles
Wastewater system capacity: 36 million gallons per day
Wastewater treatment type: Secondary
Secondary treatment method: Activated sludge
Annual systemwide purchased electricity: $174,300 (2.2 million kilowatt-hours)
Cogeneration capacity: 1.4 megawatts
Annual savings attributed to energy-efficient strategies: $611,000

How does a wastewater agency continue to operate economically and maintain high quality while serving its rapidly growing customer base? Staff at the Encina Wastewater Authority decided energy efficiency was the answer. They set into motion a comprehensive energy management program addressing every aspect of the facility's energy use, from demand control to lighting retrofits. The plant now profits from increased energy efficiency, operational savings, and a staff more attuned to methods of achieving these benefits.

---

* Adapted from CEC, *Success Story: Encina Wastewater Authority,* California Energy Commission, Sacramento, 2000 (www.energy.ca.gov/process/pubs/encina.pdf).

DID YOU KNOW?

Aeration in wastewater treatment operations involves mixing air and a liquid by one of the following methods: spraying the liquid in the air, diffusing air into the liquid, or agitating the liquid to promote surface adsorption of air.

6.2.3.1.1 Key Improvements

Staff of the Encina Wastewater Authority integrated the following measures:

- *Use cogeneration to produce onsite electricity and thermal energy.* Encina's system consists of three engine generators that run on purchased natural gas. Heat from the generators maintains a constant 96°F for digesters and is used to heat offices and run three absorption chillers that provide cooling. Although the system can produce 1425 kilowatts, emission restrictions currently allow use of only two generator engines. The facility's Cogeneration Optimization Project will improve the system's efficiency, giving Encina "qualified facility" status from the Federal Energy Regulatory Commission, which will lower costs for natural gas. Upgrades will reduce emissions by converting engines to "lean burn." Encina's cogeneration facility produces 80% of onsite power and provides heat to digesters and HVAC applications that would otherwise operate solely on purchased natural gas and electricity. As it currently operates, the system produces about 8 million kilowatt hours per year. When upgrades are completed, improvements in emissions will permit a third engine to be brought online, increasing generation capacity by 50% while continuing to meet air quality restrictions.

- *Use fine-bubble diffusers for aeration.* Because aeration constitutes as much as 50% of the energy costs of an activated sludge plant (see Figure 6.1), increased efficiency in this area is critical. When expanding their plant, Encina chose fine-bubble diffusers over the less efficient coarse-bubble versions previously used. Additionally, Encina has automated control of dissolved oxygen levels for over 10 years, using probes throughout the aeration basins to monitor and help maintain dissolved oxygen levels. Again, these automated controls maintain dissolved oxygen levels at predetermined set points. Because fine-bubble diffusers transfer more dissolved oxygen into the water than coarse versions, less oxygen needs to be introduced,

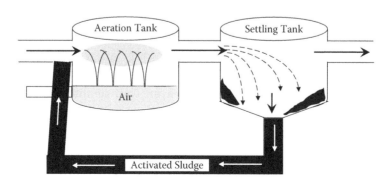

FIGURE 6.1
Activated sludge process.

thus lowering the energy required to drive dissolved oxygen compressors. The fine-bubble diffusers are estimated to save about 2,920,000 kilowatt hours per year. Some facilities are reluctant to automate dissolved oxygen controls, because they are concerned that fouled probes will hamper reliable readings.

- *Enact demand control strategies.* Encina's energy management program emphasizes off-peak pumping, enabling the facility to profit from lower utility rates. Staff manually shut down select high-demand equipment during on-peak periods. Because monthly billings are based on the highest energy demand in a 15-minute block, strict compliance is essential. Many wastewater agencies use control systems, but the Encina staff have demonstrated that expensive automated controls are not a prerequisite for success.

- *Pump water more efficiently with variable-frequency drives.*

- *Upgrade standard motors to energy-efficient motors.*

6.2.3.2 *Moulton Niguel Water District**

Service area: 37.5 square miles
Potable water system capacity: 48 million gallons per day
Wastewater system capacity: 17 million gallons per day
Wastewater treatment type: Tertiary
Secondary treatment method: Activated sludge
Annual systemwide purchased electricity: $1,310,000 (15.5 million kilowatt-hours)
Annual savings attributed to energy-efficient strategies: $332,000

For over a decade, automation and instrumentation have helped Southern California's Moulton Niguel Water District supply water and treat wastewater economically and efficiently. Facing a major rise in energy costs, the agency explored other methods to increase energy efficiency. Working closely with Southern California Edison and San Diego Gas & Electric to identify optimal rate schedules and energy-efficiency strategies, the district implemented a program in 1992 that has yielded substantial savings in the reservoir-fed branches of their distribution system. Additionally, the district plans to investigate potential improvements to their potable water systems requiring full-speed pumping to maintain system pressure.

6.2.3.2.1 *Key Improvements*
Moulton Niguel Water District staff implemented changes in the following areas:

- *Install programmable logic controllers to benefit from lower off-peak utility rates.* Moulton Niguel uses automated controls and programmable logic controllers to enable 77 district pumping stations to benefit from lower off-peak utility rates. The controls activate pumps during off-peak hours, bringing reservoirs to satisfactory levels. On-peak, pumping is halted, allowing reservoir levels to fall. Stations employing this strategy are on "reservoir duty," meaning the system is pressurized by the static head of the reservoir. All stations previously operated in closed grid mode, running pumps 24 hours a day to maintain system pressure. The programmable

---

* Adapted from CEC, *Success Story: Moulton Niguel Water District*, California Energy Commission, Sacramento, 2000 (www.energy.ca.gov/process/pubs/moulton.pdf).

logic controller's sophisticated internal clock and calendar automatically adjust for seasonal changes, consistently keeping equipment running off-peak. This strategy has decreased pumping costs, saving nearly $320,000 annually, and allowing reservoir levels to fall has improved water quality.

- *Regulate lift station wastewater levels using proportional, integral, and derivative controls to automatically transmit data to a central computer.* Moulton Niguel previously cycled constant-speed wastewater pump drives on and off to distribute wastewater. As a result, drive control was limited; pump motors were subject to starting surges, and the system shutdown left sewage sitting in pipes, producing offensive odors. They decided to replace their standard motor drives with variable-frequency drives linked to proportional, integral, and derivative controls. This system regulates wastewater levels by sending a signal to variable-frequency drive controllers to modulate wastewater flow. The new proportional, integral, and derivative/variable-frequency drive system provides a continuous, modulating flow that uses less energy, reduces motor wear and high energy demands from motor starting surges, ensures that sewage does not remain stagnant in pipes (thereby reducing odor problems), and has reduced energy costs by about 4%.
- *Install variable-frequency drives on the wastewater system to control pump speed in coordination with the proportional, integral, and derivative system, reducing costs.*
- *Specify that all motors used in new construction be 95 to 97% efficient, and replace standard-efficiency motors with energy-efficient motors (ongoing).*

6.3 Biosolids Management

Wastewater treatment unit processes remove solids, biochemical oxygen demand (BOD), and nutrients from the wastestream before the liquid effluent is discharged to its receiving waters. What remains to be disposed of is a mixture of solids and wastes called *process residuals*, more commonly referred to as *sludge* or *biosolids*. A commonly accepted name for wastewater solids is *sludge*; however, if wastewater sludge is used for beneficial reuse (e.g., as a soil amendment, fertilizer, or composting ingredient), it is commonly called *biosolids*. Moreover, because the author feels that sludge is an ugly four-letter word that is inappropriate considering its reuse value, biosolids (usually referred to in the plural) is the term of choice used in this text.

The most costly and complex aspect of wastewater treatment can be the collection, processing, and disposal of sludge. This is the case because the quantity of biosolids produced may be as high as 2% of the original volume of wastewater, depending somewhat on the treatment process being used.

DID YOU KNOW?

Biosolids treatment is generally divided into three major categories: *thickening, stabilization,* and *dewatering.* Many of these processes include complex biosolids treatment methods (e.g., heat treatment, vacuum filtration, incineration).

DID YOU KNOW?

Treatment and handling of wastewater solids account for more than half of the total costs in a typical secondary treatment plant.

Because biosolids can have a water content as high as 97% and because the cost of disposal is related to the volume of sludge being processed, one of the primary purposes or goals (along with stabilizing it so it is no longer objectionable or environmentally damaging) of sludge treatment is to separate as much of the water from the solids as possible. Sludge treatment methods may be designed to accomplish both of these purposes.

6.3.1 Biosolids: Background Information

Note that even as wastewater treatment standards have become more stringent because of increasing environmental regulations, so has the volume of wastewater biosolids increased. Note also that before biosolids can be disposed of or reused they require some form of treatment to reduce their volume, to stabilize them, and to inactivate pathogenic organisms. Biosolids form initially as a 3 to 7% suspension of solids. With each person typically generating about 4 gallons of biosolids per week, the total quantity generated each day, week, month, and year is significant. Because of the volume and nature of the material, biosolids management is a major factor in the design and operation of all water pollution control plants.

6.3.2 Sources of Biosolids

Wastewater biosolids are generated in primary, secondary, and chemical treatment processes. In primary treatment, the solids that float or settle are removed. The floatable material makes up a portion of the solid waste known as *scum*. Scum is not normally considered biosolids; however, it should be disposed of in an environmentally sound way. The settleable material that collects on the bottom of the clarifier is known as *primary biosolids*. Primary biosolids can also be referred to as *raw biosolids* because they have not undergone decomposition. Raw primary biosolids from a typical domestic facility are quite objectionable and have a high percentage of water, two characteristics that make handling difficult.

Those solids not removed in the primary clarifier are carried out of the primary unit. These solids are known as *colloidal suspended solids*. The secondary treatment system (trickling filter, activated biosolids, etc.) is designed to change those colloidal solids into settleable solids that can be removed. Once in the settleable form, these solids are removed in the secondary clarifier. The biosolids at the bottom of the secondary clarifier are referred to as *secondary*. Secondary biosolids are light and fluffy and more difficult to process than primary biosolids—in short, secondary biosolids do not dewater well.

The addition of chemicals and various organic and inorganic substances prior to sedimentation and clarification may increase the solids capture and reduce the amount of solids lost in the effluent. This *chemical addition* results in the formation of heavier solids, which trap the colloidal solids or convert dissolved solids to settleable solids. The resultant solids are known as *chemical biosolids*. As chemical usage increases, so does the quantity of biosolids that must be handled and disposed of. Chemical biosolids can be very difficult to process; they do not dewater well and contain lower percentages of solids.

TABLE 6.1

Typical Water Content of Biosolids

Water Treatment Process	% Moisture of Biosolids	lb Water/lb Biosolids Solids Generated
Primary sedimentation	95	19
Trickling filter		
Humus—low rate	93	13.3
Humus—high rate	97	32.3
Activated biosolids	99	99

Source: USEPA, *Operational Manual: Sludge Handling and Conditioning*, EPA-430/9-78-002, U.S. Environmental Protection Agency, Washington, DC, 1978.

6.3.3 Biosolids Characteristics

The composition and characteristics of sewage biosolids vary widely and can change considerably with time. Notwithstanding these facts, the basic components of wastewater biosolids remain the same. The only variations occur in quantity of the various components as the type of biosolids and the process from which they originated change. The main component of all biosolids is *water.* Prior to treatment, most biosolids contain 95 to 99% water (see Table 6.1). This high water content makes biosolids handling and processing extremely costly in terms of both money and time. Biosolids handling may represent up to 40% of the capital cost and 50% of the operation cost of a treatment plant. As a result, the importance of optimum design for handling and disposal of biosolids cannot be overemphasized. The water content of biosolids is present in a number of different forms. Some forms can be removed by several biosolids treatment processes, thus allowing some flexibility in choosing the optimum biosolids treatment and disposal method.

The various forms of water and their approximate percentages for a typical activated biosolids are shown in Table 6.2. The forms of water associated with biosolids are

- *Free water*—Water that is not attached to sludge solids in any way. This can be removed by simple gravitational settling.
- *Floc water*—Water that is trapped within the floc and travels with the floc. Its removal is possible by mechanical dewatering.

TABLE 6.2

Distribution of Water in Activated Biosolids

Water Type	Percent (%) Volume
Free water	75
Floc water	20
Capillary water	2
Particle water	2.5
Solids	0.5
Total	100

Source: USEPA, *Operational Manual: Sludge Handling and Conditioning*, EPA-430/9-78-002, U.S. Environmental Protection Agency, Washington, DC, 1978.

- *Capillary water*—Water that adheres to the individual particles and can be squeezed out of shape and compacted.
- *Particle water*—Water that is chemically bound to the individual particles and cannot be removed without inclination.

From a public health view, the second and probably more important component of biosolids is the *solids matter*. Representing from 1 to 8% of the total mixture, these solids are extremely unstable. Wastewater solids can be classified into two categories based on their origin—organic or inorganic. *Organic solids* in wastewater, simply put, are materials that are or were at one time alive and that will burn or volatilize at 550°C after 15 minutes in a muffle furnace. The percentage of organic material within the biosolids will determine how unstable they are.

The inorganic material within biosolids will determine how stable they are. The *inorganic solids* are those solids that were never alive and will not burn or volatilize at 550°C after 15 minutes in a muffle furnace. Inorganic solids are generally not subject to breakdown by biological action and are considered stable. Certain inorganic solids, however, can create problems relative to the environment—for example, heavy metals such as copper, lead, zinc, mercury, and others. These can be extremely harmful if discharged.

Organic solids may be subject to biological decomposition in either an aerobic or anaerobic environment. Decomposition of organic matter (with its production of objectionable byproducts) and the possibility of toxic organic solids within the biosolids compound the problems of biosolids disposal.

To aid in the disposal of biosolids and the reduction of biosolids handling costs, facilities incorporate sustainable biosolids treatment, transport, and end-use practices. Examples of these practices include biofuels production and landfill application (Spellman, 2007).

6.4 Operations and Maintenance: Energy- and Cost-Saving Procedures[*]

From 2008 to 2010, the U.S. Environmental Protection Agency (USEPA) developed and conducted energy workshops with over 500 water and wastewater utilities in a handful of western states. The following sections identify ten utilities that participated in the program and provide a summary of their results. Many of the projects required no additional resources outside of existing staff time and minor equipment purchases made within existing expense accounts. Some facilities focused on collecting and using renewable energy (specifically, energy generated from the force of water dropping in elevation while traveling through pipes). Other sites concentrated on reducing energy consumption by increasing energy efficiency. Some utilities reduced energy use during the day when energy costs were higher and increased energy use during times of the day when energy costs were lower. At several of the facilities, optimizing operations resulted in significant savings without requiring a large capital outlay. One step each of the ten facilities profiled below developed and implemented was an Energy Improvement Management Plan. More details on the specific projects can be found in the case studies developed by the utilities and described in the following.

---

[*] Adapted from USEPA, *2011 U.S. EPA Region 9 Energy Management Initiative for Public Wastewater and Drinking Water Utilities Facilitating Utilities toward Sustainable Energy Management*, U.S. Environmental Protection Agency, Washington, DC, 2012.

6.4.1 Chandler Municipal Utilities, Arizona

6.4.1.1 Facility Profile

Chandler Municipal Utilities is located in the City of Chandler, Arizona, within Maricopa County. The Municipal Utilities Department oversees wastewater treatment, reclaimed water, and the drinking water supply for the city. The utility selected their potable water system for the Energy Management Initiative. The Chandler potable water system serves 255,000 customers. The system treats an average of 52 million gallons per day (MGD) of groundwater and surface water at two treatment plants. The use of groundwater, from 31 wells, requires more energy than the use of surface water from the Salt River Project and Central Arizona Project. Additional energy is needed to bring groundwater to the surface for treatment and distribution.

6.4.1.2 Baseline Data

Chandler spent $2.9 million on electricity in the previous year treating potable water. The annual electricity used at the water treatment plant is 33,880,000 kWh. In 2010, the plant's energy consumption generated 22,268 metric tons of carbon dioxide equivalent (MTCO$_2$) of greenhouse gas (GHG) emissions.

6.4.1.3 Energy Improvement Management Plan

Chandler Municipal Utilities chose to reduce energy consumption by optimizing the potable water system. They did this in 2 days. First, they revised tank management practices based on hydraulic modeling and master planning to find the best configuration and operating program to reduce groundwater pumping. Second, the utility staff upgraded pumps and revised pressure zones to operate more efficiently under the new operating program. Based on this energy management approach, Chandler's goal was to reduce the number of kilowatt-hours used to produce and distribute 1 million gallons of potable water by 5% from 2010 levels. Chandler chose to develop a strong team as an area of focus for the Energy Management System.

6.4.1.4 Challenges

One of the biggest challenges Chandler faced was changing staff attitudes and long-established habits associated with operating a small groundwater-based system compared to the practicality of operating a large surface-water-dominated system. A series of small victories led to a staff-driven team approach to system optimization. The City of Chandler's potable water production and distribution system had expanded rapidly to meet the growth of the 1990s and early 2000s. Wells and mains were added so the system was able to meet all of its demands. The recent economic slowdown gave the Chandler staff the opportunity to analyze the system as a whole, rather than as a collection of separate parts. Complicating factors included a lack of consistent historic design philosophy and evolution of the system from a small groundwater-based utility to a surface-water-dominated system serving a population of 250,000 and several major industrial and commercial customers. Surface water is the most cost-effective source of water for Chandler, but due to a lack of dedicated transmission infrastructure staff had a difficult time filling tanks with surface water. Facing budgets constraints, staff revisited all aspects of the system. The results included: (1) an expanded second pressure zone, (2) consistent hydraulic grade lines for the pressure zones, (3) focused rehabilitation of key facilities, and (4) a new tank management strategy. During this time the programmable logic controllers the system used were no longer being supported by the manufacturer, so they were replaced by controllers with much better information collection capabilities. The new technology gave the operators better information and control over the system. Given the new tools, the operators developed and tested new operation strategies that have resulted in a more robust system that produces better quality water while using fewer resources.

6.4.1.5 Accomplishments

Chandler fell a bit short of the 5% goal but was able to achieve a reduction of 4.2%. They also produced 4.7% more potable water in 2011 than they did in 2010. Had Chandler not reduced the energy necessary to produce and distribute 1 million gallons they would have used an additional 1,445,000 kWh. Using the average cost per kilowatt-hour that Chandler paid for power in 2011, this amounts to an energy savings of almost $130,000 and avoids the generation of 950 MTCO$_2$ of GHG emissions. One aspect of Chandler's potable water system optimization approach involved using a higher percentage of surface water than in previous years. This resulted in substantial savings in water resource costs and a significant reduction in chlorine use. Chandler adopted a team approach to optimization that resulted in a high level of understanding of system dynamics throughout the organization, an involved staff that continually identifies ways to improve the efficiency and operation of the system, improvements in data acquisition and management, and multiple open channels of communication.

- *Annual energy savings:* 1,445,000 kWh
- *Annual cost savings:* $130,000
- *Annual GHG reductions:* 996 MTCO$_2$, equal to removing 195 passenger vehicles from the road
- *Project cost:* No additional funds required
- *Payback period:* Immediate

6.4.1.6 Future Steps

Staff are continuing to evaluate system performance and seek additional opportunities for optimization. Some lighting has been upgraded; more lighting upgrades are planned in the future. Staff are investigating the feasibility of onsite power generation using solar panels and in-pipe hydraulic power generation.

6.4.2 Airport Water Reclamation Facility, Prescott, Arizona

6.4.2.1 Facility Profile

The Airport Water Reclamation Facility (AWRF) is one of two facilities owned and operated by the City of Prescott, within Yavapai County. Prescott is positioned close to the center of Arizona between Phoenix and Flagstaff, just outside the Prescott National Forest. The original wastewater treatment plant was built in 1978 and received a major facility upgrade in 1999. The next major upgrade began in 2012. The City of Prescott also operates another wastewater treatment plant called Sundog.

6.4.2.2 Baseline Data

AWRF treats 1.1 million gallons of wastewater per day for approximately 18,000 residents. The facility spends $160,000 annually on electricity costs and uses 1.8 million kWh. Greenhouse gas (GHG) emissions for the SWTF are 1023 metric tons of carbon dioxide ($MTCO_2$).

6.4.2.3 Energy Improvement Management Plan

The City of Prescott elected to construct a hydroturbine electric generation unit as part of the Energy Improvement Management Plan at the Airport Water Reclamation Facility to conserve energy and better manage resources. The turbine will be placed at the discharge point of the recharge water pipeline to convert the potential energy in the flowing water to electricity. This hydroturbine has the potential to produce 125,000 kWh per year, which would save the City approximately $12,000 per year.

6.4.2.4 Challenges

The greatest challenge so far has been the time commitment required to accomplish the project concurrent with the design of the facility expansion. Initially, there was also some difficulty connecting with the appropriate staff at the power company; however, that has since been resolved.

6.4.2.5 Accomplishments

Energy and cost savings will result with the new hydroturbine electric generation unit installation at the AWRF. This project is estimated to produce 125,000 kWh and save $12,000 in electrical costs per year. The Energy Management Program has promoted an awareness of energy uses and potential savings associated with minor and major changes to operations. The program also highlighted many other programs and projects that have improved the City's knowledge of energy-saving considerations. The City was successful in developing and adopting an Energy Conservation Policy.

- *Annual projected GHG reductions:* 86 $MTCO_2$, equal to removing 17 passenger vehicles from the road
- *Project cost:* $25,000
- *Payback period:* 25 months

6.4.2.6 Future Steps

Go out to bid with the hydroturbine project; the major upgrade project is ready to bid, as well.

6.4.3 Somerton Municipal Water, Arizona

6.4.3.1 Facility Profile

Somerton Municipal Water is located in the City of Somerton, Arizona, within Yuma County. Yuma County is situated in the southwest corner of Arizona close to the California and Mexico borders. The water treatment plant was built in 1985. In 1998, the plant carried out a major facility upgrade by installing a new 1.2 million gallon storage tank and a new 100-hp booster pump. The drinking water treatment plant is called the Somerton Municipal Water System, and it sources water from wells 3 to 300 feet deep.

6.4.3.2 Baseline Data

The System serves a population of 14,267 residents and treats 2 million gallons of water per day (MGD). Its operations use approximately 907,000 kWh per year, and electricity costs an average of $85,000 per year. The plant generated 512.79 metric tons of carbon dioxide equivalent ($MTCO_2$) of greenhouse gas (GHG) emissions.

6.4.3.3 Energy Improvement Management Plan

Water treatment plant staff started the energy improvement process by first evaluating the energy efficiency of existing wells and pumps. Upon inspection, repairs were made to two wells and three pumps. By fixing the wells and pumps, the System will save $27,000 each year on its electricity bill. Moreover, it will save an average 250,000 kWh annually, which will result in a reduction in CO_2 emissions by an estimated 172 $MTCO_2$.

6.4.3.4 Challenges

It was difficult for staff to find the time to participate in the Energy Management sessions. No additional staff resources were available to complete the project.

6.4.3.5 Accomplishments

Somerton staff gained a better understanding of how projects are selected and increased their effectiveness in working with management. By upgrading wells and booster pumps they were able to save approximately $27,000 per year per pump on energy costs.

- *Annual projected GHG reductions:* 172 $MTCO_2$, equal to removing 34 passenger vehicles from the road
- *Project cost:* $131,203, with $33,500 covered by incentives
- *Payback period:* 4 years

6.4.3.6 Future Steps

Project is complete; work on a solar project that will produce 1.5 million kWh is being initiated.

6.4.4 Hawaii County Department of Water Supply

6.4.4.1 Facility Profile

The Hawaii County Department of Water Supply (DWS) is one of 21 departments within the County. The County encompasses the entire island of Hawaii, and the administrative offices are located in Hilo. The DWS is the public potable water distribution utility; however, the Island of Hawaii also has several other water systems that are not owned and operated by the DWS. Hawaii, the largest island in the Hawaiian chain, is 93 miles long and 76 miles wide with a land area of approximately 4030 square miles. The DWS operates and maintains 67 water sources and almost 2000 miles of water distribution pipeline. Over 90% of the water served is from a groundwater source that requires minimal treatment with chlorine for disinfection.

6.4.4.2 Baseline Data

The DWS serves 41,507 customers and produces 31.1 million gallons per day. Because of the vast area, mountainous terrain of the Island, and the many separate water sources, the energy required to pump water to the surface is significant and plant employees drive significant distances to operate the water distribution system. In 2010, the plant spent $16.5 million on energy costs, used 54,781,373 kWh for electricity, and used 95,100 gallons of gas and diesel. The Hawaii County DWS emits an estimated 31,784.35 metric tons of carbon dioxide equivalent ($MTCO_2$) of greenhouse gas (GHG) emissions.

6.4.4.3 Energy Improvement Management Plan

The main goal of the Hawaii County DWS energy reduction strategy is to reduce gas and diesel consumption. The performance target was to reduce fuel purchases by 200 gallons per month, or 2400 gallons annually. The target was met by establishing more efficient operator routes around Hilo, by using an automated SCADA system, and by installing GPS equipment in 30 vehicles. The new automated SCADA system replaced the manual system. By reducing fuel consumption and increasing efficiency, the DWS will reduce CO_2 emissions by an estimated 17.3 total CO_2 per year. This project was fully implemented December 31, 2011. The DWS also chose to develop an Energy Policy.

6.4.4.4 Challenges

Introducing a new automated SCADA system (to replace the old manual system) required programming time and duplicate systems until the new system was proven. The data being collected were all manual until the new vehicle equipment was installed. The DWS decided to purchase vehicle GPS units, so a new vehicle policy was necessary. The pilot project modified Hilo operator routes. The pilot project covers about one third of the island, and implementing route changes was met with resistance because operators lost overtime. Establishing an Energy Management Team was unsuccessful, so the project was implemented by one person, but there were many moving parts.

6.4.4.5 Accomplishments

The Hawaii County DWS created an Energy Policy.

- *Annual projected energy savings:* 1965 gallons of gasoline
- *Annual overtime savings:* $9780
- *Annual projected cost savings:* $17,670
- *Annual projected GHG reductions:* 18 $MTCO_2$, equal to removing 3.4 passenger vehicles from the road
- *Project cost:* $25,300
- *Payback period:* 1.5 years

6.4.4.6 Future Steps

Complete purchase of vehicle GPS systems.

- Fully implement the project throughout the island, which will include 100 vehicles.
- Begin wind power project to generate renewable energy.

6.4.5 Eastern Municipal Water District, California

6.4.5.1 Facility Profile

The Eastern Municipal Water District (EMWD) has over 250 operating facilities that have been constructed over the last 60 years. EMWD selected a drinking water filtration plant for the Energy Management Initiative. The Perris Water Filtration Plant is located in Perris, California, within Riverside County. Perris is situated southeast of Los Angeles along the Escondido Freeway. The plant treats water pumped from the Colorado River and from the California State Water Project.

6.4.5.2 Baseline Data for Perris Water Filtration Plant

- *Water treatment design capacity:* 24 million gallons per day (MGD)
- *Annual energy consumption:* 5.6 million kWh
- *Annual cost of energy:* $695,000
- *Annual GHG emissions:* 3862 metric tons of carbon dioxide equivalent (MTCO$_2$)
- *Design average flow (FY 2011):* 10.2 MGD

6.4.5.3 Energy Improvement Management Plan

The EMWD chose to reduce their energy consumption and greenhouse gas (GHG) emissions by producing renewable energy onsite. The municipality will install a renewable power generator (in-conduit hydro generation) at the Perris Water Filtration Plant. The renewable power generator will produce up to 290,000 kWh of energy each year. The project will take one year to complete once funding is secure and will cost approximately $350,000.

6.4.5.4 Challenges

Challenges associated with this project have included identifying hydro-generation technology capable of meeting the low head pressure and varying flow conditions existing at the Perris Water Filtration Plant. These technical challenges were eventually overcome. Also, a grant application for funding from the U.S. Bureau of Reclamation was not funded, but EMWD obtained feedback on the proposal and will fund the project without a grant.

6.4.5.5 Accomplishments

Raw water supply to EMWD's Perris Valley Water Filtration plant is controlled through on existing valve that requires frequent, and costly, replacement. The benefits of this project are that it will eliminate the need for this replacement and provide the necessary flow control capabilities combined with energy generation. EMWD completed a feasibility study that demonstrated the viability of the proposed project. The project, capable of producing nearly 300,000 kWh of electricity, has now completed the design phase. Included in the design are revised cost estimates and inclusion of hydroturbine technology that meets the unique challenges of this application (low head, with highly variable flow). EMWD has increased awareness of systematic processes for analyzing overall energy management efforts and developed structure and a strategic approach to energy management as a whole.

- *Annual projected energy savings:* 290,000 kWh
- *Annual projected cost savings:* $36,000
- *Annual projected GHG reductions:* 200 MTCO$_2$, equal to removing 39 passenger vehicles from the road
- *Project cost:* $350,000
- *Payback period:* 5 years (factoring in the need to replace an existing, non-energy-generating valve; 10 years if valve did not need to be replaced).

6.4.5.6 Future Steps

- Board approval of funding and going out to bid

6.4.6 Port Drive Water Treatment Plant, Lake Havasu, Arizona

6.4.6.1 Facility Profile

The Port Drive Water Treatment Plant (PDWTP) is located in Mohave County, Arizona, and serves the majority of the population of Lake Havasu City, which is located along the Colorado River on the eastern shores of Lake Havasu. The water treatment plant was built between 2002 and 2004 and has never received a major facility upgrade. Currently, an estimated 86% of the water treated is sourced from groundwater wells, and a small percentage comes directly from Lake Havasu.

6.4.6.2 Baseline Data

PDWTP uses a natural biological process to remove iron and manganese from 11 million gallons of water per day. The treated drinking water is then distributed to 50,000 customers. Annually, the plant uses 6,636,960 kWh of energy at a cost of $612,749 to the City each year. This equates to 4577 metric tons of carbon dioxide equivalent (MTCO$_2$) of greenhouse gas (GHG) emissions.

6.4.6.3 Energy Improvement Management Plan

Lake Havasu City Water Treatment Plant staff, encouraged by the USEPA, completed a test "change of operations" of the North and South High Service Pump Stations.

6.4.6.4 Challenges

Overall, since the 2009 recession, one major challenge to any non-core activity involved with the effort has been a lack of personnel resources. Despite these challenges, the City was able to implement small incremental changes. A major benefit in the City was the realization that those changes can and will be valuable in the future. This was seen in the demonstration project described in more detail in the Accomplishments section. The lack of staff hours prevented the City from moving forward with an Energy Improvement Plan. This also delayed the implementation of a pump study until November 2011; the study was expected to produce some energy savings in 2012.

6.4.6.5 Accomplishments

The City changed how much and when water would be released at each lead pump. The lead pump for each system was changed to operate at full water flow until it reached the turn-off level set point. Prior to this, the variable-frequency drive (VFD) controller was programmed to slow the lead pump at a nearly full tank level to provide continuous flow through the water treatment plant process. This change resulted in an average 8.7% energy reduction in pump stations for the months of March, April, and May. For the months of June, July, and August, the average energy reduction was 6% for the North pumps and 6.7% for the South pumps, compared to the same months in 2010. There was no change in the water quality or ability to supply water on demand with these changes. All light fixtures were replaced. In the last 10 to 12 years, increases in electric costs had not been passed on to users. In the previous year, there was a 24% rate hike but Lake Havasu reduced energy use by 30%. In addition to the test project, the City qualified for an energy audit at the North Regional Wastewater Treatment Plant by a consultant funded through the USEPA. A draft report was submitted to the City in September 2011. The report identified seven projects that may be scheduled in the future. Many of the suggestions in the report may be relevant to the City's other treatment plants. Approximately 4 million gallons a day of wastewater is treated by these facilities.

- *Anticipated energy savings:* Should equate to approximately 130,000 kWh annually
- *Annual projected GHG reductions:* 90 MTCO$_2$, equal to removing 18 passenger vehicles from the road
- *Project cost:* Zero
- *Payback period:* Zero

6.4.6.6 Future Steps

Monitor anticipated savings of 6 to 8.7% over the next 6 months, and implement energy audit recommendations. All parts of the plant are being examined for energy saving opportunities.

6.4.7 Truckee Meadows Water Authority, Reno, Nevada

6.4.7.1 Facility Profile

Truckee Meadows Water Authority (TMWA) chose its drinking water facility, Chalk Bluff Water Treatment Plant, for the Energy Management Initiative. The plant was built in 1994 and serves more than 330,000 customers throughout 110 square miles within Washoe County, Nevada. The Chalk Bluff Plant treats water from the Truckee River, which flows from Lake Tahoe and the Sierra Nevada Mountain range.

6.4.7.2 Baseline Data

TMWA serves 93,000 customer connections. In 2009, the water authority spent just under $7 million on electricity and $88,000 on natural gas. The Chalk Bluff Water Treatment Plant uses 13.5 gigawatt hours (GWh) and 74,452 therms per year, and spends $1.3 million for electricity. The total energy use results in estimated annual greenhouse gas (GHG) emissions of 9309 metric tons of carbon dioxide equivalent (MTCO$_2$).

6.4.7.3 Energy Improvement Management Plan

Although TMWA relies on gravity as much as possible, in a mountainous community pumping water is a reality. The Chalk Bluff Plant is TMWA's largest water producer and the highest energy use facility. The high energy use is due to pumping water uphill from the river into the plant. To reduce energy consumption at Chalk Bluff, the implementation plan consists of two parts: (1) optimizing the time-of-use operating procedures, and (2) implementing water supply capital improvements.

1. The first strategy is to optimize time-of-use operating procedures by creating a mass flow/electric cost model of the treatment and effluent pumping processes. The model will be used to predict how changes to the operating procedure will affect electricity cost. In 2010, TMWA spent $938,000 on 7.8 GWh for non-water-supply processes of the plant. This project was intended to reduce non-water-supply electric costs by 15%, or $141,000.

2. The second project involves water supply improvements to the Highland Canal, which transports 90% of Chalk Bluff's water directly to the plant using gravity. The improvement plan will allow 100% of the water to be brought to the plant using the Highland Canal and meets multiple objectives. Improvements will be made during winter months when customer water demands are lowest in order to reduce water supply pumping costs during construction. Currently, TMWA spends $60,000 on 0.5 GWh for water supply pumping at the Chalk Bluff Plant. Energy use will be zero when the project is complete. The design life of the new infrastructure is over 100 years, and it will require no energy to operate.

6.4.7.4 Challenges

Originally scheduled to begin construction during the fall of 2011, delays in obtaining highway encroachment permits postponed construction. To minimize water supply pumping costs during construction and therefore continue to reduce energy costs, this project has been delayed until the fall of 2012. TMWA attempted to use a mass balance/electric cost model to optimize time-of-use operating procedures; however, the mass balance/electrical cost model is not capable of the sophisticated decision making routinely performed by the experienced water plant operators. For this reason, the purpose of the model has shifted from generating decisions to being one of several techniques useful for improving time-of-use energy optimization at the Chalk Bluff Plant.

6.4.7.5 Accomplishments

1. TMWA began setting and tracking time-of-use electricity goals for the Chalk Bluff Plant in November of 2010. The goals depend on time of day (e.g., 200 kW on-peak, 400 kW mid-peak, and 950 kW off-peak) and vary with season, based on

the electric utility's tariffs. Water plant operators have the ability to be innovative in order to meet electricity use goals and system demands. The mass balance/ electric cost modeling effort was valuable to establish baseline energy usage by (1) formally inventorying energy-intensive unit processes, (2) establishing kilowatt draw of equipment, (3) establishing and ranking historic kilowatt-hour usage of equipment, and (4) suggesting starting point kilowatt targets for further optimization by operators. TMWA considers the time-of-use optimization project a great success due to its ability to save energy costs and will continue to optimize and track the project's results. From November 2010 through October 2011 the time-of-use optimization saved more than $225,000 (24.4%) compared to the same period the previous year. During this time electric energy usage was reduced by only 0.45 GWh (5.8%), indicating that the savings were primarily due to improved time-of-use cost management. Going through the process of identifying the energy needs of each process was eye opening. Talking to the operators and getting them to work toward the time-of-use goals proved to be educational and engaging and strengthened the team.

- *Annual projected GHG reductions:* 310 MTCO$_2$, equal to removing 61 passenger vehicles from the road
- *Project cost:* Zero
- *Payback period:* Zero

2. Design is substantially complete for the water supply improvement project, and highway encroachment permits are expected in time for the project to proceed in the fall of 2012.

- *Annual projected GHG reductions:* 345 MTCO$_2$, equal to removing 68 passenger vehicles from the road
- *Project cost:* $3,000,000
- *Payback period:* 50 years

6.4.7.6 Future Steps

(1) Complete and continue to optimize and track project results, and (2) get permits and begin construction.

6.4.8 Tucson Water, Arizona

6.4.8.1 Facility Profile

Tucson Water provides clean drinking water to a 330-square-mile service area in Tucson, Arizona. Located in Pima country, Tucson is positioned along Highway 10 about 70 miles north of the U.S./Mexico border. The potable water system serves roughly 85% of the Tucson metropolitan area, serving 228,000 customers. The potable system includes 212 production wells, 65 water storage facilities, and over 100 distribution pumps. A separate reclaimed water system serves parks, golf courses, and other turf irrigation. Tucson Water sources drinking water through groundwater and the Colorado River by recharging and recovering river water delivered through the Central Arizona Project. Tucson Water uses recycled water for its reclaimed water system.

6.4.8.2 Baseline Data

Tucson Water spends on average $12.5 million on the energy required to operate its potable water system and produces approximately 110 million gallons per day of potable water. Annually, Tucson Water uses approximately 115,000,000 kWh and 5,000,000 therms of energy to run the system. This energy use results in an estimated 83,367.43 metric tons of carbon dioxide equivalent ($MTCO_2$) of greenhouse gas (GHG) emissions.

6.4.8.3 Energy Improvement Management Plan

Tucson Water is partnering with Tucson Electric Power (TEP), a privately held regulated electric utility, to reduce peak demand. TEP contracted EnerNOC (a company that develops and provides energy management applications and services for commercial, institutional, and industrial customers, as well as electric power grid operators and utilities) to facilitate a new demand management program to reduce the energy load during peak hours. EnerNOC will pay customers to shed the load. Tucson Water will participate by identifying sites that are appropriate for load shedding. It has been estimated that the program will save Tucson Water 24,000 kWh during peak energy use and create $9000 in offsetting revenue to be put toward energy costs. As part of the City of Tucson's award under the Department of Energy's Energy Efficiency and Conservation Block Grant (funded by the American Recovery and Reinvestment Act, ARRA), Tucson Water is implementing a Water System Distribution Pump Efficiency Project. The project is designed to establish baseline data and data management tools for system booster pumps, provide energy savings recommendations for the distribution system, and implement prioritized energy-savings upgrades. In addition, training will be provided and results from the project will provide actionable information on the cost effectiveness of continuing a program without grant funding. Projected energy and cost savings for the project are 350,000 kWh and $30,000 (year one, past project). The project is scheduled to be completed in early fall of 2012.

6.4.8.4 Challenges

It was difficult to complete the project within a year. A 2-year program would have given more time to implement the project in the Energy Management Plan.

6.4.8.5 Accomplishments

In addition to the projected $9000 cost savings of the EnerNOC program, energy data will inform any plans to expand real-time energy monitoring. At the end of the grant-funded booster pump project, the utility will realize energy and cost savings and have the information necessary to scope a more permanent pump efficiency program.

- *Annul projected energy savings:* 24,000 kWh during peak energy use periods
- *Annual projected cost savings:* $9000
- *Annual projected GHG reductions:* None
- *Project cost:* Zero
- *Payback period:* Zero

6.4.8.6 Future Steps

The peak demand project is complete. Tucson will continue implementing an ARRA-funded pump efficiency project that is estimated to save $30,000 and 350,000 kWh per year.

6.4.9 Prescott–Chino Water Production Facility, Prescott, Arizona

6.4.9.1 Facility Profile

The Prescott–Chino Water Production Facility is located within Yavapai County in the town of Chino Valley, Arizona. Prescott is positioned close to the center of Arizona between Phoenix and Flagstaff, just outside the Prescott National Forest. The Facility consists of a production well field, reservoir, and booster pump facility. The Facility was built in 1947 and received its last major facility upgrade in 2004.

6.4.9.2 Baseline Data

The Facility supplies 50,000 residents with drinking water. During the winter the plant treats 4.5 million gallons of water per day and peaks at 12 million gallons during the summer. The cost of electricity for the plant is $1,600,000 for 11,000,000 kWh per year. Annual greenhouse gas (GHG) emissions are 7581 metric tons of carbon dioxide equivalent (MTCO$_2$).

6.4.9.3 Energy Improvement Management Plan

The Facility has wells that are 15 miles north and lower in elevation than the treatment plant. The challenge was to reduce the $2000/month demand charge. The Energy Management Plan for the Facility includes replacing the existing step voltage starts on three wells with soft-start units. The soft-start units will reduce the instantaneous demand on the power supply, which will reduce the demand charge on the utility bill. It will also help to reduce the power surge on the power distribution system. In addition, softer starters help to extend the life of well motors. Overall, the project saves money and electricity through reduced demand charges, energy use, and maintenance. The estimated immediate cost saving associated with the reduced demand is $1260/month, for a total savings of $15,120/year. The savings toward the maintenance and reduced strain on the electrical distribution system will be a long-term progressive savings.

6.4.9.4 Challenges

The main challenge has been the loss of two members of the City's Energy Management Team, which reduced management support and buy-in from staff. This delayed the ultimate implementation and construction of the soft-start project.

6.4.9.5 Accomplishments

As the City neared the end of the program, the project gained acceptance. The City has completed a specification package and will soon advertise for construction. The Energy Management Program has promoted an awareness of the City's energy use and potential savings associated with minor or major changes to operations. It has provided a good networking opportunity to learn, expand concepts, and consider new options. This project will reduce direct electrical costs and long-term maintenance costs and will result in unseen benefits, such as improved safety due to reduced instantaneous electrical demands. The Energy Management Program also highlighted many other programs and projects that improved the City's awareness of energy savings. The City has implemented a new Energy Conservation Policy.

- *Annual projected GHG reductions:* None
- *Project cost:* $42,000
- *Payback period:* 34 months

6.4.9.6 Future Steps

Advertise for construction and begin 2-megawatt solar generation project to offset 30% of water booster costs.

6.4.10 Somerton Municipal Wastewater Treatment Plant, Arizona

6.4.10.1 Facility Profile

The Somerton Municipal Wastewater Treatment Plant is located in the city of Somerton within Yuma County, Arizona. Somerton is positioned along Highway 95, close to the border of California and Mexico. The wastewater treatment plant was built in 1985 and received its last major facility upgrade in 2011, when it was changed from a sequencing batch reactor (SBR) facility with a capacity to treat 0.8 MGD to a modified Ludzack–Ettinger (MLE) process with a capacity of 1.8 MGD.

6.4.10.2 Baseline Data

Somerton Municipal Wastewater Treatment Plant serves a population of 14,296 residents. The wastewater treatment plant treats a daily average of 0.750 MGD of wastewater each day and uses approximately 744,480 kWh per year, generating 32 metric tons of carbon dioxide equivalent ($MTCO_2$) of greenhouse gas (GHG) emissions.

6.4.10.3 Energy Improvement Management Plan

The objective of the Somerton Municipal Wastewater Treatment plant is to save energy by running a more efficient plant. The City achieved this by replacing four old blowers used to run the four-tank aeration system with two new more efficient blowers. (The current system only requires one new blower.) The City also replaced an old diffuser system with one high-efficiency diffuser. In addition, two old blowers were replaced, one for the digester and one spare, with one high-efficiency blower. Overall, the upgrades are expected to save the wastewater treatment plant 10% on its annual electricity bill and reduce its electricity use by 6%, while doubling the capacity of the facility.

6.4.10.4 Challenges

Originally, the Energy Improvement Management Plan called for doubling the number of tanks in the existing sequencing batch reactor system. It was later determined that, by using more efficient diffusers and turbo blowers, the same number of tanks could be maintained by changing the process. This resulted in keeping the same footprint and more than doubling capacity while reducing energy use by 10%.

6.4.10.5 Accomplishments

Plant energy use awareness increased. Even though the expansion is still not complete, the plant succeeded in reducing electricity usage by 12% and costs by 11.4%, while doubling treatment capacity.

- *Annual projected cost savings:* $29,328
- *Annual projected GHG reductions:* 51 MTCO$_2$, equal to removing 10 passenger vehicles from the road
- *Project cost:* $146,640
- *Payback period:* year

6.4.10.6 Future Steps

Begin a 1.5 million-kWh solar project.

6.5 Inflow and Infiltration Control

Inflow and infiltration (I&I) in a facility's collection system are costly for a wastewater treatment plant. Increased flow in the system results in higher operational and capital costs. Energy costs are associated with processing and sorting the additional flow, and sewer lift stations (pumping stations) must pump continuously. The system runs the risk of overflows as it becomes overloaded. Inflow and infiltration problems can result in millions of gallons per day of increased flow into the wastewater treatment plant, necessitating increased discharges of treated effluent to the receiving stream. The USEPA (1999) has provided suggestions for reducing inflow, and many of these suggestions are described in detail in the following.

6.5.1 Combined Sewer Systems[*]

Combined sewer systems (CSSs) are wastewater collections systems designed to carry both sanitary sewage and stormwater runoff in a single pipe to a wastewater treatment plant. During wet weather periods, the hydraulic capacity of the CSS may become overloaded, causing overflows to receiving waters at discharge points within the CSS. These overflows are called *combined sewer overflows*, or CSOs. Inflow reduction refers to a set of control technologies used to reduce the amount of stormwater entering the CSS from surface sources. Inflow reduction can be a cost-effective way to reduce the volume of flow entering the CSS and the volume and/or number of CSOs. It is particularly applicable in CSO communities where open land is available to accommodate redirected flow for infiltration or detention, or where stormwater can be diverted to surface waters. By helping to reduce the overall flow volumes into a CSS, inflow reduction technologies help to optimize the system's storage capabilities. Maximizing storage in the collection system is one of the nine minimum controls that every CSO community is expected to implement in accordance with the USEPA's CSO Control Policy (USEPA, 1994):

1. Proper operation and regular maintenance programs for the sewer system and the CSOs
2. Maximum use of the collection system for storage

---

[*] Adapted from USEPA, *Combined Sewer Overflow Technology Fact Sheet: Inflow Reduction*, EPA 832-F-99-035, U.S. Environmental Protection Agency, Washington, DC, 1999.

3. Review and modification of pretreatment requirements to ensure that CSO impacts are minimized

4. Maximization of flow to the publicly owned treatment works for treatment

5. Prohibition of CSOs during dry weather

6. Control of solid and floatable materials in CSOs

7. Pollution prevention

8. Public notification to ensure that the public receives adequate notification of CSO occurrences and CSO impacts

9. Monitoring to effectively characterize CSO impacts and the efficacy of CSO controls

Technologies used to reduce inflow include the following:

- Roof drain redirection
- Basement sump pump redirection
- Flow restriction and flow slipping
- Stormwater infiltration sumps
- Stream diversion

All of these technologies have relatively low costs, and most require little maintenance in comparison with other CSO controls and are explained in detail below.

6.5.1.1 Roof Drain Redirection

Roof drains often convey rainfall directly from residential and commercial roofs into a CSS. Flow into the CSS can be reduced by redirecting roof drains onto lawns or into dry wells or drainfields where flows can infiltrate into the soil. Roof drain redirection works best in residential areas where homes have open yards. Telephone, mail, or door-to-door surveys are necessary to determine the extent of roof drain connections to the CSS. Because the net volume reduction per household is small, implementation must be broad and consistent throughout the service area. The cumulative reduction effects of such a program can be substantial. Redirection can be either voluntary or mandatory. Cash or other incentives can be instrumental in achieving widespread reduction.

DID YOU KNOW?

The USEPA estimates that the more than 19,000 collection systems in the United States would have a replacement value of $1 trillion dollars. Another source estimates that waste treatment and collection systems represent about 10 to 15% of the total infrastructure value in the United States. The collection systems of a single large municipality can represent an investment worth billions of dollars. Usually, the asset value of the collection system is not fully recognized and the collection system operation and maintenance programs are given low priority compared with wastewater treatment needs and other municipal responsibilities (USEPA, 2012).

DID YOU KNOW?

The feasibility of discharging downspout and basement sump discharges into a yard, a dry well, or an infiltration field depends on the soil type, the slope, and the drainage conditions around the home.

Roof drain redirection is a relatively simple task that can be done easily by individual homeowners; therefore, it is essential to prepare guidance for homeowners who elect to perform the redirection themselves. This guidance material should also include a list of contractors able to provide redirection for a reasonable cost. Homeowners should be reminded to periodically check to make sure that redirected water is infiltrating and is not contributing to other surface water problems. The community should develop a schedule and check that homeowners who have agreed to disconnect have remained disconnected. Roof drain redirection should be combined with basement sump pump redirection where possible.

The advantage of implementing roof drain redirection is that it is a relatively simple and low-cost (per unit) option for reducing stormwater inflow to a CSS. Some programs have offered rebates or incentives of from $40 to $75 to homeowners and businesses participating in voluntary redirection programs to offset typical costs for redirection. On the other hand, in order for a roof drain redirection program to be successful, the public must be educated as to the benefits and methods for implementing such a program. This can be time consuming and will most likely require some sort of rebate program or other incentive for compliance. In addition, because the effect per individual roof drain redirection is small, this program must be implemented in a wide service area to be effective.

In St. Paul, Minnesota, an estimated 20% of CSO volume came from roof drains. As a result of a $40 rebate for voluntary redirection and other innovative outreach efforts, approximately 18,000 homes redirected their roof drains over a 3-year period. Currently, 99% of all residential properties are disconnected. In addition, the city is evaluating creative funding options to assist commercial and industrial property owners to disconnect roof drains.

6.5.1.2 Basement Sump Pump Redirection

Many buildings have sump pumps to pump floodwater from basements. In the past, sump pumps were connected to the CSS. Unfortunately, this additional water often overloads the CSS lines and causes overloads and sewer backups. Redirecting this flow away from the CSS and onto lawns or dry wells or drainfields reduces the volume of stormwater entering the CSS.

Unlike roof drains, which are common to most buildings, basement sump pumps usually exist only in discrete areas. For this reason, surveys (preferably onsite) are necessary to determine where sump pumps are located, whether they discharge directly to a CSS, and whether it is feasible to redirect them into a yard, a dry well, or an infiltration field.

As with roof drain redirection, guidelines for basement sump pump self-redirection should be distributed, and the area should be checked regularly to ensure that redirected water is infiltrating and is not contributing to other surface water problems.

Advantages and disadvantages of a basement sump pump redirection program are similar to those of roof drain redirection programs. A basement sump pump redirection program can be a relatively inexpensive (typical costs for basement sump pump regulations are $300 to $500 per home) and easy way to reduce flow to CSSs. Full rebates can be used to encourage homeowners to participate; however, unlike roof drains, basement sump pumps may occur less frequently and their locations may be more difficult to determine. Also, as with roof drain redirection, a basement sump pump redirection program can be time consuming and will most likely require some sort of rebate program or other incentive for compliance. Similarly to roof drain redirection, because the effect per individual sump pump redirection is small, this program must be implemented in a wide service area to be effective.

South Portland, Maine, conducted a visual survey of the 6000 residential buildings in the city and found 380 homes with roof drains and 300 homes with sump pumps discharging directly into the CSS. The city mailed letters offering to reimburse each property owner for redirecting their roof drains and sump pumps. After the work was completed and inspected by the city, homeowners were reimbursed $75 for each redirected roof gutter and $400 for each redirected sump pump. The program has redirected more than 379 roof drains and 304 basement sump pumps. The resulting reduction is approximately 58 million gallons of water per year. Because of the variables involved (e.g., rainfall patterns, differing drainage areas), the city has not determined a direct correlation between these programs and CSO events; however, overall flow to the city's wastewater treatment plant has been reduced by 2% through these efforts.

6.5.1.3 *Flow Restriction and Flow Slipping*

Flow restriction and flow slipping methods utilize roadways and overland flow routes to temporarily store stormwater on the surface or to convey stormwater away from the CSS. Flow restriction is accomplished by installing static flow or braking devices in catch basins to limit the rate at which surface runoff can enter the CSS. Excess storm flow is retained on the surface and enters the system at a controlled rate, eliminating or reducing the chance that the system will be hydraulically overloaded and overflow. The volume of on-street storage is governed by the capacity of the static flow device, or orifice, used for restriction, as well as surface drainage patterns.

As opposed to flow restriction, where flow rates into the CSS are reduced but all storm flow eventually flows into the storm sewer system, flow slipping refers to the intentional blocking of stormwater from entering the CSS at catch basins for the purpose of routing, or "slipping," it elsewhere. Flow slipping is accomplished by partially or completely blocking the entry of surface runoff at catch basin inlets and letting the runoff follow overland flow routes. Flow restriction and flow slipping can effectively reduce inflow during peak runoff periods and can decrease CSO volume. Use of these methods must be carefully planned to ensure that sufficient surface storage overland throughput capacity exists in the drainage area. These methods are almost always used in conjunction with other practices, such as roof drain and basement sump pump redirection.

Flow restriction works best in relatively flat areas where temporary ponding and detention of water on streets is acceptable. Extensive public education and testing are required to build support and address concerns that residents and elected officials may initially have regarding on-street storage. Flow slipping is an option where opportunities for on-street storage are not available. The slipped flow can be diverted along natural drainage routes to separate receiving waters, separate storm sewer systems, or even to more optimal locations within the CSS. Flow restriction and flow slipping methods can be effective ways to manage flow in specific parts of the CSS. They are relatively inexpensive (flow-restricting orifice devices for catch basins are priced from $500 to $1200 each, depending on the size and number ordered)

DID YOU KNOW?

The use of flow restriction and flow slipping requires a detailed evaluation of the collection system and catch basins. The community must assess the potential for unsafe travel conditions, flood damage, and damage to roadways. Pilot studies and monitoring are recommended to identify impact and confirm performance.

and easy to implement. However, implementation will require a public awareness campaign to inform the public of the purpose of the flow diversions. In addition, these inflow reduction methods require large surface areas to temporarily store flow before it enters the CSS.

Experience has demonstrated that gently graded berms can be added to roads and used in conjunction with flow restriction to maximize on-street temporary storage. In Skokie, Illinois, for example, a fully developed suburb of Chicago served by a CSS that covers 8.6 square miles, an integrated program was developed that emphasized berms and flow restrictors to both increase on-street storage and to reduce the peak rate of flow entering the CSS. In order to accomplish this, 2900 flow-restricting devices were installed at catch basins, and 871 berms were constructed on streets. In addition, 100% of the roof drains previously connected to the CSS were disconnected.

6.5.1.4 Stormwater Infiltration Sumps

Stormwater infiltration sumps are below-ground structures used to collect stormwater runoff and pass it into the soil. The infiltration sumps collect runoff in standard stormwater inlet structures at the ground surface and route it to a two-chambered system consisting of a manhole structure and an attached sump chamber. The manhole chamber serves as a sedimentation basin. As this chamber fills, flow reaches an overflow point and begins to fill the second chamber, the perforated sump. The perforations allow the water to percolate outward into the soil. The sump chambers are typically 8 to 12 meters (25 to 35 feet) deep and are surrounded with granular backfill to promote infiltration.

Infiltration sumps can reduce inflow in areas where the underlying soils are moderately to highly permeable, and the water table is well below the ground surface. They are generally more applicable in residential areas that are less than 50% impervious. It is important to get the surface runoff to streets where sumps have been installed, and implementation in conjunction with roof drain and basement sump pump redirection is recommended. Due the potential for chemical contamination of groundwater, infiltration sumps are not recommended for commercial or industrial areas.

Before starting construction of a stormwater infiltration sump, the community should develop a sump management plant that address the policy, design, construction, maintenance, and public education aspects of the infiltration sump program, in addition to addressing spill response, well monitoring, stormwater runoff quality, and stormwater sampling protocols. Consideration should be given to local traffic disruptions during sump installation. Sumps will have to be cleaned every 2 to 3 years to remove material that collects in the sedimentation basin.

Although more expensive than roof drain or basement sump pump redirection (total costs range for $2 to $8 per 1000 gallons per year), stormwater infiltration sumps can be very effective in areas with highly permeable soil and little chance of groundwater contamination. Infiltration sumps are not recommended for use in areas with high groundwater tables or in soil with low permeability.

DID YOU KNOW?

Inflow reduction can be a very important part of a CSO management program. By reducing the amount of extraneous inflow to a system, CSSs may avoid CSO problems.

Infiltration sumps can be retrofitted within combined sewer areas, usually beneath the street system. In Portland, Oregon, much of the combined sewer area has highly permeable soils with a high hydraulic capacity (i.e., measure of the volume of water that a structure can pass). Portland is currently wrapping up a program that installed approximately 4000 infiltration sumps between 1994 and 1998. Combining this program with other CSO control programs, including a successful roof drain disconnection program, sewer separation, and stream diversion, is predicted to reduce total CSO volume by 3 billion gallons per year (approximately half of the total CSO volume). Portland is initiating another program to maximize storage capacity in the CSS to further reduce the overflow volume.

6.5.1.5 Stream Diversion

As cities grew during the 19th and early 20th centuries, many small streams were routed (engineered or hydromodified) into pipes to facilitate development. In communities where streams have been routed into CSSs, the surface runoff once conveyed in these streams reduces capacity in the CSS and contributes to overflows. Rerouting natural streams and surface runoff away from the CSS and back to their original watercourse or to other receiving waters can have a significant impact on CSS capacity.

Urban stream diversion is one of the more expensive inflow reduction options, as it typically requires design and construction of new storm drain lines. Stream diversion resembles CSO separation in that new alternative flow routes are required for surface runoff. It is typically employed in situations where less expensive and less disruptive options for inflow reduction are not feasible or do not provide sufficient inflow reduction. The potential amount of inflow to be diverted from the CSS must be well documented in order to assess its cost-effectiveness.

The first step in implementing a stream diversion program is to identify and evaluate alternative routes from the stream to nearby receiving waters. It may be possible to "daylight" streams by allowing them to flow in an open channel in some semblance of their natural condition. Stream daylighting can provide greenspace and can serve as an amenity to the community. Unfortunately, stream daylighting opportunities are limited in most CSS communities because the streams were originally piped in order to support property development. Diversion through separate pipe systems is also an option and may be more practical where open space is limited and the potential for flooding is high.

Stream diversion can be an expensive method for reducing stormwater inflow into the CSS. Changing a stream channel may also require numerous permits and may not meet with public approval; however, stream diversion will often cause a significant decrease in the amount of stormwater flowing into a CSS. The Ramsey Lake Wetland Project in Portland, Oregon, involved completely separating an urban stream from the CSS and creating a wetland to treat the separated stormwater flow prior to discharge into a receiving stream. The project also provided community involvement opportunities such as school field trips for the planting of wetland plants, and there are future plans to use the facility as a nature interpretive center. Portland has six additional stream diversion projects planned.

References and Recommended Reading

Bureau of Environmental Services (BES). (1994). *Combined Sewer Overflow Facilities Plan*, prepared by CH2M Hill for City of Portland, OR.

CEC. (2001). *Electrical Load Management*, California Energy Commission, Sacramento.

Hides, S. (1997). Quantity Is the Key to Improving Quality: A Common Sense Approach for Reducing Wet Weather Impacts, paper presented at the 26th Annual WEAO Technical Symposium and OPCEA Exhibitions, London, Ontario, Canada.

McKelvie, S. (1996). Flowslipping: an effective management technique for urban runoff, in *Urban Wet Weather Pollution Conference Proceedings*, Quebec City.

Spellman, F.R. (2007). *Handbook of Water and Wastewater Treatment Plant Operations*, 2nd ed., CRC Press, Boca Raton, FL.

USEPA. (1994). Combined sewer overflow (CSO) control policy; Notice, *Federal Register*, 59(75), 18687–18698.

USEPA. (1999). *Combined Sewer Overflow Technology Fact Sheet: Inflow Reduction*, EPA 832-F-99-035, U.S. Environmental Protection Agency, Washington, DC.

USEPA. (2012). *Water & Energy Efficiency in Water and Wastewater Facilities*, U.S. Environmental Protection Agency, Washington, DC (http://www.epa.gov/region9/waterinfrastructure/technology.html).

WSEO. (1993). *Improving the Energy Efficiency of Wastewater Treatment Facilities*, WSEO-192, Washington State Energy Office, Olympia.

Section III

Energy-Efficient Technology

Another world is not only possible, she is on her way. On a quiet day,
I can hear her breathing.

— **Arundhati Roy**

7

Combined Heat and Power (CHP)

> By installing a CHP system designed to meet the thermal and electrical base loads of a facility, CHP can increase operational efficiency and decrease energy costs, while reducing emissions of greenhouse gases that contribute to the risks of climate change.
>
> **—USEPA Combined Heat and Power Partnership**

> The first rule of sustainability is to align with natural forces, or at least not try to defy them.
>
> **—Paul Hawken, Environmentalist**

7.1 Introduction*

Combined heat and power (CHP), also known as *cogeneration*, is an efficient, clean, and reliable approach to generating power and thermal energy from a single fuel source. Basically, CHP is the simultaneous production of electricity and heat from a single fuel source, such as natural gas, biomass, biogas, coal, or oil. Wastewater treatment plants that have anaerobic digesters create methane gas as a byproduct of digestion of biosolids. Currently, a number of these plants release methane gas by *flaring*, converting methane to CO_2 and releasing it into the environment. Methane gas, however, is a good source of energy. By installing a CHP system designed to meet the thermal and electrical base loads of the plant, CHP can greatly increase the plant's operational efficiency and decrease energy costs. At the same time, CHP reduces the emission of greenhouse gases, which some feel may contribute to global climate change.

The plant's ability to produce its own electricity onsite has significant advantages. CHP systems have economical and environmental benefits; for example, they can reduce energy costs, offset capital costs, protect revenue streams, hedge against volatile energy prices, and reduce reliance on outside energy sources. CHP is not a single technology, but an energy system that can be modified depending on the needs of the energy end user. CHP systems consist of a number of individual components configured into an integrated whole. These components include the prime mover, generator, heat recovery equipment, and electrical interconnections. The prime mover that drives the overall system typically indentifies the CHP system. Prime movers for CHP systems include turbines, microturbines, internal combustion/reciprocating engines, steam engines/turbines, and fuel cells (USEPA, 2008). After a discussion of CHP basics, each of these technologies is discussed in detail in chapters that follow.

---

* Adapted from USEPA, *Catalog of CHP Technologies*, U.S. Environmental Protection Agency, Washington, DC, 2008.

DID YOU KNOW?

A number of wastewater treatment plants have installed cogeneration or combined heat and power. Excess methane can be captured and may be accepted by local power companies. Using the gas directly instead of converting it to electricity is an efficient application because energy is lost each time it is converted or transmitted over long distances.

7.2 CHP Key Definitions[*]

Understanding the application, efficiency, and operation of a CHP system requires an understanding of several key terms, described below:

- *CHP system* includes the unit in which fuel is consumed (e.g., turbine, boiler, engine), the electric generator, and the heat recovery unit that transforms otherwise wasted heat to useable thermal energy.

- *Total fuel input (Q_{FUEL})* is the thermal energy associated with the total fuel input. Total fuel input is the sum of all the fuel used by the CHP system. The total fuel energy input is often determined by multiplying the quantity of fuel consumed by the heating value of the fuel. Commonly accepted heating values for natural gas, coal, and diesel fuel are

  1020 Btu per cubic foot for natural gas

  10,157 Btu per pound of coal

  138,000 Btu per gallon of diesel fuel

- *Net useful power output (W_E)* is the gross power produced by the electric generator minus any parasitic electric losses—in other words, the electrical power used to support the CHP system. (An example of a parasitic electric loss is the electricity that may be used to compress the natural gas before the gas can be fired in a turbine.)

- *Net useful thermal output ($\sum Q_{TH}$)* is equal to the gross useful thermal output of the CHP system minus the thermal input. An example of thermal input is the energy of the condensate return and makeup water fed to a heat recovery steam generator (HRSG). Net useful thermal output represents the otherwise wasted thermal energy that was recovered by the CHP system.

- *Gross useful thermal output* is the thermal output of a CHP system utilized by the host facility. The term *utilized* is important here. Any thermal output that is not used should not be considered. Consider, for example, a CHP system that produces 10,000 pounds of steam per hour, with 90% of the steam used for space heating and the remaining 10% exhausted in a cooling tower. The energy content of 9000 pounds of steam per hour is the gross useful therm output.

---

[*] Adapted from USEPA, *Combined Heat and Power Partnership*, U.S. Environmental Protection Agency, Washington, DC, 2012.

- *Effective electric efficiency* is the net electric output divided by the effective fuel input. Effective fuel input is the total fuel used by the CHP system minus the fuel that would be used by an 80% efficient boiler to generate the same amount of steam as produced by the CHP system.
- *Fuel sources* in CHP systems can include natural gas, biomass, coal, biogas, or fuel oil.
- *Opportunity fuels* are materials from agricultural or industrial processes that would otherwise be wasted but could power a CHP system and are available at or in close proximity to a CHP site.
- *Prime movers* are the devices that convert fuels to electrical or mechanical energy.
- *Reliable power* refers to the ability to provide electric power that meets stringent standards for minimal power interruptions.
- *Spark spread* is the relative difference between the price of fuel and the price of power. Spark spread is highly dependent on the efficiency for conversion. For a CHP system, spark spread is the difference between the cost of fuel for the CHP system to produce power and heat on the site and the offset cost of purchased grid power.

7.3 Calculating Total CHP System Efficiency[*]

Figure 7.1 shows a typical CHP system. To produce 75 units of useful energy, conventional generation or separate heat and power systems use 154 units of energy—98 for electricity production and 56 to produce heat—resulting in an overall efficiency of 49%. However, the CHP system requires only 100 units of energy to produce the 75 units of useful energy from a single fuel source, resulting in a total system efficiency of 75%.

The most commonly used approach to determining the efficiency of a CHP system is to calculate *total system efficiency*. Also known as *thermal efficiency*, the total system efficiency (η_o) of a CHP system is the sum of the net useful power output (W_E) and net useful thermal outputs (ΣQ_{TH}) divided by the total fuel input (Q_{FUEL}), as shown below:

$$\eta_o = \frac{W_E + \Sigma Q_{TH}}{Q_{FUEL}} \tag{7.1}$$

The calculation of total system efficiency is a simple and useful method that evaluates what is produced (i.e., power and thermal output) compared to what is consumed (i.e., fuel). CHP systems with a relatively high net useful thermal output typically correspond to total system efficiencies in the range of 60 to 85%.

Note that this metric does not differentiate between the value of the power output and the thermal output; instead, it treats power output and thermal output as additive properties with the same relative value. In reality and in practice, thermal output and power output are not interchangeable because they cannot be converted easily from one to another. However, typical CHP applications have coincident power and thermal demands that must be met. It is reasonable, therefore, to consider the values of power and thermal output from a CHP system to be equal in many situations.

---

[*] Adapted from USEPA, *Efficiency Metrics for CHP Systems: Total System and Effective Electric Efficiencies*, U.S. Environmental Protection Agency, Washington, DC, 2012.

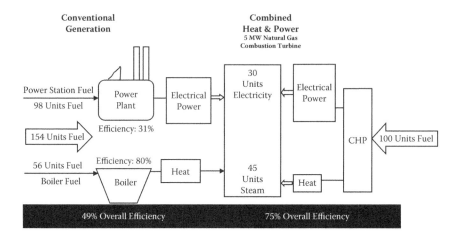

FIGURE 7.1
Conventional generation vs. CHP: overall efficiency. (Adapted from USEPA, *Efficiency Metrics for CHP Systems: Total System and Effective Electric Efficiencies*, U.S. Environmental Protection Agency, Washington, DC, 2012.)

7.4 Calculating Effective Electric Efficiency*

Effective electric efficiency calculations allow for a direct comparison of CHP to conventional power generation system performance (e.g., electricity produced from central stations, which is how the majority of the electricity is produced in the United States). Effective electric efficiency (ε_{EE}) can be calculated using Equation 7.2, where W_E is the net useful power output, $\sum Q_{TH}$ is the sum of the net useful thermal outputs, Q_{FUEL} is the total fuel input, and α equals the efficiency of the conventional technology that otherwise would be used to produce the useful thermal energy output if the CHP system did not exist:

$$\varepsilon_{EE} = \frac{W_E}{Q_{FUEL} - \sum \left(Q_{TH} / \alpha \right)} \qquad (7.2)$$

For example, if a CHP system is natural gas fired and produces steam, then α represents the efficiency of a conventional natural gas-fired boiler. Typical α values for boilers are 0.8 for a natural gas-fired boiler, 0.75 for a biomass-fired boiler, and 0.83 for a coal-fired boiler.

The calculation of effective electrical efficiency is essentially the CHP net electric output divided by the additional fuel the CHP system consumes over and above what would have been used by conventional systems to produce the thermal output for the site. In other words, this metric measures how effectively the CHP system generates power once the thermal demand of a site has been met.

Typical effective electrical efficiencies for combustion turbine-based CHP systems are in the range of 51 to 69%. Typical effective electrical efficiencies for reciprocating engine-based CHP systems are in the range of 69 to 84%.

---

* Adapted from USEPA, *Efficiency Metrics for CHP Systems: Total System and Effective Electric Efficiencies*, U.S. Environmental Protection Agency, Washington, DC, 2012.

DID YOU KNOW?

Many CHP systems are designed to meet a host site's unique power and thermal demand characteristics. As a result, a truly accurate measure of the efficiency of a CHP system may require additional information and broader examination beyond what is described in this text.

7.5 Selecting CHP Efficiency Metrics[*]

The selection of an efficiency metric depends on the purpose of calculating CHP efficiency:

- If the objective is to compare CHP system energy efficiency to the efficiency of a site's CHP options, then the *total system efficiency metric* may be the right choice. Calculation of CHP efficiency is a weighted average (based on a CHP system's net useful power output and net useful thermal output) of the efficiencies of the CHP production components. The separate power production component is typically 33% efficient grid power. The separate heat production component is typically a 75 to 85% efficient boiler.
- If CHP electrical efficiency is needed for a comparison of CHP to conventional electricity production (i.e., the grid), then the *effective electric efficiency metric* may be the right choice. Effective electric efficiency accounts for the multiple outputs of CHP and allows for a direct comparison of CHP and conventional electricity production by crediting that portion of the CHP system's fuel input allocated to thermal output.

Both the total system and effective electric efficiencies are valid metrics for evaluating CHP system efficiency. They both consider all the outputs of CHP systems and, when used properly, reflect the inherent advantages of CHP. However, because each metric measures a different performance characteristic, use of the two different metrics for a given CHP system produces different values.

7.6 Wastewater Treatment Facilities with CHP[†]

As of June 2011, wastewater treatment CHP systems were in place at 133 sites in 30 states, representing 437 megawatts (MW) of capacity (USEPA, 2011). Although the majority of facilities with CHP use digester gas as the primary fuel source, some employ CHP using fuels other than digester biogas (e.g., natural gas, fuel oil) either because they do not operate anaerobic digesters (so do not generate biogas) or because biogas is not a viable option

---

[*] Adapted from USEPA, *Efficiency Metrics for CHP Systems: Total System and Effective Electric Efficiencies*, U.S. Environmental Protection Agency, Washington, DC, 2012.

[†] Adapted from USEPA, *Opportunities for Combined Heat and Power at Wastewater Treatment Facilities: Market Analysis and Lessons from the Field*, U.S. Environmental Protection Agency, Washington, DC, 2011.

TABLE 7.1

Number of Digester Gas Wastewater CHP Systems and Total Capacity

State	Number of Sites	Capacity (MW)	State	Number of Sites	Capacity (MW)
AR	1	1.73	MT	3	1.09
AZ	1	0.29	NE	3	5.40
CA	33	62.67	NH	1	0.37
CO	2	7.07	NJ	4	8.72
CT	2	0.95	NY	6	3.01
FL	3	13.50	OH	3	16.29
IA	2	3.40	OR	10	6.42
ID	2	0.45	PA	3	1.99
IL	2	4.58	TX	1	4.20
IN	1	0.13	UT	2	2.65
MA	1	18.00	WA	5	14.18
MD	2	3.33	WI	5	2.02
MI	1	0.06	WY	1	0.03
MN	4	7.19	Total	104	189.80

Source: USEPA, *Opportunities for Combined Heat and Power at Wastewater Treatment Facilities: Market Analysis and Lessons from the Field*, U.S. Environmental Protection Agency, Washington, DC, 2011.

due to site-specific technical or economic conditions. Of the 133 wastewater treatment plants using CHP, 104 plants (78%), representing 190 MW of capacity, utilize digester gas as the primary fuel source. Note that some wastewater treatment plants blend biogas with natural gas if the volume of biogas from the digesters is not sufficient to meet a plant's thermal or electric requirement (e.g., in the winter when digester heat loads are higher). Table 7.1 shows the number of sites that use digester gas as the primary fuel source for CHP and their capacity by state. Several types of CHP prime movers can be used to generate electricity and heat at wastewater treatment plants. Table 7.2 shows the CHP prime movers currently used at wastewater treatment plants that use digester gas as the primary

TABLE 7.2

CHP Prime Movers Used at Wastewater Treatment Plants

Prime Mover	Number of Sites	Capacity (MW)
Reciprocating engine	54	85.8
Microturbine	29	5.2
Fuel cell	13	7.9
Combustion turbine	5	39.9
Steam turbine	1	23.0
Combined cycle	1	28.0
Total	104	189.8

Source: USEPA, *Opportunities for Combined Heat and Power at Wastewater Treatment Facilities: Market Analysis and Lessons from the Field*, U.S. Environmental Protection Agency, Washington, DC, 2011.

fuel source. The most commonly used prime movers at wastewater treatment plants are reciprocating engines, microturbines, and fuel cells. The power capacities of these prime movers most closely match the energy content of biogas generated by digesters at typically sized wastewater treatment plants. Opportunities for using combustion turbines, steam turbines, and combined cycle systems are typically found in the few very large wastewater treatment plants (i.e., greater than 100 MGD).

7.7 Overview of CHP Technologies

The following five chapters characterize the different CHP technologies (gas turbine, microturbines, reciprocating engines, steam turbines, and fuel cells) in detail. The chapters supply information on applications of each technology and detailed descriptions of its functionality and design characteristics, performance characteristics, emissions, and emissions control options. Table 7.3 (next page) provides a snapshot of the various CHP technologies with regard to their advantages, disadvantages, and available sizes. Again, a more detailed discussion of these technologies is presented in the five chapters that follow.

References and Recommended Reading

Bureau of Environmental Services (BES). (1994). *Combined Sewer Overflow Facilities Plan*, prepared by CH2M Hill for City of Portland, OR.

California Energy Commission. (2001). *Electric Load Management*, California Energy Commission, Sacramento (www.energy.ca.gov/process/pubs/eload.pdf).

Spellman, F.R. (2007). *Handbook of Water and Wastewater Treatment Plant Operations*, 2nd ed., CRC Press, Boca Raton, FL.

USEPA. (1999). *Combined Sewer Overflow Technology Fact Sheet: Inflow Reduction*, EPA 832-F-99-035, U.S. Environmental Protection Agency, Washington, DC.

USEPA. (2008). *Catalog of CHP Technologies*, U.S. Environmental Protection Agency, Washington, DC (www.epa.gov/chp/documents/catalog_chptech_full.pdf).

USEPA. (2011). *Opportunities for Combined Heat and Power at Wastewater Treatment Facilities: Market Analysis and Lessons from the Field*, U.S. Environmental Protection Agency, Washington, DC (www.epa.gov/chp/documents/wwtf_opportunities.pdf).

USEPA. (2012a). *Methods for Calculating Efficiency*, U.S. Environmental Protection Agency, Washington, DC (www.epa.gov/chp/basic/methods.html).

USEPA. (2012b). *Water & Energy Efficiency in Water and Wastewater Facilities*, U.S. Environmental Protection Agency, Washington, DC (www.epa.gov/region9/waterinfrastructure/technology.html).

USEPA. (2012c). *Efficiency Metrics for CHP Systems: Total System and Effective Electric Efficiencies*, U.S. Environmental Protection Agency, Washington, DC.

USEPA. (2012d). *Waste Heat to Power Systems*, U.S. Environmental Protection Agency, Washington, DC.

WSEO. (1993). *Improving the Energy Efficiency of Wastewater Treatment Facilities*, WSEO-192, Washington State Energy Office, Olympia.

TABLE 7.3

Summary of CHP Technologies

CHP System	Advantages	Disadvantages	Available Sizes
Gas turbine	High reliability Low emissions High-grade heat available No cooling required	Requires high-pressure gas or in-house gas compressor Poor efficiency at low loading Output falls as ambient temperature rises	500 kW to 250 MW
Microturbine	Small number of moving parts Compact size and light weight Low emissions No cooling required	High costs Relatively low mechanical efficiency Limited to lower temperature cogeneration applications	30 to 250 kW
Reciprocating engine (spark ignition, compression ignition with dual fuel pilot ignition)	High power efficiency with part-load operational flexibility Fast start-up Relatively low investment cost Can be used in island mode and have good load following capability Can be overhauled onsite with normal operators Operates on low-pressure gas	High maintenance costs Limited to lower temperature cogeneration applications Relatively high air emissions Must be cooled even if recovered heat is not used High levels of low-frequency noise	Distributed generation, <5 MW High speed (1200 rpm), ≤4 MW Low speed (102–514 rpm), 4 to 75 MW
Steam turbine	High overall efficiency Any type of fuel may be used Ability to meet more than one site heat grade requirement Long working life and high reliability Power-to-heat ratio can be varied	Slow start-up Low power-to-heat ratio	50 kW to 250 MW
Fuel cells	Low emissions and low noise High efficiency over load range Modular design	High costs Low durability Fuels require processing unless pure hydrogen is used	5 kW to 2 MW

Source: USEPA, *Catalog of CHP Technologies*, U.S. Environmental Protection Agency, Washington, DC, 2008.

8

Gas Turbines*

Imagination is more important than knowledge.

—**Albert Einstein**

For us there is no past, no future. We live in the present and are full.

—**John Muir**

8.1 Introduction

Figure 8.1 shows the primary components of a simple-cycle gas turbine. Engineering advancements pioneered the development of the simple-cycle gas turbine in the early 1900s, and turbines began to be used for stationary electric power generation in the later 1930s. Turbines revolutionized airplane propulsion in the 1940s, and in the 1990s through today they have been a popular choice for new power generation plants in the United States. Available in sizes ranging from 500 kilowatts (kW) to 150 megawatts (MW), gas turbines can be used in power-only generation or in combined heat and power (CHP) systems. The most efficient commercial technology for central station power-only generation is the gas turbine–steam turbine combined-cycle plant, with efficiencies approaching 60% (lower heating value, LHV). Simple-cycle gas turbines for power-only generation are available with efficiencies approaching 40% (LHV). Gas turbines have long been used by utilities for peaking capacity; however, with changes in the power industry and advancements in technology, the gas turbine is now being increasingly used for base-load power.

DID YOU KNOW?

Most of the efficiencies reported in this book are based on higher heating value (HHV), which includes the heat of condensation of the water vapor in the combustion process. In engineering and scientific literature concerning heat engine efficiencies, the lower heating value (LHV), which does not include the heat of condensation of the water vapor in the combustion products, is usually used. The HHV is greater than the LHV by approximately 10% with natural gas as the fuel (e.g., 50% LHV is equivalent to 55% HHV). HHV efficiencies are about 8% greater for oil (liquid petroleum products) and 5% for coal.

---

* Material in this chapter is adapted from USEPA, *Catalog of CHP Technologies*, U.S. Environmental Protection Agency, Washington, DC, 2008.

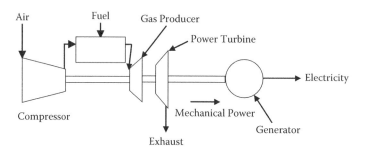

FIGURE 8.1
Components of a simple-cycle gas turbine.

Gas turbines produce high-quality exhaust heat that can be used for CHP configurations to reach overall system efficiencies (electricity and useful thermal energy) of 70 to 80%. By the early 1980s, the efficiency and reliability of smaller gas turbines (1 to 40 MW) had progressed sufficiently to be an attractive choice for industrial and large institutional uses for CHP applications.

Gas turbines are one of the cleanest means of generating electricity. Emissions of oxides of nitrogen (NO_x) from some large turbines are in the single-digit parts per million (ppm) range, with either catalytic exhaust cleanup or lean premixed combustion. (For comparative purposes, we say that 1 ppm is analogous to a full shotglass of water sitting in the bottom of a full standard-size swimming pool.) Because of their relatively high efficiency and reliance on natural gas as the primary fuel, gas turbines emit substantially less carbon dioxide (CO_2) per kilowatt-hour (kWh) generated than any other fossil technology in general commercial use.

8.2 Applications

Process industries use gas turbines to drive compressors and other large mechanical equipment. The oil and gas industry commonly uses them to drive pumps and compressors. Many industrial and institutional facilities use turbines to generate electricity for use onsite. When used to generate power onsite, gas turbines are often used in combined heat and power mode where energy in the turbine exhaust provides thermal energy to the facility (USEPA, 2008a).

There is a significant amount of gas turbine-based CHP capacity in the United States located at industrial and institutional facilities. Much of this capacity is concentrated in large combined-cycle CHP systems that maximize power production for sale to the grid. However, a significant number of simple-cycle, gas turbine-based CHP systems are in operation for a variety of applications, including oil recovery, chemicals, paper production, food processing, and universities. Simple-cycle CHP applications are most prevalent in smaller installations, typically less than 40 MV.

Gas turbines are ideally suited for CHP applications because their high-temperature exhaust can be used to generate process steam at conditions as high as 1200 pounds per square inch gauge (psig) and 900°F or used directly in industrial processes for heating or drying. A typical industrial CHP application for gas turbines is a chemicals plant with a 25-MW, simple-cycle gas turbine supplying base-load power to the plant with an unfired heat recovery steam generator on the exhaust. Approximately 29 MW thermal (MWth) of steam is produced for process use within the plant.

A typical commercial/institutional CHP application for gas turbines is a college or university campus with a 5-MW, simple-cycle gas turbine. Approximately 8 MWth of 150- to 400-psig steam (or hot water) is produced in an unfired heat recovery steam generator and sent into a central thermal loop for campus space heating during winter months or to single-effect absorption chiller to provide cooling during the summer.

While the recovery of thermal energy provides compelling economics for gas turbine CHP, smaller gas turbines supply prime power in certain applications. Large industrial facilities install simple-cycle gas turbines without heat recovery to provide peaking power in capacity-constrained areas, and utilities often place gas turbines in the 5- to 40-MW size range at substations to provide incremental capacity and grid support. A number of turbine manufacturers and packagers offer mobile turbine generator units in this size range that can be used in one location during a period of peak demand and then trucked to another location for the following season.

8.3 Gas Turbine Technology

With gas turbines systems, it is all about the Brayton cycle, which consists of the following processes:

- Adiabatic process (compression)
- Isobaric process (heat addition)
- Adiabatic process (expansion)
- Isobaric process (heat rejection)

In the Brayton cycle, atmosphere is compressed, heated, and then expanded. The excess power produced by the turbine (also called the *expander*) over that consumed by the compressor is used for power generation. The power produced by an expansion turbine and consumed by the compressor is proportional to the absolute temperature of the gas passing through the device. Consequently, it is advantageous to operate the expansion turbine at the highest practical temperature consistent with economic materials and internal blade cooling technology and to operate the compressor with inlet air flow at as low a temperature as possible. As technology advances permit higher turbine inlet temperature, the optimum pressure ratio also increases (USEPA, 2008b).

Higher temperature and pressure ratios result in higher efficiency and specific power; thus, the general trend in gas turbine advancement has been toward a combination of higher temperatures and pressures. Although such advancements increase the manufacturing cost of the machine, the higher value, in terms of greater power output and higher efficiency, provides net economic benefits. The industrial gas turbine is a balance between performance and cost that results in the most economic machine for both the user and manufacturer.

8.3.1 Modes of Operation

Figure 8.1 shows the primary components of a simple cycle gas turbine, but what the figure does not show is the several variations of the Brayton cycle in use today. Fuel consumption may be decreased by preheating the compressed air with heat from the turbine exhaust using a recuperator or regenerator; the compressor work may be reduced and net power increased by using intercooling or precooling; or the exhaust may be used to raise steam in a boiler and to generate additional power in a combined cycle. For smaller industrial turbines, gas turbine exhaust is quite hot, up to 800 to 900°F, and it can rise to 1100°F for some new, large central station utility machines and aeroderivative turbines. Such high exhaust temperatures allow direct use of the exhaust. With the addition of a heat recovery steam generator (HRSG), the exhaust heat can produce steam or hot water. A portion of all the steam generated by the HRSG may be used to generate additional electricity through a steam turbine in a combined cycle configuration. A gas turbine-based system is operating in combined heat and power mode when the waste heat generated by the turbine is applied in an end-use. For example, a simple-cycle gas turbine using the exhaust in a direct heating process is a CHP system, whereas a system that features all of the turbine exhaust feeding a HRSG and all of the steam output going to produce electricity in a combined-cycle steam turbine is not.

8.3.2 Design Characteristics

- *Thermal output*—Gas turbines produce high-quality (high-temperature) thermal output suitable for most combined heat and power applications. High-pressure steam can be generated or exhaust can be used directly for process drying and heating.

- *Fuel flexibility*—Gas turbines operate on natural gas, synthetic gas, landfill gas, and fuel oils. Plants typically operate on gaseous fuel with a stored liquid fuel for backup to obtain the less expensive interruptible rate for natural gas.

- *Reliability and life*—Modern gas turbines have proven to be reliable power generators given proper maintenance. Time to overhaul is typically 25,000 to 50,000 hours.

- *Size range*—Gas turbines are available in sizes from 500 kW to 250 MW.

- *Emissions*—Many gas turbines burning gaseous fuels (mainly natural gas) feature lean premixed burners (also called dry low-NO_x combustors) that produce NO_x emissions below 25 ppm, with laboratory data down to 9 ppm, and simultaneous low CO emissions in the range of 10 to 50 ppm. Selective catalytic reduction (SCR) or catalytic combustion further reduces NO_x emissions. Many gas turbines sited in locales with stringent emission regulations use SCR after-treatment to achieve single-digit (below 9 ppm) NO_x emissions.

- *Part-load operation*—Because gas turbines reduce power output by reducing combustion temperatures, efficiency at part load can be substantially below that of full-power efficiency.

DID YOU KNOW?

Aeroderivative gas turbines for stationary power are adapted from their jet and turboshaft aircraft engine counterparts. These turbines are lightweight and thermally efficient, but they are usually more expensive than products designed and built exclusively for stationary applications.

DID YOU KNOW?

Gas turbines have high oxygen content in their exhaust because they burn fuel with high excess air to limit combustion temperatures to levels that the turbine blades, combustion chamber, and transition section can handle without compromising system life. Consequently, emissions from gas turbines are evaluated at a reference condition of 15% oxygen. For comparison, boilers use 3% oxygen as the reference conditions for emissions, because they can minimize excess air and thus waste less heat in the stack exhaust. Note that due to the different amount of diluent gases in the combustion products, the mass of NO_x measured as 9 ppm at 15% oxygen is approximately 27 ppm at 3% oxygen, the condition used for boiler NO_x regulations.

8.3.3 Performance Characteristics

8.3.3.1 Electrical Efficiency

The thermal efficiency of the Brayton cycle is a function of pressure ratio, ambient air temperature, turbine inlet air temperature, efficiency of the compressor and turbine elements, turbine blade cooling requirements, and any performance enhancements (e.g., recuperation, intercooling, inlet air cooling, reheat, steam injection, recombined cycle). All of these parameters, along with gas turbine internal mechanical design features, have been improving with time; therefore, new machines are usually more efficient than older ones of the same size and general type. The performance of a gas turbine is also appreciably influenced by the purpose for which it is intended. Emergency power units generally have lower efficiency and lower capital costs, while turbines intended for prime power, compressor stations, and similar applications with high annual capacity factors have higher efficiency and higher capital costs. Emergency power units are permitted for a maximum number of hours per year and allowed to have considerably higher emissions than turbines permitted for continuous duty.

8.3.3.2 Fuel Supply Pressure

Gas turbines require a minimum gas pressure of about 100 psig for the smallest turbines with substantially higher pressures for larger turbines and aeroderivative machines. Depending on the supply pressure of the gas being delivered to the site, the cost and power consumption of the fuel gas compressor can be a significant consideration.

8.3.3.3 Part-Load Performance

When less than full power is required from a gas turbine, the output is reduced by lowering the turbine inlet temperature. In addition to reducing power, this change in operating conditions also reduces efficiency. Emissions are generally increased at part load conditions, especially at half loaded and below.

8.3.3.4 Effects of Ambient Conditions on Performance

The ambient conditions under which a gas turbine operates have a noticeable effect on both the power output and efficiency. At elevated inlet air temperatures, both the power and efficiency decrease. The power decreases due to the decreased air flow mass rate (the

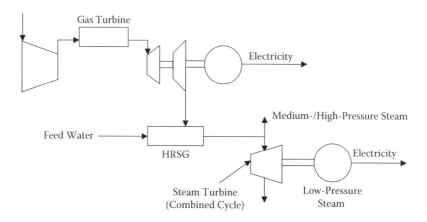

FIGURE 8.2
Heat recovery from a gas turbine/HRSG system.

density of air declines as temperature increases), and the efficiency decreases because the compressor requires more power to compress air of higher temperature. Conversely, the power and efficiency increase when the inlet air temperature is reduced.

8.3.3.5 Heat Recovery

The economics of gas turbines in process applications often depend on effective use of the thermal energy contained in the exhaust gas, which generally represents 60 to 70% of the inlet fuel energy. The most common use of this energy is for steam generation in unfired or supplementary fired heat recovery steam generators. The gas turbine exhaust gases can also be used as a source of direct process energy, for unfired or fired process fluid heaters, or as preheated combustion air for power boilers. Figure 8.2 shows a typical gas turbine/HRSG configuration. An unfired HRSG is the simplest steam CHP configuration and can generate steam at conditions ranging from 150 psig to approximately 1200 psig.

8.3.3.6 CHP System Efficiency

Overall or total efficiency of a CHP system is a function of the amount of energy recovered from the turbine exhaust. The two most important factors influencing the amount of energy available for steam generation are gas turbine exhaust temperature and HRSG stack temperature. Turbine firing temperature and turbine pressure ratio combine to determine gas turbine exhaust temperature. Typically, aeroderivative gas turbines have higher firing temperatures than do industrial gas turbines, but when the higher pressure ratio of aeroderivative gas turbines is recognized, the turbine discharge temperatures of the two turbine types remain somewhat close, typically in the range of 850 to 950°F. For the same HRSG exit temperature, higher turbine exhaust temperature (higher HRSG gas inlet temperature) results in greater available thermal energy and increased HRSG output. Similarly, the lower the HRSG stack temperature, the greater the amount of energy recovered and the higher the total-system efficiency. HRSG stack temperature is a function of steam conditions and fuel type. Saturated steam temperatures increase with increasing steam pressure. Because of pinch point considerations within the HRSG, higher steam pressures result in higher HRSG exhaust stack temperatures, less utilization of available

DID YOU KNOW?

Because very little of the available oxygen in the turbine air flow is used in the combustion process, the oxygen content in the gas turbine exhaust allows supplementary fuel firing ahead of the HRSG to increase steam production relative to an unfired unit. Supplementary firing can raise the exhaust gas temperate entering the HRSG up to 1800°F and increase the amount of steam produced by the unit by a factor of two. Moreover, because the turbine exhaust gas is essentially preheated combustion air, the fuel consumed in supplementary firing is less than that required for a stand-alone boiler providing the same increment in steam generation. The HHV efficiency of incremental steam production from supplementary firing above that of an unfired HRSG is often 85% or more when firing natural gas.

thermal energy, and a reduction in total CHP system efficiency. In general, minimum stack temperatures of about 300°F are recommended for sulfur-bearing fuels. Generally, unfired HRSGs can be designed to economically recover approximately 95% of the available energy in the turbine exhaust (the energy released in going from turbine exhaust temperature to HRSG exhaust temperature). Overall CHP efficiency generally remains high under part-load conditions. The decrease in electric efficiency from the gas turbine under part-load conditions results in a relative increase in heat available for recovery under those conditions. This can be a significant operating advantage for applications in which the economics are driven by high steam demand.

8.3.3.7 Performance and Efficiency Enhancements

Several technologies that increase the output power or efficiency of gas turbines have been developed and put into limited commercial service.

8.3.3.7.1 Recuperators

Fuel use can be reduced (and efficiency improved) by using a heat exchanger called a *recuperator*, which uses the hot turbine exhaust to heat the compressed air entering the combustor.

8.3.3.7.2 Intercoolers

Intercoolers are used to increase gas turbine power by dividing the compressor into two sections and cooling the compressed air exiting the first section before it enters the second compressor section. Intercoolers reduce the power consumption in the second section of the compressor, thereby adding to the net power delivered by the combination of the turbine and compressor.

8.3.3.7.3 Inlet Air Cooling

Decreased power and efficiency of gas turbines at high ambient temperatures mean that gas turbine performance is at its lowest at times when power is often in greatest demand and most valued. Cooling the air entering the turbine by 40 to 50°F on a hot day can increase power output by 15 to 20%. The decreased power and efficiency resulting from high ambient air temperatures can be mitigated by any of several approaches to inlet-air cooling, including refrigeration, evaporative cooling, and thermo-energy storage using off-peak cooling.

8.3.3.8 Capital Cost

A gas turbine CHP plant is a complex process with many interrelated subsystems. The basic package consists of the gas turbine, gearbox, electric generator inlet and exhaust ducting, inlet air filtration, lubrication and cooling systems, standard starting system, and exhaust silencing. The basic package cost does not include extra systems such as the fuel-gas compressor, heat-recovery system, water-treatment system, or emissions-control systems such as selective catalytic reduction (SCR) or a continuous emission monitoring system (CEMS). Not all of these systems are required at every site. The cost of the basic turbine package plus the costs for added systems required for the particular application comprise the total equipment cost. The total plant cost consists of total equipment cost plus installation labor and materials (including site work), engineering, project management (including licensing, insurance, commissioning, and start-up), and financial carrying costs during the 6- to 18-month construction period.

8.3.3.9 Maintenance

Daily maintenance includes visual inspection by site personnel of filters and general site conditions. Routine inspections are required every 4000 hours to ensure that the turbine is free of excessive vibration due to worn bearings, rotors, and damaged blade tips. Inspections generally include onsite hot gas path boroscope inspections and nondestructive component testing using dye penetrant and magnetic particle techniques to ensure the integrity of components. The combustion path is inspected for fuel nozzle cleanliness and wear, along with the integrity of other hot gas path components. A gas turbine overhaul is needed every 25,000 to 50,000 hours, depending on service, and typically includes a complete inspection and rebuild of components to restore the gas turbine to nearly original or current (upgraded) performance standards. A typical overhaul consists of dimensional inspections, product upgrades and testing of the turbine and compressor, rotor removal, inspection of thrust and journal bearings, blade inspection and clearances, and setting packing seals. Gas turbine maintenance costs can vary significantly depending on the quality and diligence of the preventive maintenance program and operating conditions. Although gas turbines can be cycled, cycling every hour triples maintenance costs compared to a turbine that operates for intervals of 1000 hours or more. In addition, operating the turbine over the rated capacity for significant periods of time will dramatically increase the number of hot path inspections and overhauls. Gas turbines that operate for extended periods on liquid fuels will experience higher than average overhaul intervals.

8.3.3.10 Fuels

All gas turbines intended for service as stationary power generators in the United States are available with combustors equipped to handle natural gas fuel. A typical range of heating values of gaseous fuel acceptable to gas turbines is 900 to 1100 Btu per standard

DID YOU KNOW?

A *boroscope* (or *borescope*) is an optical device consisting of a rigid or flexible tube with an eyepiece on one end and an objective lens on the other linked together by a relay optical system in between. It is used to inspect internal turbine components.

DID YOU KNOW?

The pollutants referred to as nitrogen oxides (NO_x) are a mixture of mostly NO and NO_2 in variable composition. In emissions measurement, NO_x is reported as parts per million by volume in which both species count equally. NO_x is formed by three mechanisms: thermal NO_x, prompt NO_x, and fuel-bound NO_x. The predominant NO_x formation mechanism associated with gas turbines is thermal NO_x (USEPA, 2008a).

cubic foot (SCF), which covers the range of pipeline quality natural gas. Clean liquid fuels are also suitable for use in gas turbines. Special combustors developed by some gas turbine manufacturers are capable of handling cleaned gasified solid and liquid fuels. Burners have been developed for medium-Btu fuel (in the range of 400 to 500 Btu/SCF), which is produced with oxygen-blown gasifiers, and for low-Btu fuel (90 to 125 Btu/SCF), which is produced by air-blown gasifiers. These burners for gasified fuels exist for large gas turbines but are not available for small gas turbines. Contaminants in fuel such as ash, alkalis (sodium and potassium), and sulfur result in alkali sulfate deposits, which impede flow, decrease performance, and cause corrosion in the turbine hot section. Fuels must have only low levels of specified contaminants in them (typically, less than 10 ppm total alkalis and single-digit ppm of sulfur). Liquid fuels require their own pumps, flow control, nozzles, and mixing systems. Many gas turbines are available with either gas or liquid firing capability. In general, gas turbines can convert from one fuel to another quickly. Several gas turbines are equipped with dual firing and can switch fuels with minimum or no interruptions.

8.3.4 Emissions

Gas turbines are among the cleanest fossil-fueled power generation equipment commercially available. Gas turbine emission control technologies continue to evolve, with older technologies gradually being phased out as new technologies are developed and commercialized. The primary pollutants from gas turbines are oxides of nitrogen (NO_x), carbon monoxide (CO), and volatile organic compounds (VOCs). Other pollutants such as oxides of sulfur (SO_x) and particulate matter (PM) are primarily dependent on the fuel used. The sulfur content of the fuel determines emissions of sulfur compounds, primarily SO_2. Gas turbines operating on desulfized natural gas or distillate oil emit relatively insignificant levels of SO_x. In general, SO_x emissions are greater when heavy oils are fired in the turbine. SO_x control is thus a fuel purchasing issue rather than a gas turbine technology issue. Particulate matter is a marginally significant pollutant for gas turbines using liquid fuels. Ash and metallic additives in the fuel may contribute to PM in the exhaust.

It is important to note that the gas turbine operating load has a significant effect on the emissions levels of the primary pollutants of NO_x, CO, and VOCs. Gas turbines typically operate at high loads; consequently, gas turbines are designed to achieve maximum efficiency and optimum combustion conditions at high loads. Controlling all pollutants simultaneously at all load conditions is difficult. At higher loads, higher NO_x emissions occur due to peak flame temperatures. At lower loads, lower thermal efficiencies and more incomplete combustion occur, resulting in higher emission of CO and VOCs.

References and Recommended Reading

Spellman, F.R. (2007). *Handbook of Water and Wastewater Treatment Plant Operations*, 2nd ed., CRC Press, Boca Raton, FL.

USEPA. (2008a). *Catalog of CHP Technologies*, U.S. Environmental Protection Agency, Washington, DC (www.epa.gov/chp/documents/catalog_chptech_full.pdf).

USEPA. (2008b). *Technology Characterization: Gas Turbines*, U.S. Environmental Protection Agency, Washington, DC (www.epa.gov/chp/documents/catalog_chptech_gas_turbines.pdf).

USEPA. (2011). *Opportunities for Combined Heat and Power at Wastewater Treatment Facilities: Market Analysis and Lessons from the Field*, U.S. Environmental Protection Agency, Washington, DC (www.epa.gov/chp/documents/wwtf_opportunities.pdf).

USEPA. (2012). *Methods for Calculating Efficiency*, U.S. Environmental Protection Agency, Washington, DC (www.epa.gov/chp/basic/methods.html).

WSEO. (1993). *Improving the Energy Efficiency of Wastewater Treatment Facilities*, WSEO-192, Washington State Energy Office, Olympia.

9

Microturbines*

Microturbines are a new, innovative technology based on jet engines (more specifically the turbo charger equipment found in jet engines) that use rotational energy to generate power....Microturbines can run on bio-gas, natural gas, propane, diesel, kerosene, methane, and other fuel sources, making them suitable for a variety of applications.... From an environmental standpoint, these new machines take up less space, have higher efficiencies, and generate lower emissions than reciprocating engines....If operated from a natural gas pipeline, no onsite gas storage is needed, thus reducing safety concerns.

—USEPA (2006a,b)

9.1 Introduction

This chapter describes the use of microturbines as auxiliary and supplemental power sources (ASPSs) for wastewater treatment plants (WWTPs). Microturbines are a new, innovative technology based on jet engines. They are small prime movers that burn gaseous and liquid fuels to create the high-speed rotation of electrical generators. The size range for microturbines available and in development is from 30 to 250 kilowatts (kW), while conventional gas turbine sizes range from 500 kW to 250 megawatts (MW). Microturbines run at high speeds and, like larger gas turbines, can be used in power-only generation or in combined heat and power (CHP) systems. They are able to operate on a variety of fuels including natural gas, sour gases (high sulfur, low Btu content), and liquid fuels such as gasoline, kerosene, and diesel fuel/distillate heating oil. In resource recovery applications, they burn waste gases that would otherwise be flared or released directly into the atmosphere.

9.2 Microturbine Applications

Microturbines are ideally suited for distributed generation applications because of their flexibility in connection methods, ability to be stacked in parallel to serve large loads, ability to provide stable and reliable power, and low emissions. From an economic standpoint, the microturbine generators are less expensive to build and run when compared to larger conventional gas- or diesel-powered generators. The technology is well understood and has been implemented in many applications throughout the United States. Moreover,

---

* Material in this chapter is adapted from USEPA, *Catalog of CHP Technologies*, U.S. Environmental Protection Agency, Washington, DC, 2008.

microturbines are relatively inexpensive and easy to manufacture, and they have few moving parts. These power plants can use various types of fuel and can provide reliable unattended operation; this is important because some locations are remote from the grid. Types of applications include the following:

- Peak shaving and base-load power (grid parallel)
- Combined heat and power
- Stand-alone power
- Backup/standby power
- Ride-through connection
- Primary power with grid as backup
- Microgrid
- Resource recovery

In CHP applications, the waste heat from the microturbine is used to produce hot water, to heat building space, to drive absorption cooling or desiccant dehumidification equipment, and to supply other thermal energy needs in a building or industrial process.

9.3 Microturbine Technology

Microturbines are small gas turbines, most of which feature an internal heat exchanger called a *recuperator*. In a microturbine, a radial flow (centrifugal) compressor compresses the inlet air, which is then preheated in the recuperator using heat from the turbine exhaust. Next, the heated air from the recuperator mixes with fuel in the combustor, and hot combustion gas expands through the expansion and power turbines. The expansion turbine turns the compressor and, in single-shaft models, turns the generator as well. Two-shaft models use the compressor-driven turbine exhaust power as a second turbine that drives the generator. Finally, the recuperator uses the exhaust of the power turbine to preheat the air from the compressor.

Single-shaft models generally operate at speeds over 60,000 revolutions per minute (rpm) and generate electrical power of high frequency and of variable frequency (alternating current, AC). The power is rectified to direct current (DC) and then inverted to 60 hertz (Hz) for U.S. commercial use. In the two-shaft version, the power turbine connects via a gearbox to a generator that produces power at 60 Hz. Some manufacturers offer units producing 50 Hz for use in countries where 50 Hz is standard, such as in Europe and parts of Asia.

Keep in mind that microturbines operate on the same thermodynamic cycle, known as the *Brayton cycle* (see Chapter 8), as larger gas turbines. In this cycle, atmospheric air is compressed, heated, and then expanded, with the excess power produced by the expander (also called the turbine) over that consumed by the compressor being used for power generation. The power produced by an expansion turbine and consumed by a compressor is proportional to the absolute temperature of the gas passing through those devices. Consequently, it is advantageous to operate the expansion turbine at the highest practical temperature consistent with economic materials and to operate the compressor with inlet airflow at as low a temperature as possible. As technological advances permit higher turbine inlet temperatures, the optimum pressure ratio also increases. Higher temperature and pressure ratios result in higher efficiency and specific power. Thus, the general trend in gas turbine advancement has been toward a combination of higher temperatures and pressure. However, microturbine inlet temperatures are generally limited to 1800°F or below to enable the use of relatively inexpensive materials for the turbine wheel and to maintain pressure ratios at a comparatively low 3.5 to 4.0.

9.3.1 Basic Components

9.3.1.1 Turbocompressor Package

The basic components of a microturbine are the compressor, turbine generator, and recuperator (Figure 9.1). The heart of the microturbine is the compressor–turbine package, which is commonly mounted on a single shaft along with the electric generator. Two bearings support the single shaft. The single moving part of the one-shaft design has the potential for reducing maintenance needs and enhancing overall reliability. There are also two-shaft versions, in which the turbine on the first shaft directly drives the compressor while a power turbine on the second shaft drives a gearbox and conventional electrical generator producing 60-Hz power. The two-shaft design features more moving parts but does not require complicated power electronics to convert high-frequency AC power output to 60 Hz.

Moderate- to large-size gas turbines use multistage axial flow turbines and compressors, in which the gas flows along the axis of the shaft and is compressed and expanded in multiple stages. However, microturbine turbomachinery is based on single-stage radial flow compressors and turbines. Radial flow turbomachinery handles the small volumetric flows of air and combustion products with reasonably high component efficiency. (*Note:* With axial flow turbomachinery, blade height would be too small to be practical.) Large-size axial flow turbines and compressors are typically more efficient than radial flow components; however, in the size range of microturbines—0.5 to 5 lb/sec of air/gas flow—radial flow components offer minimum surface and end wall losses and provide the highest efficiency.

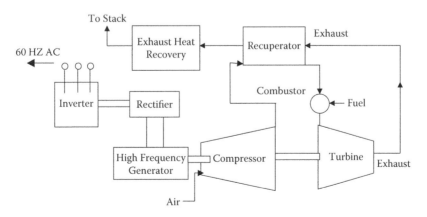

FIGURE 9.1
Microturbine-based CHP system (single-shaft design).

In microturbines, the turbocompressor shaft generally turns at high rotational speed, about 96,000 rpm in the case of a 30-kW machine and about 80,000 rpm in a 75-kW machine. One 45-kW model on the market turns at 116,000 rpm. There is no single rotational speed–power size rule, as the specific turbine and compressor design characteristics strongly influence the physical size of components and consequently rotational speed. For a specific aerodynamic design, as the power rating decreases the shaft speed increases, hence the high shaft speed of the small microturbines.

The radial flow turbine-driven compressor is quite similar in terms of design and volumetric flow to automobile, truck, and other small reciprocating engine turbochargers. Superchargers and turbochargers have been used for almost 80 years to increase the power of reciprocating engines by compressing the inlet air to the engine. Today's world market for small automobile and truck turbochargers is around 2 million units per year. Small gas turbines, of the size and power rating of microturbines, serve as auxiliary power systems on airplanes. Cabin cooling (air-conditioning) systems of airplanes use this same size and design family of compressors and turbines. The decades of experience with these applications provide the basis for the engineering and manufacturing technology of microturbine components.

9.3.1.2 Generator

The microturbine produces electrical power either via a high-speed generator turning on the single turbo-compressor shaft or with a separate power turbine driving a gearbox and conventional 3600-rpm generator. The high-speed generator of the single-shaft design employs a permanent magnet (typically samarium–cobalt) alternator, and it requires that the high-frequency AC output (about 1600 Hz for a 30-kW machine) be converted to 60 Hz for general use. This power conditioning involves rectifying the high-frequency AC to DC and then inverting the DC to 60-Hz AC. Power conversion comes with an efficiency penalty (approximately 5%). To start up a single-shaft design, the generator acts as a motor turning the turbo-compressor shaft until sufficient rpm is reached to start the combustor. A power storage unit (typically a battery-supplied uninterruptible power supply, or UPS) is used to power the generator for startup if the system is operating independent of the grid. This is referred to as *black starting*, which is restoring power to the operation without relying on the external electrical power transmission network.

9.3.1.3 Recuperators

Recuperators are heat exchangers that use the hot turbine exhaust gas (typically around 1200°F) to preheat the compressed air (typically around 300°F) going into the combustor, thereby reducing the fuel required to heat the compressed air to turbine inlet temperature. Depending on microturbine operating parameters, recuperators can more than double machine efficiency; however, because of the increased pressure drop in both the compressed air and turbine exhaust sides of the recuperator, power output typically declines 10 to 15% from that attainable without the recuperator. Recuperators also lower the temperature of the microturbine exhaust, reducing the effectiveness of the microturbine in CHP applications.

9.3.1.4 Bearings

Microturbines operate on either oil-lubricated or air bearings, which support the shafts. *Oil-lubricated bearings* are mechanical bearings that come in three forms: (1) high-speed metal roller, (2) floating sleeve, and (3) ceramic surface. The latter type typically offers the most attractive benefits in terms of life, operating temperature, and lubricant flow. Although they are a well-established technology, they require an oil pump, oil filtering system, and liquid cooling, all of which add to microturbine cost and maintenance. In addition, the exhaust from machines featuring oil-lubricated bearings may not be usable for direct space heating in cogeneration configurations due to the potential for contamination. Because the oil never comes in direct contact with hot combustion products, as is the case in small reciprocating engines, it is believed that the reliability of such a lubrication system is more typical of ship propulsion diesel systems (which have separate bearings and cylinder lubrication systems) and automotive transmissions than cylinder lubrication in automotive engines.

Air bearings have been in service on airplane cabin cooling systems for many years. They allow the turbine to spin on a thin layer of air, so friction is low and rpm is high. No oil or oil pump is needed. Air bearings offer simplicity of operation without the cost, reliability concerns, maintenance requirements, or power drain of an oil supply and filtering system. Concern does exist for the reliability of air bearings under numerous and repeated starts due to metal-on-metal friction during startup, shutdown, and load changes. Reliability depends largely on the individual manufacturer's quality control methodology more than on design engineering and will only be proven after significant experience with substantial numbers of units with long numbers of operating hours and on/off cycles.

9.3.1.5 Power Electronics

Single-shaft microturbines feature digital power controllers to convert the high-frequency AC power produced by the generator into usable electricity. The high-frequency AC is rectified to DC, inverted back to 60- or 50-Hz AC, and then filtered to reduce harmonic distortion. This is a critical component in the single-shaft microturbine design and represents significant design challenges, specifically in matching turbine output to the required load. To allow for transients and voltage spikes, power electronics designs are generally able to handle seven times the nominal voltage. Most microturbine power electronics are generating three-phase electricity. Electronic components also direct all of the operating and startup functions. Microturbines are generally equipped with controls that allow the unit to be operated in parallel or independent of the grid and internally incorporate many of the grid and system protection features required of interconnect. The controls also allow for remote monitoring and operation.

9.3.2 CHP Operation

In CHP operation, a second heat exchanger, the *exhaust gas heat exchanger*, transfers the remaining energy from the microturbine exhaust to a hot water system. Exhaust heat can be used for a number of different applications, including potable water heating, driving absorption cooling and desiccant dehumidification equipment, space heating, process heating, and other building or site uses. Some microturbine-based CHP applications do not use recuperators. With these microturbines, the temperature of the exhaust is higher and thus more heat is available for recovery (see Figure 9.1).

9.4 Design Characteristics

- *Thermal output*—Microturbines produce thermal output at temperatures in the range of 400 to 600°F, suitable for supplying a variety of building thermal needs.

- *Fuel flexibility*—Microturbines can operate using a number of different fuels: natural gas, sour gases (high sulfur, lower Btu content), and liquid fuels such as gasoline, kerosene, and diesel fuel/heating oil.

- *Reliability and life*—Design life is estimated to be in the range of 40,000 to 80,000 hours. Although units have demonstrated reliability, they have not been in commercial service long enough to provide definitive life data.

- *Size range*—Microturbines available and under development are sized from 30 to 250 kW.

- *Emissions*—Low inlet temperatures and high fuel-to-air ratios result in NO_x emissions of less than 10 parts per million (ppm) when running on natural gas.

- *Modularity*—Units may be connected in parallel to serve larger loads and provide power reliability.

- *Part-load operation*—Because microturbines reduce power output by reducing mass flow and combustion temperature, efficiency at part load can be below that of full-power efficiency.

- *Dimensions*—About 12 cubic feet.

9.5 Microturbine Performance Characteristics

Microturbines are more complex than conventional simple-cycle gas turbines, as the addition of the recuperator both reduces fuel consumption (thereby substantially increasing efficiency) and introduces additional internal pressure losses that moderately lower efficiency and power. As the recuperator has four connections—to the compressor discharge, the expansion turbine discharge, the combustor inlet, and the system exhaust—it becomes a challenge to the microturbine product designer to make all of the connections in a manner that minimizes pressure loss, keeps manufacturing costs low, and entails the least compromise of system reliability. Various manufacturers' models have evolved in unique ways.

The addition of a recuperator opens numerous design parameters to performance–cost tradeoffs. In addition to selecting the pressure ratio for high efficiency and best business opportunity (high power for lower price), the recuperator has two performance parameters, effectiveness and pressure drop, that also have to be selected for the combination of efficiency and cost that creates the best business conditions. Higher effectiveness recuperation requires greater recuperator surface area, which both increases cost and incurs additional pressure drop. Such increased internal pressure drops reduce net power production and consequently increase microturbine cost per kilowatt.

Microturbine performance, in terms of both efficiency and specific power (i.e., power produced by the machine per unit of mass flow though the machine), is highly sensitive to small variations in component performance and internal losses. This is because the high-efficiency recuperated cycle processes a much larger amount of air and combustion products flow per kilowatt of net power delivered than is the case for high-pressure-ratio, simple-cycle machines. When the net output is the small difference between two large numbers (the compressor and expansion turbine work per unit of mass flow), small losses in component efficiency, internal pressure losses, and recuperator effectiveness have large impacts on net efficiency and net power per unit of mass flow. For these reasons, it is advisable to focus on trends and comparisons in considering performance, while relying on manufacturers' guarantees for precise values.

9.5.1 Effects of Ambient Conditions on Performance

The ambient conditions under which a microturbine operates have a noticeable effect on both the power output and efficiency. At elevated inlet air temperatures, both the power and efficiency decrease. The power decreases due to the decreased airflow mass rate (because the density of air declines as temperature increases), and the efficiency decreases because the compressor requires more power to compress air of higher temperature. Conversely, the power and efficiency increase with reduced inlet air temperature.

9.5.2 Heat Recovery

Microturbine system economics are improved by effective use of the thermal energy contained in the exhaust gas. Exhaust heat can be recovered and used in a variety of ways, including water heating, space heating, and driving thermally activated equipment such as an absorption chiller or a desiccant dehumidifier. Microturbine CHP system efficiency is a function of exhaust heat temperature. Recuperator effectiveness strongly influences the microturbine exhaust temperature. Consequently, the various microturbine CHP systems have substantially different CHP efficiency and net heat rate chargeable to power. These variations in CHP efficiency and net heat rate are mostly due to the mechanical design and manufacturing cost of the recuperators and their resulting impact on system cost, rather than being due to differences in system size.

9.5.3 Performance and Efficiency Enhancements

9.5.3.1 Recuperators

Most microturbines include built-in recuperators. The inclusion of a high-effectiveness (90%) recuperator essentially doubles the efficiency of a microturbine with a pressure ratio of 3.2, from about 14% to about 29%, depending on component details. Without a recuperator, such a machine would be suitable only for emergency, backup, or possibly peaking power operation. With the addition of the recuperator, a microturbine can be suitable for intermediate duty or price-sensitive baseload service.

While recuperators previously in use on industrial gas turbines developed leaks attributable to the consequences of differential thermal expansion accompanying thermal transients, microturbine recuperators have proven quite durable in testing to date. This

---

DID YOU KNOW?

Effectiveness is the technical term in the heat exchanger industry for the ratio of the actual heat transferred to the maximum achievable.

durability has resulted from using higher strength alloys and higher quality welding along with engineering design to avoid the internal differential expansion that causes internal stresses and leakage. Such practical improvements result in recuperators being of appreciable cost, which detracts from the economic attractiveness of the microturbine. The cost of a recuperator becomes easier to justify as the number of full-power operational hours per year increases.

Incorporation of a recuperator into the microturbine results in pressure losses in the recuperator itself and in the ducting that connects it to other components. Typically, these pressure losses result in 10 to 15% less power being produced by the microturbine and a corresponding loss of a few points in efficiency. The pressure loss parameter in gas turbines that is the measure of lost power is $\delta p/p$. As $\delta p/p$ increases, the net pressure ratio available for power generation decreases, and hence the power capability of the expansion process diminishes as well.

9.5.3.2 Firing Temperature

With firing temperatures well above those of the metallurgical limit of the best gas turbine alloys, large turbines (25 to 2000 lb/sec of mass flow) are usually equipped with internal cooling capability to permit optimum operation. Indeed, progress to higher and higher gas turbine efficiency, via higher firing temperatures, has occurred more through the development and advancement of blade and vane internal cooling technology than through improvement of the high-temperature capabilities of gas turbine alloys. Unfortunately for microturbine development, the nature of the three-dimensional shape of radial inflow turbines has not yet lent itself to the development of a manufacturing method that can produce internal cooling. Consequently, microturbines are limited to firing temperatures within the capabilities of gas turbine alloys. An ongoing program at the U.S. Department of Energy (DOE) Office of Energy Efficiency seeks to apply the technology of ceramic radial inflow turbines (previously advanced for the purpose of developing automotive gas turbines) to microturbines to increase their efficiency to 36% (higher heating value, HHV). The design and materials technology from the previous efforts are applicable, because the automotive gas turbines were in the same size range, and the same general geometry, as those used in microturbines.

9.5.3.3 Inlet Air Cooling

The decreased power and efficiency of microturbines at high ambient temperatures means that microturbine performance is at its lowest at the times power is often in greatest demand and most valued. The use of inlet air cooling can mitigate the decreased power and efficiency resulting from high ambient air temperatures. Although inlet air cooling is not a feature of today's microturbines, cooling techniques now entering the market on large gas turbines can be expected to work their way to progressively smaller equipment sizes and, at some future date, to be used with microturbines.

Evaporative cooling, a relatively low capital cost technique, is the most likely to be applied to microturbines. It disperses a very fine spray of water directly into the inlet air stream. Evaporation of the water reduces the temperature of the air. Because cooling is limited to the wet-bulb air temperature, evaporative cooling is most effective when the wet-bulb temperature is appreciably below the dry-bulb (ordinary) temperature. In most locales with high daytime dry-bulb temperatures, the wet-bulb temperature is often 20°F lower. This affords an opportunity for substantial evaporative cooling; however, evaporative cooling can consume large quantities of water, making it difficult to operate in arid climates.

It is also technically feasible to use refrigeration cooling in microturbines. In refrigeration cooling, a compression-driven or thermally activated (absorption) refrigeration cycle cools the inlet air through a heat exchanger. The heat exchanger in the inlet air stream causes an additional pressure drop in the air entering the compression, thereby slightly lowering cycle power and efficiency. Because the inlet air is now substantially cooler than the ambient air, however, there is a significant net gain in power and efficiency. Electric motor compression refrigeration requires a substantial parasitic power loss. Thermally activated absorption cooling can use waste heat from the microturbine, reducing the direct parasitic loss. The relative complexity and cost of these approaches, in comparison with evaporative cooling, render them less likely.

Finally, it is also technically feasible to use thermal energy storage systems—typically ice, chilled water, or low-temperature fluids—to cool inlet air. These systems eliminate most parasitic losses from the augmented power capacity. Thermal energy storage is a viable option if on-peak power pricing occurs only a few hours a day. In that case, the shorter time of energy storage discharge and longer time for daily charging allow for a smaller and less expensive thermal energy storage system.

9.5.4 Maintenance

Normal maintenance includes periodic air and fuel filter inspections and changes, igniter and fuel injector replacement, and major overhauls of the turbine itself. A typical microturbine maintenance schedule includes

- 8000 hours—Replace air and fuel filters
- 16,000 to 20,000 hours—Inspect/replace fuel injectors, igniters, and thermocouples
- 20,000 hours—Battery replacement (standalone units)
- 40,000 hours—Major overhaul, core turbine replacement

In addition to the microturbine itself, the fuel compressor also requires periodic inspection and maintenance. The actual level of compressor maintenance depends on the inlet pressure and site conditions.

A major overhaul is required every 30,000 to 40,000 turbine run hours depending on technology and service. A typical overhaul consists of replacing the main shaft with the compressor and turbine attached, and inspecting and if necessary replacing the combustor. At the time of the overhaul, other components are examined to determine if wear has occurred, with replacements made as required.

Maintenance requirements can be affected by fuel type and site conditions. Waste gas and liquid fuel applications may require more frequent inspections and component replacement than natural gas systems. Microturbines operating in dusty or dirty environments require more frequent inspections and filter replacements.

DID YOU KNOW?

Most manufacturers offer service contracts that cover scheduled and unscheduled events. The cost of full service contracts covers the inspections and component replacements listed above, including the major overhaul.

9.5.5 Fuels

Microturbines have been designed to use natural gas as their primary fuel; however, they are able to operate on a variety of fuels, including the following:

- *Liquefied petroleum gas (LPG)*—Propane and butane mixtures
- *Sour gas*—Unprocessed natural gas as it comes directly from the gas well
- *Biogas*—Combustible gas produced from biological degradation of organic wastes, such as landfill gas, sewage digester gas, and animal waste digester gas
- *Industrial waste gases*—Flare gases and process off-gases from refineries, chemical plants, and steel mills
- *Manufactured gases*—Typically low- and medium-Btu gas produced as products of gasification of pyrolysis processes

With some waste fuels, contaminants are a concern. This is especially the case with acid gas components (H_2S, halogen acids, hydrogen cyanide; ammonia; salts and metal-containing compounds; organic halogen-, sulfur-, nitrogen-, and silicon-containing compounds) and oils. In combustion, halogen and sulfur compounds form halogen acids, SO_2, some SO_3, and possibly H_2SO_4 emissions. The acids can also corrode downstream equipment. A substantial fraction of any fuel nitrogen oxidizes into NO_x in combustion. Solid particulates must be kept to low concentrations to prevent corrosion and erosion of components. Various fuel scrubbing, droplet separation, and filtration steps will be required if any fuel contaminant levels exceed manufacturer specifications. Landfill gas in particular often contains chlorine compounds, sulfur compounds, organic acids, and silicon compounds that dictate pretreatment.

9.5.6 Availability

Generally, systems in the field have shown a high level of availability and reliability. The basic design and low number of moving parts hold the potential for systems of high availability; manufacturers have targeted availabilities of 98 to 99%. The use of multiple units or backup units at a site can further increase the availability of the overall facility.

9.5.7 Disadvantages

As mentioned earlier, microturbines have many advantages for use in WWTPs; however, one disadvantage of microturbines is a limit on the number of times they can be turned on. Microturbines also run at very high speeds and high temperatures, causing noise pollution for nearby residents and potential risks for operators and maintenance staff. It may also take several microturbines set in a series to provide enough energy to power a small WWTP (USEPA, 2006a).

9.6 Emissions

When properly operated and maintained, microturbines have the potential for extremely low emissions. All microturbines operating on gaseous fuels feature lean premixed (dry low NO_x, or DLN) combustor technology, which was developed relatively recently in the history of gas turbines and is not universally featured on larger gas turbines. All of the example commercial units have been certified to meet extremely stringent standards in Southern California of less than 4 to 5 ppmvd (parts per million, volumetric dry) of NO_x (15% O_2). Carbon monoxide and volatile organic compound emissions are at the same level. "Non-California" versions have NO_x emissions of less than 9 ppmvd.

CASE STUDY 9.1. SAN ELIJO JOINT POWER WATER RECLAMATION FACILITY

In 2000, microturbines were installed at San Elijo Joint Power Water Reclamation Facility. San Elijo is a small (3 MGD) WWTP. Instead of just burning the excess gas, San Elijo now uses its bio gas from its digesters to fuel the microturbines. Three microturbines were installed, producing 80 kW of energy. The system produces approximately 15% of the plant's demand. The exhaust from the microturbines was captured and used to heat water at the reclamation facility, a process known as combined heat and power (CHP). One generator system uses one fuel source to yield two usable energy outputs with very high fuel efficiency. The plant experienced a decline in electricity costs estimated at $4000 per month and expects a payback on their investment in 3 to 4 years. The microturbine exhaust is also lower in methane and NO_x than emissions from flaring digester gas and substantially less than those from conventional reciprocating engine-driven generator sets. In addition to realizing savings in peak energy demand charges, the San Elijo Joint Power Authority also received a $76,000 rebate check from San Diego Gas and Electric's Self-Generation Incentive Program (USEPA, 2006a).

References and Recommended Reading

Spellman, F.R. (2007). *Handbook of Water and Wastewater Treatment Plant Operations*, 2nd ed., CRC Press, Boca Raton, FL.

USEPA. (2006a). *Auxiliary and Supplemental Power Fact Sheet: Microturbines*, EPA 832-F-05-014, U.S. Environmental Protection Agency, Washington, DC.

USEPA. (2006b). *Auxiliary and Supplemental Power Fact Sheet: Viable Sources*, EPA 832-F-05-009, U.S. Environmental Protection Agency, Washington, DC.

USEPA. (2008a). *Technology Characterization: Microturbines*, U.S. Environmental Protection Agency, Washington, DC (www.epa.gov/chp/documents/catalog_chptech_microturbines.pdf).

USEPA. (2008b). *Catalog of CHP Technologies*, U.S. Environmental Protection Agency, Washington, DC (www.epa.gov/chp/documents/catalog_chptech_full.pdf).

WSEO. (1993). *Improving the Energy Efficiency of Wastewater Treatment Facilities*, WSEO-192, Washington State Energy Office, Olympia.

10

Reciprocating Engines[*]

Industrial man—a sentient reciprocating engine having a fluctuating output, coupled to an iron wheel revolving with uniform velocity. And then we wonder why this should be the golden age of revolution and mental derangement.

—Aldous Huxley (*Time Must Have a Stop*, 1944)

Electric generators can be furnished with engines that can run on diesel fuel, natural gas, or bio-gas. In many cases the engine can be provided with duel fuel capability. All of the engines currently being manufactured are required to meet Clean Air Act (CAA) emissions requirements as stated in sections 89-90, Chapter 40 of the Code of Federal Regulations.

—USEPA (2006b)

10.1 Introduction

Reciprocating internal combustion engine technology is widespread and well known. North American production exceeds 35 million units per year for automobiles, trucks, construction and mining equipment, marine propulsion, lawn care, and a diverse set of power generation applications. Various stationary engine products are available for a range of power generation market applications and duty cycles, including standby and emergency power peaking service, intermediate and base-load power, and combined heat and power (CHP). Reciprocating engines are available for power generation applications in sizes ranging from a few kilowatts to over 5 MW.

The two basic types of reciprocating engines are *spark ignition* (SI) and *compression ignition* (CI). Spark ignition engines for power generation use natural gas as the preferred fuel, although they can be set up to run on propane, gasoline, or landfill gas. Compression ignition engines (often called diesel engines) operate on diesel fuel or heavy oil, or they can be set up to run in a dual-fuel configuration that burns primarily natural gas with a small amount of diesel pilot fuel.

Diesel engines have historically been the most popular type of reciprocating engine for both small and large power generation applications; however, in the United States and other industrialized nations, diesel engines are increasingly restricted to emergency standby or limited duty-cycle service because of air emission concerns. Consequently, the natural gas-fueled SI engine is now the engine of choice for the higher-duty-cycle stationary power market (over 500 hr/yr) and is the primary focus of this chapter.

---

[*] Material in this chapter is adapted from USEPA, *Catalog of CHP Technologies*, U.S. Environmental Protection Agency, Washington, DC, 2008.

DID YOU KNOW?

Stoichiometric ratio is the chemically correct ratio of fuel to air for complete combustion; that is, there is no unused fuel or oxygen after combustion.

Current-generation natural gas engines offer low first cost, fast startup, proven reliability when properly maintained, excellent load-following characteristics, and significant heat recovery potential. Electric efficiencies of natural gas engines range from 30% (lower heating value, LHV) for small stoichiometric engines (<100 kW) to over 40% (LHV) for large lean burn engines (>3 MW). Waste heat recovered from the hot engine exhaust and from the engine cooling systems produces either hot water or low-pressure steam for CHP applications. Overall CHP system efficiencies (electricity and useful thermal energy) of 65 to 80% are routinely achieved with natural gas engine systems.

Reciprocating engine technology has improved dramatically over the past three decades, driven by economic and environmental pressures for power density improvements (more output per unit of engine displacement), increased fuel efficiency, and reduced emissions. Computer systems have greatly advanced reciprocating engine design and control by accelerating advanced engine designs and making possible more precise control and diagnostic monitoring of the engine process. Stationary engine manufacturers and worldwide engine research and development (R&D) firms continue to drive advanced engine technology, including accelerating the diffusion of technology and concepts from the automotive market to the stationary market. The emissions signature of natural gas SI engines in particular has improved significantly in the last decade through better design and control of the combustion process and through the use of exhaust catalysts. Advanced lean-burn natural gas engines are available that produce NO_x levels as low as 50 ppmv at 15% O_2 (dry basis).

10.2 Applications

Reciprocating engines are well suited to a variety of distributed generation applications. These varied applications include industrial, commercial, and institutional facilities in the United States and Europe for power generation and CHP. Reciprocating engines start quickly, follow load well, have good part-load efficiencies, and generally have high reliabilities. In many cases, multiple reciprocating engine units further increase overall plant capacity and availability. Reciprocating engines have higher electrical efficiencies than gas turbines of comparable size and thus low fuel-related operating costs. In addition, the first costs of reciprocating engine gensets (generator sets) are generally lower than gas turbine gensets up to 3 to 5 MW in size. Reciprocating engine maintenance costs are generally higher than comparable gas turbines, but the maintenance can often be handled by in-house staff or provided by local service organizations.

Potential distributed generation applications for reciprocating engines include standby, peak shaving, grid support, and CHP applications in which hot water, lower pressure steam, or waste-heat-fired absorption chillers are required. Reciprocating engines are also used extensively as direct mechanical drives in applications such as water pumping, air and gas compression, and chilling/refrigeration.

10.2.1 Combined Heat and Power

The use of reciprocating engines for various distributed generation applications is expected to grow; the most prevalent onsite generation application for natural gas SI engines has traditionally been CHP, and this trend is likely to continue. The economics of natural gas engines in onsite generation applications are enhanced by effective use of the thermal energy contained in the exhaust gas and cooling systems, which generally represents 60 to 70% of the inlet fuel energy.

The four sources of usable waste heat from a reciprocating engine are exhaust gas, engine jacket cooling water, lube oil cooling water, and turbocharger cooling. Recovered heat is in the form of hot water or low-pressure steam (<30 psig). The high-temperature exhaust can generate medium-pressure steam (up to about 150 psig), but the hot exhaust gas contains only about one half of the available thermal energy from a reciprocating engine. Some industrial CHP applications use the engine exhaust gas directly for process drying. Generally, the hot water and low-pressure steam produced by reciprocating engine CHP systems are appropriate for low-temperature process needs, space heating, potable water heating, and driving absorption chillers that provide cold water, air conditioning, or refrigeration.

There are many engine-based CHP systems operating in the United States in a variety of applications, including universities, hospitals, water treatment facilities, industrial facilities, and commercial and residential buildings. Facility capacities range from 30 kW to 30 MW, and many larger facilities are comprised of multiple units. Spark ignition engines fueled by natural gas or other gaseous fuels represent 84% of the installed reciprocating engine CHP capacity.

Thermal loads most amenable to engine-driven CHP systems in commercial/institutional buildings are space heating and hot water requirements. The simplest thermal load to supply is hot water. The primary applications for CHP in the commercial/institutional and residential sectors are those building types with relatively high and coincident electric and hot water demand such as colleges and universities, hospitals and nursing homes, multifamily residential buildings, and lodging. If space heating needs are incorporated, office buildings and certain warehousing and mercantile/service applications can be economic applications for CHP. Technology development efforts targeted at heat-activated cooling/refrigeration and thermally regenerated desiccants expand the application of engine-driven CHP by increasing the thermal energy loads in certain building types. Use of CHP thermal output for absorption cooling or desiccant dehumidification could increase the size and improve the economics of CHP systems in existing CHP markets such as schools, multifamily residential buildings, lodging, nursing homes, and hospitals. Use of these advanced technologies in applications such as restaurants, supermarkets, and refrigerated warehouses provides a base thermal load that opens these applications to CHP.

A typical commercial application for reciprocating engine CHP is a hospital or healthcare facility with a 1-MW CHP system comprised of multiple 200- to 300-kW natural gas engine gensets. The system is designed to satisfy the baseload electric needs of the facility.

DID YOU KNOW?

Peak shaving is a process whereby electrical demand is shifted from peak times (e.g., noon) to times with lower demand (e.g., night), thus shaving the peak.

Approximately 1.6 MW therm (MWth) of hot water is recovered from engine exhaust and engine cooling systems to provide space heating and domestic hot water to the facility and to drive absorption chillers for space conditioning during summer months. Overall efficiency of this type of CHP can exceed 70%.

Industry also uses engine-driven CHP in a variety of industrial applications where hot water or low-pressure steam is required. A typical industrial application for engine CHP would be a food processing plant with a 2-MW natural gas engine-driven CHP system comprised of multiple 500- to 800-kW engine gensets. The system provides baseload power to the facility and approximately 2.2 MWth low-pressure steam for process heating and washdown. Overall efficiency for a CHP system of this type approaches 75%.

With regard to wastewater treatment plants (WWTPs), most have electric power connections to at least two independent power substations, such that if power from one substation fails (e.g., due to a localized storm, power plant malfunction, reactor shutdown, downing of a local power line), the WWTP could receive power from the other substation. If the entire grid fails (such as it did for much of the northeast and the Great Lakes states in August 2003), having power feeds from separate substations that are all connected to the same main grid will not meet the auxiliary power needs to keep many WWTPs operating during such a failure. Without an adequate reliable auxiliary power source, many WWTPs will discharge (i.e., bypass treatment unit processes) untreated sewage into the receiving waters (USEPA, 2006b).

10.3 Reciprocating Engine Technology

The two primary reciprocating engine designs relevant to stationary power generation applications are the spark ignition Otto-cycle engine and the compression ignition diesel-cycle engine. The essential mechanical components of the Otto-cycle and diesel-cycle engines are the same. Both use a cylindrical combustion chamber in which a close-fitting piston travels the length of the cylinder. The piston connects to a crankshaft that transforms the linear motion of the piston into the rotary motion of the crankshaft. Most engines have multiple cylinders that power a single crankshaft.

The primary difference between the Otto and diesel cycles is the method of igniting the fuel. Spark ignition (Otto-cycle) engines use a spark plug to ignite a premixed air–fuel mixture introduced into the cylinder. Compression ignition (diesel-cycle) engines compress the air introduced into the cylinder to a high pressure, raising its temperature to the auto-ignition temperature of the fuel that is injected at high pressure.

Engines are further categorized by crankshaft speed (rpm), operating cycle (two- or four-stroke), and whether turbocharging is used. Reciprocating engines are also categorized by their original design purpose—automotive, truck, industrial, locomotive, and marine. Hundreds of small-scale stationary power, CHP, irrigation, and chiller applications use automotive engine models. These are generally low-priced engines due to large production volumes; however, unless conservatively rated, these engines have limited durability. Truck engines have the cost benefit of production volume and are designed for reasonably long life (e.g., 1 million miles). Several truck engines are available as stationary engines. Engines intended for industrial use are designed for durability and for a wide range of mechanical drive and electric power applications. Their sizes range from 20 kW up to 6 MW, including industrialized truck engines in the range of 200 to 600 kW and industrially applied marine and locomotive engines above 1 MW.

Both the spark ignition and the diesel four-stroke engines most relevant to stationary power generation applications complete a power cycle in four strokes of the piston within the cylinder:

1. *Intake stroke*—Introduction of air (diesel) or air–fuel mixture (spark ignition) into the cylinder.
2. *Compression stroke*—Compression of air or an air–fuel mixture within the cylinder. In diesel engines, the fuel is injected at or near the end of the compression stroke (top dead center, or TDC) and ignited by the elevated temperature of the compressed air in the cylinder. In spark ignition engines, the compressed air–fuel mixture is ignited by an ignition source at or near TDC.
3. *Power stroke*—Acceleration of the piston by expansion of the hot, high-pressure combustion gases.
4. *Exhaust stroke*—Expulsion of combustion products from the cylinder through the exhaust port.

10.4 Design Characteristics

Several features have made reciprocating engines a leading prime mover for CHP and other distributed generation applications, including the following:

- *Size range*—Reciprocating engines are available in sizes from 10 kW to over 5 MW.
- *Thermal output*—Reciprocating engines can produce hot water and low-pressure steam.
- *Fast start-up*—The fast start-up capability of reciprocating engines allows timely resumption of the system following a maintenance procedure. In peaking or emergency power applications, reciprocating engines can quickly supply electricity on demand.
- *Black-start capability*—In the event of an electric utility outage, reciprocating engines require minimal auxiliary power requirements. Generally only batteries are required.
- *Availability*—Reciprocating engines have typically demonstrated availability in excess of 95% in stationary power generation applications.
- *Part-load operation*—The high part-load efficiency of reciprocating engines ensures economical operation in electric load following applications.
- *Reliability and life*—Reciprocating engines have proven to be reliable power generators, given proper maintenance.
- *Emissions*—Diesel engines have relatively high emissions levels of NO_x and particulates; however, natural gas spark ignition engines have improved emissions profiles.

10.5 Performance Characteristics

10.5.1 Electrical Efficiency

Approximately 50 to 60% of the waste heat from an engine system is recovered from jacket cooling water and lube oil cooling systems at a temperature too low to produce steam. This feature is generally less critical in commercial/institutional applications where it is more common to have hot-water thermal loads. Steam can be produced from the exhaust heat if required (maximum pressure of 150 psig), but if no hot water is needed then the amount of heat recovered from the engine is reduced and total CHP system efficiency drops accordingly.

10.5.2 Load Performance

In power generation and CHP applications, reciprocating engines generally drive synchronous generators at constant speed to produce steady alternating current (AC) power. As load is reduced, the heat rate of spark ignition engines increases and efficiency decreases.

10.5.3 Heat Recovery

The economics of engines in onsite power generation applications often depend on effective use of the thermal energy contained in the exhaust gas and cooling systems, which generally represents 60 to 70% of the inlet fuel energy. Most of the waste heat is available in the engine exhaust and jacket coolant, while smaller amounts can be recovered from the lube oil cooler and the turbocharger's intercooler and aftercooler (if so equipped). The most common use of this heat is to generate hot water or low-pressure steam for process use or for space heating, process needs, and domestic hot water or absorption cooling. However, the engine exhaust gases can be used as a source of direct energy for drying or other direct heat processes.

10.5.3.1 Closed-Loop Cooling Systems

Figure 10.1 depicts the closed-loop cooling system, the most common method of recovering engine heat. These systems are designed to cool the engine by forced circulation of a coolant through engine passages and an external heat exchanger. An excess heat exchanger transfers engine heat to a cooling tower or radiator when there is excess heat generated. Closed-loop water cooling systems can be operated at coolant temperatures

DID YOU KNOW?

Electrical efficiency increases as engine size becomes larger. As electrical efficiency increases, the absolute quantity of thermal energy available to produce useful thermal energy decreases per unit of power output, and the ratio of power to heat for the CHP system generally increases. A changing ratio of power to heat impacts project economics and may affect the decisions that customers make in terms of CHP acceptance, sizing, and the desirability of selling power.

DID YOU KNOW?

Heat in the engine jacket coolant accounts for up to 30% of the energy input and is capable of producing 200 to 210°F hot water. Some engines, such as those with high-pressure or ebullient (i.e., boiling coolant) cooling systems, can operate with water jacket temperatures up to 265°F. Engine exhaust heat represents from 30 to 50% of the available waste heat. Exhaust temperatures of 850 to 1200°F are typical. By recovering heat in the cooling systems and exhaust, approximately 70 to 80% of the fuel's energy can be effectively utilized to produce both power and useful thermal energy.

from 190 to 250°F. Depending on the engine and requirements of the CHP system, the lube oil cooling and turbocharger aftercooling may be either separate or part of the jacket cooling system.

10.5.3.2 Ebullient Cooling Systems

Ebullient cooling systems cool the engine by natural circulation of a boiling coolant through the engine. This type of cooling system is typically used in conjunction with exhaust heat recovery for production of low-pressure steam. Cooling water is introduced at the bottom of the engine where the transferred heat boils the coolant, generating two-phase flow. The formation of bubbles lowers the density of the coolant, causing a natural circulation to the top of the engine. The coolant at the engine outlet is maintained at saturated steam conditions and is usually limited to 250°F and a maximum of 15 psig. Inlet cooling water is also near saturation conditions and is generally 2 to 3°F below the outlet temperature. The uniform temperature throughout the coolant circuit extends engine life and contributes to improved combustion efficiencies.

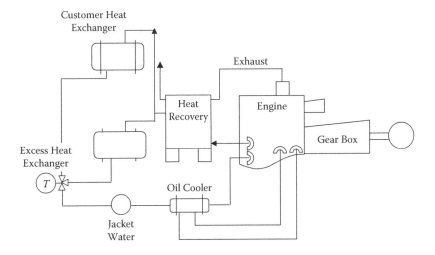

FIGURE 10.1
Closed-loop heat recovery system.

10.5.3.3 Exhaust Heat Recovery

Exhaust heat is typically used to generate hot water to about 230°F or low-pressure steam (up to 150 psig). Only a portion of the exhaust heat can be recovered, as exhaust gas temperatures are generally kept above temperature thresholds to prevent the corrosive effects of condensation in the exhaust piping. For this reason, most heat recovery units are designed for a 250 to 350°F exhaust outlet temperature. Exhaust heat recovery can be independent of the engine cooling system or coupled with it; for example, hot water from the engine cooling can be used as feedwater or feedwater preheat to the exhaust recovery unit. In a typical district heating system, jacket cooling, lube oil cooling, single-stage aftercooling, and exhaust gas heat recovery are all integrated for steam production.

10.5.4 Performance and Efficiency Enhancements

10.5.4.1 BMEP and Engine Speed

Engine power is related to engine speed and the brake mean effective pressure (BMEP) during the power stroke. BMEP can be regarded as an average cylinder pressure on the piston during the power stroke and is a measure of the effectiveness of engine power output or mechanical efficiency. Engine manufacturers often include BMEP values in their product specifications. Typical BMEP values are as high as 230 psig for large natural gas engines and 350 psig for diesel engines. Corresponding peak combustion pressures are about 1750 psig and 2600 psig, respectively. High BMEP levels increase power output, improve efficiency, and result in lower specific costs ($/kW). The BMEP can be increased by raising combustion cylinder air pressure through increased turbocharging, improved aftercooling, and reduced pressure losses through improved air passage design. These factors all increase air charge density and raise peak combustion pressures, translating into high BMEP levels. However, higher BMEP increases thermal and pneumatic stresses within the engine, and proper design and testing are required to ensure continued engine durability and reliability.

10.5.4.2 Turbocharging

Essentially all modern engines above 300 kW are turbocharged to achieve higher power densities. A turbocharger is basically a turbine-driven intake air compressor. The hot, high-velocity exhaust gases leaving the engine cylinders power the turbine. Very large engines typically are equipped with two turbochargers. On a carbureted engine, turbocharging forces more air and fuel into the cylinders, increasing engine output. On a fuel-injected engine, the mass of fuel injected must be increased in proportion to the increased air input. Cylinder pressure and temperature normally increase as a result of turbocharging, increasing the tendency for detonation for both spark ignition and dual-fuel engines and requiring a careful balance between compression ratio and turbocharger boost level. Turbochargers normally boost inlet air pressure on a 3:1 to 4:1 ratio. A wide range of turbocharger designs

---

DID YOU KNOW?

Brake mean effective pressure (BMEP) is the average (mean) pressure that, if imposed on the pistons uniformly from the top to the bottom of each power stroke, would produce the measured (brake) power output.

DID YOU KNOW?

Predictive maintenance is a proactive method of establishing baseline performance data, monitoring performance criteria over a period of time, and observing changes in performance so that failure can be predicted and maintenance can be performed on a planned, scheduled basis.

and models are used. Heat exchangers (aftercoolers or intercoolers) are often used on the discharge air from the turbocharger to keep the temperature of the air to the engine under a specified limit. Intercooling on forced induction engines improves volumetric efficiency by increasing the density of intake air to the engine; that is, the cold air charge from intercooling provides denser air for combustion, thus allowing more fuel and air to be combusted per engine stroke and increasing the output of the engine.

10.5.5 Maintenance

The type, speed, size, and numbers of engine cylinders influence costs of maintenance and typically include

- Maintenance labor
- Engine parts and materials, such as oil filters, air filters, spark plugs, gaskets, valves, piston rings, electronic components, etc., and consumables such as oil
- Minor and major overhauls

Maintenance can be either done by in-house personnel or contracted out to manufacturers, distributors, or dealers under service contracts. Full maintenance contracts (covering all recommended service) generally cost 0.7 to 2.0 cents/kWh depending on engine size. Many service contracts now include remote monitoring of engine performance and condition and allow for predictive maintenance. Service contract rates typically are all inclusive and include the travel time of technicians on service calls.

Recommended service is comprised of routine, short-interval inspections and adjustments and periodic replacement of engine oil and filters, coolant, and spark plugs (typically 500 to 2000 hours). An oil analysis is part of most preventative maintenance programs to monitor engine wear. A top-end overhaul is generally recommended between 8000 and 30,000 hours of operation and entails a cylinder head and turbocharger rebuild. A major overhaul is performed after 30,000 to 72,000 hours of operation and involves piston/liner replacement, crankshaft inspection, and inspection of bearings and seals.

10.5.6 Fuels

Spark ignition engines operate on a variety of alternative gaseous fuels:

- *Liquefied petroleum gas (LPG)*—Propane and butane mixtures
- *Sour gas*—Unprocessed natural gas as it comes directly from the gas well
- *Biogas*—Any of the combustible gases produced from biological degradation of organic waste, such as landfill gas, sewage digester gas, and animal waste digester gas

- *Industrial waste gases*—Flare gases and process off-gases from refineries, chemical plants, and steel mills
- *Manufactured gases*—Typically low- and medium-Btu gas produced as products of gasification or pyrolysis processes

Factors that impact the operation of a spark ignition engine with alternative gaseous fuels include the following:

- Because engine fuel is delivered on a volume basis, fuel volume into the engine increases as heating value decreases, requiring engine derating on fuels with very low Btu content. Derating is more pronounced with naturally aspirated engines and, depending on air requirements, turbocharging partially or totally compensates.
- Autoignition characteristics and detonation tendency vary among types of fuel.
- Contaminants may impact engine component life or engine maintenance or result in air pollutant emissions that require additional control measures.
- Hydrogen-containing fuels may require special measures (generally if hydrogen content by volume is greater than 5%) because of the unique flammability and explosion characteristics of hydrogen.

Contaminants are a concern with many waste fuels, specifically acid components (hydrogen sulfide, halogen acids, hydrogen cyanide); ammonia; salts and metal-containing compounds; organic halogen-, sulfur-, nitrogen-, and silicon-containing compounds; and oils. In combustion, halogen and sulfur compounds form halogen acids, SO_2, some SO_3, and possibly H_2SO_4 emissions. The acids can also corrode downstream equipment. A substantial fraction of any fuel nitrogen oxidizes NO_x in combustion. To prevent corrosion and erosion of components, solid particulates must be kept to very low concentrations. Various fuel scrubbing, droplet separation, and filtration steps will be required if any fuel contaminant levels exceed manufacturer specifications. Landfill gas in particular often contains chlorine compounds, sulfur compounds, organic acids, and silicon compounds, which dictate pretreatment. Once treated and acceptable for use in the engine, alternative fuels have emissions performance profiles similar to those for natural gas engine performance. Specifically, the low emissions rates of lean-burn engines can usually be maintained on alternative fuels.

10.5.6.1 Liquefied Petroleum Gas

Liquefied petroleum gas (LPG) is composed primarily of propane or butane. Propane used in natural gas engines requires retarding of ignition timing and other appropriate adjustments. LPG often serves as a backup fuel where there is a possibility of interruption in the natural gas supply. LPG is delivered as a vapor to the engine. The use of LPG is limited

DID YOU KNOW?

The high butane content of LPG is recommended only for low-compression, naturally aspirated engines. Significantly retarded timing avoids detonation.

in high-compression engines because of its relatively low octane number. In general, LPG for engines contains 95% propane by volume with a higher heating value (HHV) of 2500 Btu/scf; the remaining 5% is lighter than butane. Off-spec LPG may require cooling to condense out large volumes of butane or heavier hydrocarbons.

10.5.6.2 Field Gas

Field gas often contains more than 5% by volume of heavy ends (butane and heavier), as well as water, salts, and hydrogen sulfide, and usually requires some scrubbing before use in natural gas engines. Cooling may be required to reduce the concentrations of butane and heavier components. Field gas usually contains some propane and normally is used in low-compression engines (both naturally aspirated and turbocharged). Retarded ignition timing eliminates detonation.

10.5.6.3 Biogas

Biogases (landfill gas and digester gas) are predominately mixtures of methane and carbon dioxide, with a HHV in the range of 300 to 700 Btu/scf. Landfill gas also contains a variety of contaminants as discussed earlier. Biogases are produced essentially at atmospheric pressure so must be compressed for delivery to the engine. After compression, cooling and scrubbing or filtration are required to remove compressor oil, condensate, and any particulates that may have been entrained in the original gas. Scrubbing with a caustic solution may be required if acid gases are present. Because of the additional requirements for raw gas treatment, biogas-powered engine facilities are more costly to build and operate than natural gas-based systems.

10.5.6.4 Industrial Waste Gases

Industrial waste gases that are common reciprocating engine fuels include refinery gases and process off-gases. Refinery gases typically contain components such as hydrogen, carbon monoxide, light hydrocarbons, hydrogen sulfide, and ammonia, as well as carbon dioxide and nitrogen. Process off-gases include a wide variety of compositions. Generally, waste gases are medium- to low-Btu content. Medium-Btu gases generally do not require significant engine derating; low-Btu gases usually require derating.

10.6 Emissions

The primary environmental concern with reciprocating engines are exhaust emissions. The primary pollutants are oxides of nitrogen (NO_x), carbon monoxide (CO), and volatile organic compounds (VOCs)—unburned, non-methane hydrocarbons. Other pollutants such as oxides of sulfur (SO_x) and particulate matter (PM) are primarily dependent on the fuel used. The sulfur content of the fuel determines emissions of sulfur compounds, primarily SO_2. Engines operating on natural gas or desulfurized distillate oil emit insignificant levels of SO_x. In general, SO_x emissions are an issue only in large, slow-speed diesels firing heavy oils. Particulate matter can be an important pollutant for engines using liquid fuels. Ash and metallic additives in the fuel contribute to PM in the exhaust.

DID YOU KNOW?

Depending on their origin and contaminants, industrial gases sometimes require pretreatment comparable to that applied to raw landfill gas. Particulates (e.g., catalyst dust), oils, condensable gases, water, C_4^+ hydrocarbons, and acid gases may all have to be removed.

10.6.1 Nitrogen Oxides (NO_x)

NO_x emissions are usually the primary concern with natural gas engines and are a mixture of (mostly) NO and NO_2 in variable composition. In measurement, NO_x is reported as parts per million by volume in which both species count equally (e.g., ppmv at 15% O_2, dry). Other common units for reporting NO_x in reciprocating engines are g/hp-hr and g/kWh, or as an output rate such as lb/hr. Among natural gas engine options, lean-burn natural gas engines produce the lowest NO_x emissions directly from the engine; however, rich-burn engines can more effectively make use of three-way catalysts to produce very low emissions. If lean-burn engines must meet extremely low emissions levels, as in the California Air Resources Board (CARB) 2007 standards of 0.07 lb/MWh, then selective catalytic reduction must be added. Rich-burn engines would qualify for this standard by taking a CHP credit for avoided boiler emissions. In addition, a commercial rich-burn engine with cold exhaust gas recirculation and a three-way catalyst has been tested below the CARB 2007 standard without the CHP credit. Operation at this ultra-low emissions level still in a commercial installation requires further development and refinement of control systems.

For any engine there are generally trade-offs between low NO_x emissions and high efficiency. There are also trade-offs between low NO_x emissions and emissions of the products of incomplete combustion (carbon monoxide and unburned hydrocarbons). Three main approaches to these trade-offs come into play depending on regulations and economics. One approach is to control for lowest NO_x by accepting a fuel efficiency penalty and possibly higher carbon monoxide and hydrocarbon emissions. A second option is finding an optimal balance between emissions and efficiency. A third option is to design for highest efficiency and use post-combustion exhaust treatment.

10.6.2 Carbon Monoxide

Carbon monoxide (CO) and VOCs both result from incomplete combustion. CO emissions result when there is inadequate oxygen or insufficient residence time at high temperature. Cooling at the combustion chamber walls and reaction quenching in the exhaust process also contribute to incomplete combustion and increased CO emissions. Excessively lean conditions can lead to incomplete and unstable combustion and high CO levels.

10.6.3 Unburned Hydrocarbons

Volatile hydrocarbons, also called volatile organic compounds (VOCs), can encompass a wide range of compounds, some of which are hazardous air pollutants. These compounds are discharged into the atmosphere when some portion of the fuel remains unburned or just partially burned. Some organics are carried over as unreacted trace constituents of

DID YOU KNOW?

Three mechanisms form NO_x: (1) thermal NO_x, (2) prompt NO_x, and (3) fuel-bound NO_x. The predominant NO_x formation mechanism associated with reciprocating engines is thermal NO_x. Thermal NO_x is the fixation of atmospheric oxygen and nitrogen, which occurs at high combustion temperatures. Flame temperature and residence time are the primary variables that affect thermal NO_x levels. The rate of thermal NO_x formation increases rapidly with flame temperature.

the fuel, while others may be pyrolysis products of the heavier hydrocarbons in the gas. Volatile hydrocarbon emissions from reciprocating engines are normally reported as non-methane hydrocarbons (NMHCs). Methane is not a significant precursor to ozone creation and smog formation and is not currently regulated.

10.6.4 Carbon Dioxide

Although not considered a pollutant in the ordinary sense of directly affecting health, emissions of carbon dioxide (CO_2) are of concern due to its contribution to global warming. Atmospheric warming occurs because solar radiation readily penetrates to the surface of the planet but infrared (thermal) radiation from the surface is absorbed by the CO_2 (and other polyatomic gases, such as methane, unburned hydrocarbons, refrigerants, and volatile chemicals) in the atmosphere, with resultant increases in temperature of the atmosphere. The amount of CO_2 emitted is a function of both fuel carbon content and system efficiency. The fuel carbon content of natural gas is 34 lb carbon per MMBtu. Oil is 48 lb carbon per MMBtu, and (ash-free) coal is 66 lb carbon per MMBtu.

References and Recommended Reading

Spellman, F.R. (2007). *Handbook of Water and Wastewater Treatment Plant Operations*, 2nd ed., CRC Press, Boca Raton, FL.

USEPA. (2006a). *Auxiliary and Supplemental Power Fact Sheet: Microturbines*, EPA 832-F-05-014, U.S. Environmental Protection Agency, Washington, DC.

USEPA. (2006b). *Auxiliary and Supplemental Power Fact Sheet: Viable Sources*, EPA 832-F-05-009, U.S. Environmental Protection Agency, Washington, DC.

USEPA. (2008). *Catalog of CHP Technologies*, U.S. Environmental Protection Agency, Washington, DC (www.epa.gov/chp/documents/catalog_chptech_full.pdf).

11

Steam Turbines[*]

Steam turbines have been generating power in America for many years. Power generated by steam turbines has lit the first light bulbs and propelled ships for over 100 years. In fact, the first power plant (run by Thomas Edison using his dynamos and located at Pearl Street in New York City) was a CHP plant that generated power using a steam turbine. The excess steam was used to heat homes. Today, most of the electricity produced in the U.S. is done so by steam turbines. It is safe to say the steam turbines are a known, well understood and proven technology.

—**Rutgers University,** *Steam Turbines Used for Combined Heat and Power*

11.1 Introduction

Steam turbines are one of the most versatile and oldest prime mover technologies still in general production and used to drive a generator or mechanical machinery. As noted above, steam turbine power generation has been in use for about 100 years, when the turbines replaced reciprocating steam engines due to their higher efficiencies and lower costs. Most of the electricity produced in the United States today is generated by conventional steam turbine power plants. The capacity of steam turbines can range from 50 kW to several hundred megawatts for large utility power plants. Steam turbines are widely used for combined heat and power (CHP) applications in the United States and Europe.

Unlike gas turbine and reciprocating engine CHP systems where heat is a byproduct of power generation, steam turbines generate electricity as a byproduct of heat (steam) generation. A steam turbine is captive to a separate heat source and does not directly convert fuel to electric energy. The energy is transferred from the boiler to the turbine through high-pressure steam that in turn powers the turbine and generator. This separation of functions enables steam turbines to operate with an enormous variety of fuels varying from clean natural gas to solid waste, including all types of coal, wood, wood waste, and agricultural byproducts (e.g., sugar cane bagasse, fruit pits, rice hulls). In CHP applications, steam at lower pressure is extracted from the steam turbine and used directly in a process or for district heating (i.e., a system for distributing heat generated in a centralized location for residential and commercial heating requirements such as space heating and water heating) or it can be converted to other forms of thermal energy, including hot or chilled water.

Steam turbines offer a wide array of designs and complexity to match the desired application and performance specifications. Steam turbines for utility service may have several pressure casings and elaborate design features, all designed to maximize the efficiency

---

[*] Material in this chapter is adapted from USEPA, *Catalog of CHP Technologies*, U.S. Environmental Protection Agency, Washington, DC, 2008.

of the power plant. For industrial applications, steam turbines are generally of simpler, single-casing design and less complicated for reliability and cost reasons. CHP can be adapted to both utility and industrial steam turbine designs.

11.2 Applications

Although steam turbines themselves are competitively priced compared to other prime movers, the costs of complete boiler/steam turbine CHP systems are relatively high on a per kW of capacity basis because of their low power-to-heat ratio; the costs of the boiler, fuel handling, and overall steam systems; and the custom nature of most installations. Thus, steam turbines are well suited to medium- and large-scale industrial and institutional applications where inexpensive fuels, such as coal, biomass, various solid wastes and byproducts (e.g., wood chips), refinery residual oil, and refinery off gases are available. Because of the relatively high cost of the system, including boiler, fuel handling system, condenser, cooling tower, and stack gas cleanup, high annual capacity factors are required to enable a reasonable recovery of invested capital.

However, retrofit applications of steam turbines in existing boiler/steam systems can be competitive options for a wide variety of users depending on the pressure and temperature of the steam exiting the boiler, the thermal needs of the site, and the condition of the existing boiler and steam system. In such situations, the decision involves only the added capital cost of the steam turbine and its generator, controls, and electrical interconnection, with the balance of the plant already in place. Similarly, many facilities that are faced with replacing or upgrading existing boilers and steam systems often consider the addition of steam turbines, especially if steam requirements are relatively large compared to power needs within the facility.

In general, steam turbine applications are driven by balancing lower cost fuel or avoiding disposal costs for the waste fuel with the high capital cost and (usually high) annual capacity factor for the steam plant and the combined energy plant–process plant application. For these reasons, steam turbines are not normally direct competitors of gas turbines and reciprocating engines.

11.2.1 Industrial and CHP Applications

Combined heat and power technology exists in a wide variety of energy-intensive facility types and sizes nationwide, including the following (USEPA, 2012):

- *Industrial manufacturers*—Chemical, refining, ethanol, pulp and paper, food processing, glass manufacturing
- *Institutions*—Colleges and universities, hospitals, prisons, military bases
- *Commercial buildings*—Hotels and casinos, airports, high-tech campuses, large office buildings, nursing homes
- *Municipal*—District energy systems, wastewater treatment plants, K-12 schools
- *Residential*—Multifamily housing, planned communities

In CHP applications, steam is extracted from the steam turbine and used directly in a process or for district heating, or it can be converted to other forms of thermal energy including hot water or chilled water. The turbine may drive an electric generator or equipment such as boiler feedwater pumps, process pumps, air compressors, and refrigeration chillers. Turbines as industrial drivers are almost always a single-casing machine, either single stage or multistage, condensing or non-condensing, depending on steam conditions and the value of the steam. Steam turbines can operate at a single speed to drive an electric generator or over a speed range to drive a refrigeration compressor. For non-condensing applications, steam is exhausted from the turbine at a pressure and temperature sufficient for the CHP heating applications.

Steam turbine systems are very commonly found in paper mills as there is usually a variety of waste fuels from hog fuel to black liquor recovery. Chemical plants are the next most common industrial user of steam turbines followed by primary metals. Various other industrial applications include the food industry, particularly sugar mills. There are commercial applications as well. Many universities have coal-powered CHP generating power with steam turbines. Some of these facilities are blending biomass to reduce their environmental impact.

11.2.2 Combined-Cycle Power Plants

The trend in power plant design is the combined cycle, which incorporates a steam turbine in a cycle with a gas turbine. Steam generated in the heat recovery steam generator (HRSG) of the gas turbine is used to drive a steam turbine to yield additional electricity and improve cycle efficiency. An extraction-condensing type of steam turbine can be used in combined cycles and be designed for CHP applications. Many large independent combined-cycle power plants operating on natural gas provide power to the electric grid and steam to one or more industrial customers.

11.3 Steam Turbine: Basic Process and Components

The thermodynamic cycle for the steam turbine is the *Rankine cycle*, which is the basis for conventional power generating stations and consists of a heat source (boiler) that converts water to high-pressure steam. In the steam cycle, water is first pumped to elevated pressure, which is medium to high pressure depending on the size of the unit and the temperature to which the steam is eventually heated. It is then heated to the boiling temperature corresponding to the pressure, boiled (heated for liquid to vapor), and then most frequently superheated (heated to a temperature above that of boiling). The pressurized steam is expanded to lower pressure in a multistage turbine, then exhausted to condensate at vacuum conditions or into an intermediate temperature steam distribution system that delivers the steam to the industrial or commercial application. The condensate from the condenser or from the industrial steam utilization system is returned to the feedwater pump for continuation of the cycle. Primary components of a boiler/steam turbine system are shown in Figure 11.1.

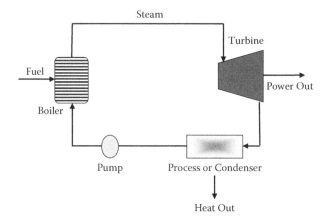

FIGURE 11.1
Components of a boiler/steam turbine system. (From USEPA, *Catalog of CHP Technologies*, U.S. Environmental Protection Agency, Washington, DC, 2008.)

The steam turbine itself consists of a stationary set of blades (called *nozzles*) and a moving set of adjacent blades (called *buckets* or *rotor blades*) installed within a casing. The two sets of blades work together such that the steam turns the shaft of the turbine and the connected load. The stationary nozzles accelerate the steam to high velocity by expanding it to lower pressure. A rotating bladed disc changes the direction of the steam flow, thereby creating a force on the blades that, because of the wheel geometry, manifests itself as torque on the shaft on which the bladed wheel is mounted. The combination of torque and speed is the output power of the turbine.

The internal flow passages of a steam turbine are very similar to those of the expansion section of a gas turbine (indeed, gas turbine engineering came directly from steam turbine design around 100 years ago). The main differences are the different gas density, molecular weight, isentropic expansion coefficient, and to a lesser extent viscosity of the two fluids.

Compared to reciprocating steam engines of comparable size, steam turbines rotate at much higher rotational speeds, which contribute to their lower cost per unit of power developed. The absence of inlet and exhaust valves that somewhat throttle (reduce pressure without generating power) and other design features enable steam turbines to be more efficient than reciprocating steam engines operating from steam at the same inlet conditions and exhausting into the same steam exhaust systems. In some steam turbine designs, part of the decrease in pressure and acceleration is accomplished in the blade row. These distinctions are known as *impulse* and *reaction turbine designs,* respectively. The competitive merits of these designs are the subject of business competition, as both designs have been sold successfully for well over 75 years.

The connection between the steam supply and power generation is the steam, and return feedwater, lines. There are numerous options in the steam supply, pressure, temperature, and extent, if any, for reheating steam that has been partially expanded from high pressure. Steam systems vary from low-pressure lines used primarily for space heating and food preparation, to medium pressure and temperature used in industrial processes and cogeneration, to high pressure and temperature used in utility power generation Generally, as the system gets larger the economics favor higher pressures and temperatures with their associated heavier walled boiler tubes and more expensive alloys.

DID YOU KNOW?

Steam turbines can be modified to fit any CHP system; therefore, the steam turbine can be fitted to match a facility's pressure and temperature requirements.

In general, utility applications involve raising steam for the exclusive purpose of power generation. Such systems also exhaust the steam from the turbine at the lowest practical pressure through the use of a water-cooled condenser. Some utility turbines utilizing dual-use power generation and steam delivery to district heating systems deliver steam at higher pressure into district heating systems or to neighboring industrial plants at pressure and consequently do not have condensers. These plants are actually large cogeneration/CHP plants.

11.3.1 Boilers

Steam turbines differ slightly from reciprocating engines and gas turbines in that the fuel is burned in a piece of equipment, the boiler, which is separate from the power generation equipment, the steam turbogenerator. The energy is transferred from the boiler to the turbine by an intermediate medium, steam under pressure. This separation of functions enables steam turbines to operate with an enormous variety of fuels. Horizontal industrial boilers are built for sizes up to approximately 40 MW. This enables them to be shipped via rail car, with considerable cost savings and improved quality, as factory labor is usually both lower in cost and greater in quality than field labor. Large shop-assembled boilers are typically capable of firing only gas or distillate oil, as there is inadequate residence time for complete combustion of most solid and residual fuels in such designs. Large, field-erected industrial boilers firing solid and residual fuels bear a resemblance to utility boilers except for the actual solid fuel injection. Large boilers usually burn pulverized coal; however, intermediate and small boilers burning coal and solid fuel employ various types of solids feeders.

11.3.2 Types of Steam Turbines

The three types of steam turbines are condensing, non-condensing (back-pressure), and extraction. Condensing turbines are not used for CHP applications and therefore will not be addressed here.

11.3.2.1 Non-Condensing (Back-Pressure) Turbine

The non-condensing turbine (also referred to as a back-pressure turbine) exhausts its entire flow of steam to the industrial process or facility steam mains at conditions close to the process heat requirements; that is, steam is expanded over a turbine and the exhaust steam is used to meet the steam needs of a plant, as shown in Figure 11.2.

DID YOU KNOW?

Water and steam are very well understood. Using standardized steam tables, we can know exactly the properties of our working fluids at certain temperatures and pressures. For this reason, steam is very predictable.

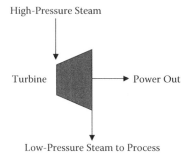

FIGURE 11.2
Non-condensing (back-pressure) steam turbine. (From USEPA, *Catalog of CHP Technologies*, U.S. Environmental Protection Agency, Washington, DC, 2008.)

11.3.2.2 Extraction Turbine

The extraction turbine is the other type of steam turbine used in CHP applications. This turbine type has openings in its casing for extraction of a portion of the steam at some intermediate pressure. The extracted steam may be used for process purposes in a CHP plant or for feedwater heating, as is the case for most utility power plants. The rest of the steam is condensed, as illustrated in Figure 11.3.

11.3.3 Design Characteristics

- *Custom design*—Steam turbines can be designed to match CHP design pressure and temperature requirements. The steam turbine can be designed to maximize electric efficiency while providing the desired thermal output.
- *Thermal output*—Steam turbines are capable of operating over a very broad range of steam pressures. Utility steam turbines operate with inlet steam pressures up to 3500 psig and exhaust vacuum conditions as low as 1 inch of Hg (absolute). Steam turbines can be custom designed to deliver the thermal requirements of the CHP applications through the use of back-pressure or extraction steam at appropriate pressures and temperatures.

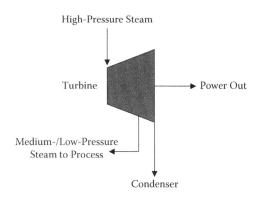

FIGURE 11.3
Extraction steam turbine. (From USEPA, *Catalog of CHP Technologies*, U.S. Environmental Protection Agency, Washington, DC, 2008.)

- *Fuel flexibility*—Steam turbines offer a wide range of fuel flexibility using a variety of fuel sources in the associated boiler or other heat source, including coal, oil, natural gas, wood, and waste products.

- *Reliability and life*—Steam turbine life is extremely long. Some steam turbines have been in service for over 50 years. Overhaul intervals are measured in years. When properly operated and maintained (including proper control of boiler water chemistry), steam turbines are extremely reliable. They require controlled thermal transients, as the massive casing heats up slowly and differential expansion of the parts must be minimized.

- *Size range*—Steam turbines are available in sizes from under 100 kW to over 250 MW. In the multi-megawatt size range, industrial and utility steam turbine designations merge, such that the same turbine (high-pressure section) is able to serve both industrial and small utility applications.

- *Emissions*—Emissions are dependent upon the fuel used by the boiler or other steam source, boiler furnace combustion section design and operation, and built-in and add-on boiler exhaust cleanup systems.

11.4 Performance Characteristics

11.4.1 Electrical Efficiency

The electrical generating efficiency of steam turbine power plants varies from a high of 36% (higher heating value, or HHV) for large, electric utility plants designed for the highest practical annual capacity factor, to under 10% (HHV) for small, simple plants that make electricity as a byproduct of delivering steam to industrial processes or district heating systems for colleges, industrial parks, and building complexes. Steam turbine *thermodynamic efficiency* (*isentropic efficiency*) refers to the ratio of power actually generated from the turbine to what would be generated by a perfect turbine with no internal losses using steam at the same inlet conditions and discharging to the same downstream pressure. Turbine thermodynamic efficiency is not to be confused with electrical generating efficiency, which is the ratio of net power generated to total fuel input to the cycle. Steam turbine thermodynamic efficiency is a measure of how efficiently the turbine extracts power from the steam itself and is useful in identifying the conditions of the steam as it exhausts from the turbine and in comparing the performance of various steam turbines. Multistage (moderate to high pressure ratio) steam turbines have thermodynamic efficiencies that vary from 65% for very small (under 1000 kW) units to over 90% for large industrial- and utility-sized units. Small, single-stage steam turbines can have efficiencies as low as 50%.

11.4.2 Operating Characteristics

Steam turbines, especially smaller units, leak steam around blade rows and out the end seals. When an end is at a lower pressure, as is the case with condensing steam turbines, air can also leak into the system. The leakages cause less power to be produced than expected, and the makeup water has to be treated to avoid boiler and turbine material problems. Air that has leaked must be removed, which is usually done by a compressor removing noncondensable gases from the condenser.

Because of the high pressures used in steam turbines, the casing is quite thick; consequently, steam turbines exhibit large thermal inertia. Steam turbines must be warmed up and cooled down slowly to minimize the differential expansion between the rotating blades and the stationary parts. Large steam turbines can take over 10 hours to warm up. Although smaller units have more rapid start-up times, steam turbines differ appreciably from reciprocating engines, which start up rapidly, and from gas turbines, which can start up in a moderate amount of time and load follow with reasonable rapidity.

Steam turbine applications usually operate continuously for extended periods of time, even though the steam fed to the unit and the power delivered may vary (slowly) during such periods of continuous operation. As most steam turbines are selected for applications with high duty factors, the nature of their application often takes care of the need to have only slow temperature changes during operation, and long start-up times can be tolerated. Steam boilers similarly have long startup times.

11.4.3 Process Steam and Performance Trade-Offs

Heat recovery methods from a steam turbine use back-pressure exhaust or extraction steam; however, the term is somewhat misleading. In the case of steam turbines, the steam turbine itself can be defined as the heat recovery device. The amount and quality of recovered heat are a function of the entering steam conditions and the design of the steam turbine. Exhaust steam from the turbine can be used directly in a process or for district heating. It can also be converted to other forms of thermal energy, including hot or chilled water. Steam discharged or extracted from a steam turbine can be used in a single- or double-effect absorption chiller. The steam turbine can also be used as a mechanical drive for a centrifugal chiller.

11.4.4 CHP System Efficiency

Steam turbine CHP systems are generally characterized by very low power-to-heat ratios, typically in the 0.05 to 0.2 range. This is because electricity is a byproduct of heat generation, with the system optimized for steam production. Hence, although steam turbine CHP system electrical efficiency (net power output/total fuel input into the system) may seem very low, it is because the primary objective is to produce large amounts of steam. The effective electrical efficiency of steam turbine systems:

$$\frac{\text{Steam turbine electric power output}}{\text{Total fuel into boiler} - \left(\text{Steam to process/Boiler efficiency}\right)}$$

is generally very high, however, because almost all of the energy difference between the high pressure boiler output and the lower pressure turbine output is converted to electricity. This means that total CHP system efficiencies (net power and steam generated divided by total fuel input) are generally very high and approach the boiler efficiency level. Steam boiler efficiencies range from 70 to 85% (HHV) depending on boiler type and age, fuel, duty cycle, application, and steam conditions.

11.4.5 Performance and Efficiency Enhancements

In industrial steam turbine systems, business conditions determine the requirements and relative values of electric power and process, or heating, steam. Plant system engineers then decide the extent of efficiency-enhancing options to incorporate in terms of their incremental effects on performance and plant cost and select appropriate steam turbine inlet and exhaust conditions. Often the steam turbine is going into a system that already exists and is being modified, so that a number of steam system design parameters are already determined by previous decisions, which exist as system hardware characteristics. As the stack temperature of the boiler exhaust combustion products still contain some heat, trade-offs are made regarding the extent of investment in heat reclamation equipment for the sake of efficiency improvement. Often the stack exhaust temperature is set at a level where further heat recovery would result in condensation of corrosive chemical species in the stack, with consequential deleterious effects on stack life and safety.

11.4.6 Steam Reheat

To increase power generation efficiency in large utility (and industrial) systems, higher pressures and steam reheat are commonly used. The higher the pressure ratio (the ratio of the steam inlet pressure to the steam exit pressure) across the steam turbine, and the higher the steam inlet temperature, the more power it will produce per unit of mass flow, provided of course that the turbine can handle the pressure ratio and that the turbine is not compromised by excessive condensation within the last expansion stage. To avoid condensation the inlet steam temperature is increased, until the economic practical limit of materials capability is reached. This limit is now generally in the range of 800 to 900°F for small industrial steam turbines. When the economically practical limit of temperature is reached, the expanding steam can reach a condition of temperature and pressure where condensation to (liquid) water begins. Small amounts of water droplets can be tolerated in the last stage of a steam turbine provided that the droplets are not too large or numerous. At pressures higher than that point, the steam is returned to the boiler, reheated, and then returned to the expansion steam turbine for further expansion. When returned to the next stage of the turbine, the steam can be further expanded without condensation.

11.4.7 Combustion Air Preheating

In large industrial systems, air preheaters recover heat from the boiler exhaust gas steam and use it to preheat the combustion air, thereby reducing fuel consumption. Boiler combustion air preheaters are large versions of the heat wheels used for the same purpose on industrial furnaces.

11.4.8 Maintenance

Even though steam turbines are very rugged units, with operation life often exceeding 50 years, they do require, as with many other major machines with moving parts, routine maintenance. Unlike maintenance requirements for many other complicated industrial machines, steam turbine maintenance is simple; it is comprised mainly of making sure that all fluids (steam flowing through the turbine and oil for the bearing) are always clean and at the proper temperature. The oil lubrication system must be clean and at the correct operating temperature and level to maintain proper performance. Other items include inspecting auxiliaries such as lubricating-oil pumps, coolers, and oil strainers and checking safety devices such as the operation of overspeed trips. In order to obtain reliable service, steam turbines require long warm-up periods to reduce thermal expansion stress and wear concerns. Steam turbine maintenance costs are quite low, typically around $0.005 per kWh. Boilers and any associated solid-fuel processing and handling equipment that are part of the boiler/steam turbine plant require their own types of main-tenance. One maintenance issue related to steam turbine operation is solids carryover from the boiler that deposits on turbine nozzles and other internal parts and degrades turbine efficiency and power output. Some of these solids are water soluble but others are not. Three methods are employed to remove such deposits: (1) manual removal, (2) cracking off deposits by shutting the turbine off and allowing it to cool, and (3) water washing while the turbine is running, for water-soluble deposits.

11.4.9 Fuels

Industrial boilers operate on a wide variety of fuels, including wood, coal, natural gas, oils (including residual oil, the material left over when the valuable distillates have been separated for separate sale), municipal solid waste, and sludges. The fuel handling, stor-age, and preparation equipment required for solid fuels adds considerably to the cost of an installation. Thus, such fuels are used only when a high annual capacity factor is expected of the facility, or when the solid material has to be disposed of to avoid an environmental or space occupancy problem.

11.4.10 Availability

Steam turbines are generally considered to have 99%+ availability, with longer than one year between shutdowns for maintenance and inspections. This high level of availability applies only to the steam turbine, not the boiler of the HRSG that is supplying the steam.

11.5 Emissions

Emissions associated with a steam turbine are dependent on the source of the steam. Steam turbines can be used with a boiler firing any one or a combination of a large variety of fuel sources, or they can be used with a gas turbine in a combined cycle configuration. Boiler emissions vary depending on fuel type and environmental conditions. Boiler emis-sions include nitrogen oxide (NO_x), sulfur oxides (SO_x), particulate matter (PM), carbon monoxide (CO), and carbon dioxide (CO_2).

References and Recommended Reading

Goldstein, L., Hedman, B., Knowles, D. et al. (2003). *Gas-Fired Distributed Energy Resource Technology Characterizations*, NREL/TP-620-34783, Gas Research Institute and National Renewable Energy Laboratory, Washington, DC (www.nrel.gov/docs/fy04osti/34783.pdf).

Spellman, F.R. (2007). *Handbook of Water and Wastewater Treatment Plant Operations*, 2nd ed., CRC Press, Boca Raton, FL.

USEPA. (2006a). *Auxiliary and Supplemental Power Fact Sheet: Microturbines*, EPA 832-F-05-014, U.S. Environmental Protection Agency, Washington, DC.

USEPA. (2006b). *Auxiliary and Supplemental Power Fact Sheet: Viable Sources*, EPA 832-F-05-009, U.S. Environmental Protection Agency, Washington, DC.

USEPA. (2008). *Catalog of CHP Technologies*, U.S. Environmental Protection Agency, Washington, DC (www.epa.gov/chp/documents/catalog_chptech_full.pdf).

USEPA. (2012). *Combined Heat and Power Partnership: Basic Information*, U.S. Environmental Protection Agency, Washington, DC (http://www.epa.gov/chp/basic/index.html).

12

Fuel Cells

Sustainable development is not a "fixed state of harmony." Rather, it is an ongoing process of evolution in which people take action leading to development that meets their current needs without compromising the ability of future generations to meet their own needs.

—Hardi and Zdan (1997)

Fuel cell systems employ an entirely different approach to the production of electricity than traditional prime mover technologies. Fuel cells are similar to batteries in that both produce a direct current (DC) through an electromechanical process without direct combustion of a fuel source. However, whereas a battery delivers power from a finite amount of stored energy [hence, the name "storage battery"], fuel cells can operate indefinitely provided the availability of a continuous fuel source.

—USEPA (2008)

12.1 Introduction[*]

A fuel cell is an electrochemical device similar to a battery. Although both batteries and fuel cells generate power through an internal chemical reaction, a fuel cell differs from a battery in that it uses an external supply that continuously replenishes the reactants in the fuel cell. A battery, on the other hand, has a fixed internal supply of reactants. A fuel cell can supply power continuously as long as the reactants are replenished, whereas a battery can generate only limited power before it must be recharged or replaced. Most types of fuel cells can operate on a wide variety of fuels including hydrogen, digester gas, natural gas, propane, landfill gas, diesel, or other combustible gas. In some cases, such as in a wastewater treatment plant (WWTP), methane (biosolids gas) from anaerobic digesters can be reused in the fuel cell instead of flaring off the excess gas. Other advantages of fuel cells include few moving parts, modular design, and negligible emission of pollutants.

This chapter describes the basics of fuel cell technology and the use of fuel cells as auxiliary and supplemental power sources (ASPSs) for wastewater treatment plants. Fuel cells have been a popular choice as an ASPS in recent years, because they are clean, quiet, and an efficient type of power generation. Fuel cells are highly efficient and emissions free. Because the fuel is not combusted, but instead reacts electrochemically, virtually no air pollution is associated with its use. Although there are many different types of fuel cells, each of which uses its own specific set of chemicals to produce power, only molten carbonate fuel cells (MCFCs), phosphoric acid fuel cells (PAFCs), and solid-oxide fuel cells (SOFCs) can generate enough energy to power a typical WWTP. Each of these types of fuel cell is appropriate for use as either a supplemental power source or an auxiliary power source.

---

[*] Adapted from USEPA, *Auxiliary and Supplemental Power Fact Sheet: Fuel Cells*, EPA 832-F-05-012, U.S. Environmental Protection Agency, Washington, DC, 2006.

DID YOU KNOW?

Sir William Grove, who demonstrated a hydrogen fuel cell in London in the 1830s, invented what is known today as fuel cell technology. Grove's technology was set aside for more than 100 years without practical application. Fuel cells returned to the laboratory in the 1950s, when power systems were being developed for the U.S. space program. Today, the topic of fuel cells encompasses a broad range of different technologies, technical issues, and market dynamics. Significant amounts of public and private investment are being applied to the development of fuel cells for applications in the power-generation and automotive industries.

12.2 Fuel Cells: The Basics[*]

Depending on your education or experience level or social, cultural, or economic background, the term *cell* may conjure up images as diverse in variety as the colors, sizes, and shapes of lightning bolts. Some may think of plant cells, animal cells, cell structure, cell biology, cell diagrams, cell membranes, human memory, cell theory, cell walls, cell parts, cell functions, honeycomb cells, prison cells, electrolytic cells (for producing electrolysis), aeronautic gas cells (contained in a balloon), ecclesiastical cells, or currently and definitely more commonly cell phones. Before moving on to our basic discussion of fuel cells and their associated terminology and applications, it is important to point out that although fuel cells are not topics of discussion anywhere near as common as these other cells (e.g., cell phones), we predict that the day is coming when we will refer to our fuel cells just as commonly as we mention our cell phones—for it will be the fuel cell (of some composition, size, shape, and type or variety) that will power our lives, just as the cell phone powers and transmits our communication.

12.2.1 Open Cells vs. Closed Cells

Batteries store electrical energy chemically (contrary to popular belief, they do not make electrical energy), so they are considered to be thermodynamically *open* systems. By contrast, fuel cells differ from conventional electrochemical cell batteries in that they consume reactant from an external source, which must be replenished; thus, they are thermodynamically *closed* systems (Anon., 2007). A fuel cell is an electrochemical cell that converts a source fuel, such as hydrogen gas or a hydrogen-rich liquid fuel (in addition to hydrocarbons, alcohols, chlorine, chlorine dioxide, and others) into an electrical current (Meibuhr, 1966). The hydrogen gas or hydrogen-rich fuel is reacted or triggered in the presence of an electrolyte (an oxidant), usually oxygen from the air, to produce electrons, heat, and water. The reactants flow into the cell and the reaction products flow out of it, while the electrolyte remains within the cell. Fuel cells can operate continuously as long as the reactant and oxidant flows are maintained. Figure 12.1 provides a generalized representation of the fuel cell process. Keep in mind that fuel cells come in many varieties (see Table 12.1); however, they all work in the same general manner. The principles of hydrogen-type fuel cell operation are described in the next section.

---

[*] Adapted from Spellman, F.R. and Bieber, R., *The Science of Renewable Energy*, CRC Press, Boca Raton, FL, 2011.

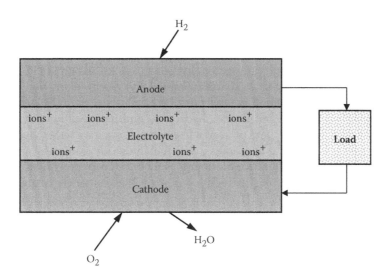

FIGURE 12.1
Generalized block diagram of the fuel cell process.

TABLE 12.1

Types of Fuel Cells

Fuel Cell Type	Electrolyte	Current Uses
Metal hydride	Aqueous alkaline solution	Commercial/research
Electrogalvanic	Aqueous alkaline solution	Commercial/research
Direct formic acid	Polymer membrane (ionomer)	Commercial/research
Zinc–air battery	Aqueous alkaline solution	Mass production
Microbial	Polymer membrane/humic acid	Research
Upflow microbial	—	Research
Regenerative	Polymer membrane (ionomer)	Commercial/research
Direct borohydride	Aqueous alkaline solution	Commercial/research
Alkaline	Aqueous alkaline solution	Commercial/research
Direct methanol	Polymer membrane (ionomer)	Commercial/research
Reformed methanol	Polymer membrane (ionomer)	Commercial/research
Direct-ethanol	Polymer membrane (ionomer)	Research
Proton exchange membrane	Polymer membrane (ionomer)	Commercial/research
RFC–redox	Liquid electrolytes with redox shuttle and polymer membrane (ionomer)	Research
Phosphoric acid (H_3PO_4)	Molten phosphoric acid	Commercial/research
Molten carbonate	Molten alkaline carbonate	Commercial/research
Tubular solid oxide	O^{2-}-conducting ceramic oxide	Commercial/research
Protonic ceramic	H^+-conducting ceramic oxide	Research
Direct carbon	Several different	Commercial/research
Planar solid oxide	O^{2-}-conducting ceramic oxide	Commercial/research
Enzymatic biofuel	Any that will not denature the enzyme	Research
Magnesium air	Saltwater	Commercial/research

12.3 Hydrogen Fuel Cells: A Realistic View[*]

Containing only one electron and one proton, hydrogen (chemical symbol H) is the simplest element on Earth. Hydrogen gas is a diatomic molecule, which means that each molecule has two atoms of hydrogen (which is why pure hydrogen is commonly expressed as H_2). Although abundant on Earth as an element, hydrogen combines readily with other elements and is almost always found as part of another substance, such as water, hydrocarbons, or alcohols. Hydrogen is also found in biomass, which includes all plants and animals.

Hydrogen is an energy carrier, not an energy source. Hydrogen can store and deliver usable energy, but because it typically does not exist by itself in nature it must be produced from compounds that contain it. Hydrogen can be produced using diverse, domestic resources, including nuclear, natural gas and coal, and biomass, as well as other renewables such as solar, wind, hydroelectric, or geothermal energy. This diversity of domestic energy sources makes hydrogen a promising energy carrier and important to our nation's energy security. It is expected and desirable for hydrogen to be produced using a variety of resources and process technologies (or pathways). The U.S. Department of Energy is focusing on energy-efficient hydrogen production technologies that result in near-zero net greenhouse gas emissions and use renewable energy sources, nuclear energy, and coal (when combined with carbon sequestration).

Hydrogen can be produced via various process technologies, including *thermal* (natural gas reforming, renewable liquid and bio-oil processing, and biomass and coal gasification), *electrolytic* (water splitting using a variety of energy resources), and *photolytic* (splitting water using sunlight via biological and electrochemical materials). Hydrogen can be produced in large, central facilities 50 to 300 miles from point of use or in smaller facilities located within 25 to 100 miles of use.

For hydrogen to be successful in the marketplace, it must be cost competitive with the available alternatives. In the light-duty vehicle transportation market, this competitive requirement means that hydrogen must be available untaxed at $2 to $3/GGE (gasoline gallon equivalent). At this price, hydrogen fuel cell vehicles would cost the same on a cost-per-mile-driven basis compared to conventional vehicles. Some hydrogen production technologies are further along in development than others; some promise to be cost competitive for the transition period beginning in 2015, but others are considered long-term technologies that will not be cost competitive until after 2030.

Infrastructure is required to move hydrogen from the location where it is produced to the dispenser at a refueling station or stationary power site. Infrastructure includes the pipelines, trucks, railcars, ships, and barges that deliver fuel, as well as the facilities and equipment needed to load and unload them. Some of the infrastructure is already in place because hydrogen has long been used in industrial applications, but it is not sufficient to support widespread consumer use of hydrogen as an energy carrier. Because hydrogen has a relatively low volumetric energy density, its transportation, storage, and final delivery to the point of use represent significant costs and result in some of the energy inefficiencies associated with using it as an energy carrier.

Options and trade-offs for hydrogen delivery from the production facilities to the point of use are complex. The choice of a hydrogen production strategy greatly affects the cost and method of delivery; for example, larger, centralized facilities can produce hydrogen

---

[*] Adapted from USDOE, *Hydrogen Production*, Energy Efficiency & Renewable Energy, U.S. Department of Energy, Washington, DC, 2008 (http://www1.eere.energy.gov/hydrogenandfuelcells/production/basics.html).

at relatively low costs due to economies of scale, but the delivery costs are higher than for smaller, localized production facilities. Although these smaller production facilities would have relatively lower delivery costs, their hydrogen production costs are likely to be higher, as lower volume production means higher equipment costs on a per-unit-of-hydrogen basis.

Key challenges to hydrogen delivery include reducing delivery costs, increasing energy efficiency, maintaining hydrogen purity, and minimizing hydrogen leakage. Further research is needed to analyze the trade-offs between the hydrogen production options and the hydrogen delivery options taken together as a system. Building a national hydrogen delivery infrastructure is a big challenge. Such an infrastructure will take time to develop and will likely include combinations of various technologies. Infrastructure options will evolve as the demand for hydrogen grows and as delivery technologies develop and improve.

12.3.1 Hydrogen Storage

For hydrogen to be a competitive fuel for vehicles, the hydrogen vehicle must be able to travel a distance comparable to that of conventional hydrocarbon-fueled vehicles. Storing enough hydrogen on board a vehicle to achieve a driving range of greater than 300 miles is a significant challenge. On a weight basis, hydrogen has nearly three times the energy content of gasoline (120 MJ/kg for hydrogen vs. 44 MJ/kg for gasoline); however, on a volume basis, the situation is reversed (8 MJ/L for liquid hydrogen vs. 32 MJ/L for gasoline). On-board hydrogen storage in the range of 5 to 13 kg is required to encompass the full platform of light-duty vehicles. Hydrogen can be stored as either a gas or a liquid. Storage as a gas typically requires high-pressure tanks (5000- to 10,000-psi tank pressure). Storage of hydrogen as a liquid requires cryogenic temperatures, because the boiling point of hydrogen at 1 atmosphere pressure is –252.8°C. Hydrogen can also be stored on the surfaces of solids (by adsorption) or within solids (by absorption). In adsorption, hydrogen is attached to the surface of a material as either hydrogen molecules or hydrogen atoms. In absorption, hydrogen is dissociated into H atoms, and then the hydrogen atoms are incorporated into the solid lattice framework.

Hydrogen storage in solids may make it possible to store large quantities of hydrogen in smaller volumes at low pressures and at temperatures close to room temperature. It is also possible to achieve volumetric storage densities greater than those of liquid hydrogen because the hydrogen molecule is dissociated into atomic hydrogen within a metal hydride lattice structure. Finally, hydrogen can be stored through the reaction of hydrogen-containing materials with water (or other compound such as alcohols). In this case, the hydrogen is effectively stored in both the material and the water. The term *chemical hydrogen storage* or *chemical hydride* is used to describe this form of hydrogen storage. It is also possible to store hydrogen in the chemical structures of liquids and solids.

12.3.2 How a Hydrogen Fuel Cell Works

A fuel cell uses the chemical energy of hydrogen to cleanly and efficiently produce electricity. Fuel cells have a variety of potential applications; for example, they can provide energy for systems as large as utility power stations or as small as laptop computers. Fuel cells offer several benefits over conventional combustion-based technologies currently used in many power plants and passenger vehicles. They produce much smaller quantities of greenhouse gases and none of the air pollutants that create smog and cause health problems. If pure hydrogen is used as a fuel, fuel cells emit only heat and water as byproducts.

A hydrogen fuel cell is a device that uses hydrogen (or hydrogen-rich fuel) and oxygen to create electricity through an electrochemical process. A single fuel cell consists of an electrolyte and two catalyst-coated electrodes (a porous anode and cathode). The various types of fuel cells all work similarly:

- Hydrogen (or hydrogen-rich fuel) is fed to an anode, where a catalyst separates the negatively charged hydrogen electrons from positively charged ions.
- At the cathode, oxygen combines with electrons and, in some cases, with species such as protons or water, resulting in water or hydroxide ions, respectively.
- For polymer electrolyte membrane and phosphoric acid fuel cells, protons move through the electrolyte to the cathode, where they combine with oxygen and electrons to produce water and heat.
- For alkaline, molten carbonate, and solid oxide fuel cells, negative ions travel through the electrolyte to the anode, where they combine with hydrogen to generate water and electrons.
- The electrons from the anode cannot pass through the electrolyte to the positively charged cathode; they must travel around it via an electrical circuit to reach the other side of the cell. This movement of electrons is an electrical current.

12.4 CHP Applications[*]

Fuel cells are either available or being developed for a number of stationary and vehicle applications. The power applications include commercial and industrial CHP systems (200 to 1200 kW), residential and commercial CHP systems (3 to 10 kW), and backup and portable power systems (0.5 to 5 kW). CHP applications combine onsite power generation with the recovery and use of byproduct heat. Continuous base-load operation and effective use of the thermal energy contained in the exhaust gas and cooling subsystem enhance the economics of on-site generation applications.

CASE STUDY 12.1[†]

King County Washington Treatment Division in Renton, Washington, installed a 1-MW molten carbonated fuel cell (MCFC) power plant to reduce energy costs to the treatment plant. The output is tied to a transformer to step up voltage to 13,000 V. The fuel cell system was chosen because of its high efficiency and low emissions. This cell is operated using methane from the anaerobic digesters. King County uses the electricity produced by the fuel cells to supplement its energy needs, which has also reduced the facility's power costs by 15%. The estimated installed cost for the MCFC system was approximately $22.8 million, including the waste heat recovery system. The waste heat recovery unit for the exhaust is sized for 1.7 million Btu per hour of waste heat. At 45% electrical efficiency and at rated heat recovery, the net thermal efficiency of the plant is expected to be around 68%. The waste heat can also be returned to the digester loop.

---

[*] Adapted from USEPA, *Catalog of CHP Technologies*, U.S. Environmental Protection Agency, Washington, DC, 2008.

[†] Adapted from USEPA, *Auxiliary and Supplemental Power Fact Sheet: Fuel Cells*, EPA 832-F-05-012, U.S. Environmental Protection Agency, Washington, DC, 2006.

References and Recommended Reading

Anon. (2007). *Science Trackers Bullets Online: Batteries, Supercapacitors, and Fuel Cells*, Science Reference Services, Library of Congress, Washington, DC (http://www.loc.gov/rr/scitech/tracer-bullets/batteriestb.html#top).

Barbir, F. (2005). *PEM Fuel Cells: Theory and Practice*, Elsevier, New York.

EG&G Technical Services, Inc. (2004). *Fuel Cell Handbook,* 7th ed., U.S. Department of Energy, National Energy Technology Laboratory, Morgantown, WV.

Eisenberg, A. and Kim, J.S. (1998). *Introduction to Ionomers*, Wiley, New York.

Goldstein, L., Hedman, B., Knowles, D. et al. (2003). *Gas-Fired Distributed Energy Resource Technology Characterizations*, NREL/TP-620-34783, Gas Research Institute and National Renewable Energy Laboratory, Washington, DC (www.nrel.gov/docs/fy04osti/34783.pdf).

Hardi, P. and Zdan, T., Eds. (1997). *Assessing Sustainable Development: Principles in Practice*, International Institute for Sustainable Development, Winnipeg.

Hoogers, G., Ed. (2003). *Fuel Cell Technology Handbook*, CRC Press, Boca Raton, FL.

Larminie, J. and Dicks A. (2003). *Fuel Cell Systems Explained*, 2nd ed., John Wiley & Sons, New York.

Meibuhr, S.G. (1966). *Electrochim. Acta*, 11, 1301.

NREL. (2009). *Ultracapacitors*, National Renewable Energy Laboratory, Washington, DC (http://www.nrel.gov/vehiclesandfuels/energystorage/ultracapacitors.html?print).

Singhal, S.C. and Kendall, K. (2003). *High-Temperature Solid Oxide Fuel Cells: Fundamentals, Design and Applications*, Elsevier, New York.

Spellman, F.R. (2007). *Handbook of Water and Wastewater Treatment Plant Operations*, 2nd ed., CRC Press, Boca Raton, FL.

Spellman, F.R. and Bieber, R. (2011). *The Science of Renewable Energy*, CRC Press, Boca Raton, FL.

USDOE. (2008a). *Hydrogen Production*, Energy Efficiency & Renewable Energy, U.S. Department of Energy, Washington, DC (http://www1.eere.energy.gov/hydrogenandfuelcells/production/basics.html).

USDOE. (2008b). *Fuel Cell Technologies Program: Hydrogen Delivery*, Energy Efficiency & Renewable Energy, U.S. Department of Energy, Washington, DC (http://www1.eere.energy.gov/hydrogenandfuelcells/delivery/m/basics.html).

USDOE. (2008c). *Fuel Cell Technologies Program: Hydrogen Storage*, Energy Efficiency & Renewable Energy, U.S. Department of Energy, Washington, DC (http://www1.eere.energy.gov/hydrogenandfuelcells/storage/).

USDOE. (2010). *Fuel Cell Technologies Program: Fuel Cells*, Energy Efficiency & Renewable Energy, U.S. Department of Energy, Washington, DC (http://www1.eere.energy.gov/hydrogenandfuelcells/fuelcells/).

USEPA. (2006a). *Auxiliary and Supplemental Power Fact Sheet: Microturbines*, EPA 832-F-05-014, U.S. Environmental Protection Agency, Washington, DC.

USEPA. (2006b). *Auxiliary and Supplemental Power Fact Sheet: Fuel Cells*, EPA 832-F-05-012, U.S. Environmental Protection Agency, Washington, DC.

USEPA. (2008). *Catalog of CHP Technologies*, U.S. Environmental Protection Agency, Washington, DC (www.epa.gov/chp/documents/catalog_chptech_full.pdf).

Vielstich, W. et al., Eds. (2009). *Handbook of Fuel Cells: Advances in Electrocatalysis, Materials, Diagnostics and Durability*, Wiley, Hoboken, NJ.

Section IV

Biomass Power and Heat Generation

13

CHP and Wastewater Biogas

"Waste" heat produced during electricity generation or industrial processes can be used to heat water and make steam. The steam can be distributed through pipes and used to heat buildings or whole communities. In 1984, Copenhagen built a system that supplies hot water to 97 percent of the city by harvesting the heat from local clean-burning biomass plants.

—USEPA WaterSense®

13.1 Grasshopper Generation[*]

In an article in *The New York Times*, Thomas Friedman (2010) stated that, "The fat lady has sung." Specifically, Friedman was speaking about America's transition from the "Greatest Generation" to what Kurt Anderson referred to as the "Grasshopper Generation." According to Friedman, we are "going from the age of government handouts to the age of citizen givebacks, from the age of companions fly free to the age of paying for each bag." Friedman goes on to say that we all accept that our parents were the greatest generation, but it is us that we are concerned about and that it is the "we" that comprise the Grasshopper Generation: "We have been eating through the prosperity that was bequeathed us like hungry locusts."

Emphasizing again the major theme of this text, we must develop and utilize alternative and renewable sources of energy, because what we are eating through, among other things, is our readily available, relatively inexpensive source of energy. The point is we can, like the grasshopper, gobble it all up until it is all gone or we can find alternatives—sustainable alternatives of energy.

A promising source of energy is bioenergy. *Bioenergy* is a general term that refers to energy derived from materials such as straw, wood, or animal wastes, which, in contrast to fossil fuels, were living matter relatively recently. Such materials can be burned directly as solids (*biomass*) to produce heat or power, but they can also be converted into liquid biofuels. Interest is growing in *biofuels*, liquid fuels (biodiesel and bioethanol) that can be used for transport. At the moment, transport has taken center stage in our search for renewable, alternative, and sustainable fuels to eventually replace hydrocarbon fuels. Unlike biofuels, *solid biomass fuel* is used primarily for electricity generation or heat supply.

Even though bioenergy is a promising source of energy for the future, it is rather ironic when the experts (or anyone else, for that matter) frequently make this point without qualification. The qualification? The reality? Simply, keep in mind that it was only a little over 100 ago that our economy was based primarily on bioenergy from biomass, or carbohydrates, rather than from hydrocarbons. In the late 1800s, the largest selling chemicals were alcohols made from wood and grain, the first plastics were produced from cotton, and about 65% of the nation's energy came from wood (USDOE, 2004).

---

[*] Adapted from Spellman, F.R. and Bieber, R., *The Science of Renewable Energy*, CRC Press, Boca Raton, FL, 2011.

By the 1920s, the economy began to shift toward the use of fossil resources, and after World War II this trend accelerated as technology breakthroughs were made. By the 1970s, fossil energy was established as the backbone of the U.S. economy, and all but a small portion of the carbohydrate economy remained (Morris, 2002). In 1989, in the industrial sector, plants accounted for about 16% of input, compared with 35% in 1925 (USDOE, 2006a).

Processing costs and the availability of inexpensive fossil energy resources continue to be driving factors in the dominance of hydrocarbon resources. In many cases, it is still more economical to produce goods from petroleum or natural gas than from plant matter. This trend is about to shift dramatically, though, as we reach peak oil and as the world continues to demand unprecedented amounts of petroleum supplies from an ever-dwindling supply.

Assisting in this trend shift are technological advances being made in the biological sciences and engineering fields, political changes, and concern for the environment. These factors have begun to swing the economy back toward the use of carbohydrates on a number of fronts. Consumption of biofuels in vehicles, for example, rose from 0 in 1977 to nearly 1.5 billion gallons in 1999. The use of inks produced from soybeans in the United States increased by fourfold between 1989 and 2000 and is now at more than 22% of total use (Morris, 2002).

Technological advances are also beginning to make an impact on reducing the cost of producing industrial products and fuels from biomass, making them more competitive with those produced from petroleum-based hydrocarbons. Developments in pyrolysis, ultracentrifuges, membranes, and the use of enzymes and microbes as biological factories are enabling the extraction of valuable components from plants at a much lower cost. As a result, industry is investing in the development of new bioproducts that are steadily gaining a share of current markets (USDOE, 2004).

New technology is enabling the chemical and food processing industries to develop new processes for more cost-effective production of all kinds of industrial products from biomass. One example is a plastic polymer derived from corn that is now being produced at a 300-million-pound-per-year plant in Nebraska, a joint venture between Cargill, the largest grain merchant, and Dow Chemical, the largest chemical producer (Fahey, 2001).

Other chemical companies are exploring the use of low-cost biomass processes to make chemicals and plastics that are now made from more expensive petrochemical processes (USDOE, 2006a). In this regard, new innovative processes such as *biorefineries* may become the foundation of the new bioindustry. A biorefinery is similar in concept to the petroleum refinery, except that it is based on conversion of biomass feedstocks rather than crude oil. Biorefineries in theory would use multiple forms of biomass to produce a flexible mix of products, including fuels, power, heat, chemicals, and materials. In a biorefinery, biomass would be converted into high-value chemical products and fuels (both gas and liquid). Byproducts and residues, as well as some portion of the fuels produced, would be used to fuel onsite power generation or cogeneration facilities producing heat and power. The biorefinery concept has already proven successful in the U.S. agricultural and forest products industries, where such facilities now produce food, feed fiber, or chemicals, as well as heat and electricity, to run plant operations. Biorefineries offer the most potential for realizing the ultimate opportunities of the bioenergy industry.

As the title suggests, this chapter discusses biogas production in wastewater treatment plants in plant combined heat and power (CHP) applications. Specifically, the anaerobic digestion process will be described as it applies to producing methane for process use within the plant site. Before discussing anaerobic digestion for power and heat generation, we begin our discussion with a basic overview of biomass.

13.2 Biomass

Biomass (all Earth's living matter) consists of the energy from plants and plant-derived organic-based materials; it is essentially stored energy from the sun. Biomass can be biochemically processed to extract sugars, thermochemically processed to produce biofuels or biomaterials, or combusted to produce heat or electricity. Biomass is also an input into other end-use markets, such as forestry products (pulpwood) and other industrial applications. This complicates the economics of biomass feedstock and requires that we differentiate between what is technically possible from what is economically feasible, taking into account relative prices and intermarket competition.

Biomass has been used since people began burning wood to cook food and keep warm. Trees have been the principal fuel for almost every society for over 5000 years, from the Bronze Age until the middle of the 19th century (Perlin, 2005). Wood is still the largest biomass energy resource today, but other sources of biomass can also be used. These include food crops, grassy and woody plants, residues from agriculture or forestry, and the organic component of municipal and industrial wastes. Even the fumes from landfills (which are methane, a natural gas) can be used as a biomass energy source. This category excludes organic material that has been transformed by geological processes into substances such as coal or petroleum. The biomass industry is one of the fastest growing industries in the United States.

13.2.1 Feedstock Types

A variety of biomass feedstocks can be used to produce transportation fuels, biobased products, and power. Feedstocks refer to the crops or products, such as waste vegetable oil, that can be used as or converted into biofuels and bioenergy. With regard to the advantages or disadvantages of one type of feedstock as compared to another, this is gauged in terms of how much usable material they yield, where they can grow, and how energy and water intensive they are. Feedstock types are categorized as first-generation or second-generation feedstocks. First-generation feedstocks include those that are already widely grown and used for some form of bioenergy or biofuel production, which means that food vs. fuel conflicts could arise. First-generation feedstocks include sugars (sugar beets, sugar cane, sugar palm, sweet sorghum, and *Nypa* palm), starches (cassava, corn, milo, sorghum, sweet potato, and wheat), waste feedstocks such as whey and citrus peels, and oils and fats (coconut oil, oil palm, rapeseed, soy beans sunflower seed, castor beans, jatropha, jojoba, karanj, waste vegetable oil, and animal fat). Second-generation feedstocks include crops that offer high potential yields of biofuels but are not widely cultivated or not cultivated as an energy crop. Examples are cellulosic feedstocks or conventional crops such as *Miscanthus* grasses, prairie grass and switchgrass, and willow and hybrid poplar trees. Algae and halophytes (saltwater plants) are other second-generation feedstocks.

Currently, a majority of the ethanol produced in the United States is made from corn or other starch-based crops. The current focus, however, is on the development of cellulosic feedstocks—non-grain, non-food-based feedstocks such as switchgrass, corn stover, and wood material—and on technologies to convert cellulosic material into transportation fuels and other products. Using cellulosic feedstocks not only can alleviate the potential concern of diverting food crops to produce fuel but also offers a variety of environmental benefits (EERE, 2008). Because such a wide variety of cellulosic feedstocks can be used for energy production, potential feedstocks are grouped into categories—or pathways.

13.2.2 Composition of Biomass

The ease with which biomass can be converted to useful products or intermediates is determined by the composition of the biomass feedstock. Biomass contains a variety of components, some of which are readily accessible and others that are much more difficult and costly to extract. The composition and subsequent conversion issues for current and potential biomass feedstock compounds are listed and described below:

- *Starch* (glucose) is readily recovered and converted from grain (corn, wheat, rice) into products. Starch from corn grain provides the primary feedstock for today's existing and emerging sugar-based bioproducts, such as polylactide, as well as the entire fuel ethanol industry. Corn grain serves as the primary feedstock for starch used to manufacture today's biobased products. Corn wet mills use a multistep process to separate starch from the germ, gluten (protein), and fiber components of corn grain. The starch streams generated by wet milling are very pure, and acid or enzymatic hydrolysis is used to break the glycosidic linkages of starch to yield glucose. Glucose is then converted into a multitude of useful products.

- *Lignocellulosic biomass*, the non-grain portion of biomass (e.g., cobs, stalks), is often referred to as agricultural stover or residues. Energy crops such as switchgrass also have valuable components, but they are not as readily accessible as starch. These lignocellulosic biomass resources (also called *cellulosic*) are comprised of cellulose, hemicellulose, and lignin. Generally, lignocellulosic material contains 30 to 50% cellulose, 20 to 30% hemicellulose, and 20 to 30% lignin. Some exceptions to this are cotton (98% cellulose) and flax (80% cellulose). Lignocellulosic biomass is perceived as a valuable and largely untapped resource for the future bioindustry; however, recovering the components in a cost-effective way is a significant technical challenge.

- *Cellulose* is one of nature's polymers and is composed of glucose, a six-carbon sugar. The glucose molecules are joined by glycosidic linkages, which allow the glucose chains to assume an extended ribbon conformation. Hydrogen bonding between chains leads to the formation of the flat sheets that lie on top of one another in a staggered fashion, similar to the way staggered bricks add strength and stability to a wall. As a result, cellulose is very chemically stable and insoluble and serves as a structural component in plant walls.

- *Hemicellulose* is a polymer containing primarily 5-carbon sugars such as xylose and arabinose, with some glucose and mannose dispersed throughout. It forms a short-chain polymer that interacts with cellulose and lignin to form a matrix in the plant wall, strengthening it. Hemicellulose is more easily hydrolyzed than cellulose. Much of the hemicellulose in lignocellulosic material is solubilized and hydrolyzed to pentose and hexose sugars.

- *Lignin* helps bind the cellulosic/hemicellulose matrix while adding flexibility to the mix. The molecular structure of lignin polymers is very random and disorganized and consists primarily of carbon ring structures (benzene rings with methoxyl, hydroxyl, and propyl groups) interconnected by polysaccharides (sugar polymers). The ring structures of lignin have great potential as valuable chemical intermediates; however, separation and recovery of the lignin is difficult.

- *Oils* and *proteins* are obtained from the seeds of certain plants (e.g., soybeans, castor beans) and have great potential for bioproducts. These oils and proteins can be extracted in a variety of ways. Plants raised for this purpose include soy, corn, sunflower, safflower, and rapeseed, among others. A large portion of the oils and proteins recovered from oilseeds and corn is processed for human or animal consumption, but they can also serve as raw materials for lubricants, hydraulic fluids, polymers, and a host of other products.

- *Vegetable oils* are composed primarily of triglycerides, also referred to as triacylglycerols. Triglycerides contain a glycerol molecule as the backbone with three fatty acids attached to glycerol's hydroxyl groups.

- *Proteins* are natural polymers with amino acids as the monomer unit. They are incredibly complex materials and their functional properties depend on molecular structure. There are 20 amino acids, each differentiated by their side chain or R-group, and they can be classified as nonpolar and hydrophobic, polar uncharged, and ionizable. The interactions among the side chains, the amide protons, and the carbonyl oxygen help create the three-dimensional shape of the protein.

13.3 Biomass for Power and Heat Generation

As part of its efforts to reduce the environmental impacts of energy production and use, the U.S. Environmental Protection Agency (USEPA) has engaged in outreach and providing technical assistance to improve understanding and use of highly efficient combined heat and power (CHP) applications. Recently, market and policy forces have driven strong interest and early implementation of new biomass-fuel CHP projects by potential users and those concerned with clean and sustainable energy sources (USEPA, 2007b).

The use of biomass offers many potential advantages compared to the use of fossil fuels to meet our energy needs. Specific benefits depend upon the intended use and fuel source but often include reductions in greenhouse gases (particularly carbon dioxide) and other air pollutants, energy cost savings, local economic development, waste reduction, and the security of having a domestic fuel supply. In addition, biomass is more flexible (e.g., can generate both power and heat) and reliable (as a non-intermittent resource) as an energy option than many other sources of renewable energy.

Biomass fuels are typically used most efficiently and beneficially to generate both power and heat through CHP. CHP, also known as cogeneration, is the simultaneous production of electricity and heat from a single fuel source, such as biomass/biogas, natural gas, coal, or oil. CHP offers the following advantages:

- *Distributed generation* of electrical and/or mechanical power
- *Waste-heat recovery* for heating, cooling, or process applications
- *Seamless system integration* of a variety of technologies, thermal applications, and fuel types into existing building infrastructure

As mentioned earlier, CHP is not single technology, but an integrated energy system that can be modified depending on the needs of the energy end-user. The hallmark of all well-designed CHP systems is an increase in the efficiency of fuel use. By using waste heat recovery technology to capture a significant proportion of heat created as a byproduct in electricity generation, CHP systems typically achieve total system efficiencies of 60 to 80% of producing electricity and thermal energy. These efficiency gains improve the economics of using biomass fuels and offer other environmental benefits. More than 60% of current biomass-powered electricity generation in the United States is in the form of CHP (EEA, 2006).

Currently, both steam or hot water and electricity are being produced from biomass in CHP facilities in the paper, chemical, wood products, and food-processing industries. These industries are major users of biomass fuels; utilizing the heat and steam in their processes can improve energy efficiency by more that 35%. The biggest industrial user of bioenergy is the forest products industry, which consumes 85% of all wood waste used for energy in the United States. Manufacturing plants that utilize forest products can typically generate more than half of their own energy from woody waste products and other renewable sources of fuel (e.g., wood chips, black liquor).

Most of the electricity, heat, and steam produced by industrial facilities are consumed onsite; however, some manufacturers that produce more electricity than they need onsite sell excess power to the grid. Wider use of biomass resources will directly benefit many companies that generate more residues (e.g., wood or processing wastes) than they can use internally. New markets for these excess materials may support business expansion as the residues are purchased for energy generation purposes, or new profit centers of renewable energy production may diversify and support the core business of these companies.

13.4 Biogas (Methane, CH_4)

Right up front it is important to point out that pure methane is not biogas and biogas is not pure methane. Instead, biogas is the gaseous emissions from anaerobic digestion or degradation of organic matter (from plants or animals) by a consortium of bacteria and is principally a mixture of methane (CH_4) and carbon dioxide (CO_2); biogas also contains other trace gases. Moreover, it is also important to point out that biogas is not the same as biofuel. Instead, biogas is only one of many types of biofuels, which include solid, liquid, or gaseous fuels from biomass.

Methane gas, the primary component of natural gas (98%), makes up 55 to 90% by volume of biogas, depending on the source of organic matter and conditions of degradation. In natural environments, biogas is produced when degradable organic material and low levels of oxygen (O_2) are present. These natural sources of biogas include aquatic sediments, wet soils, burning organic matter, animal and insect (arthropods) digestive tracts, and the cores of some trees. Human activities create additional sources, such as landfills, waste lagoons, and waste storage structures. Atmospheric emissions of biogas from natural and manmade sources contribute to climate change due to the potent greenhouse gas properties of methane. Biogas technology allows recovery of biogas from the anaerobic digestion of organic matter using sealed vessels (digesters) and makes the biogas available for use as fuel for direct heating, electrical generation, or mechanical power and other uses. Biogas is often made from wastes but also can be made from biomass feedstocks. Before discussing wastewater-treatment-derived biogas (the focus of this chapter) it is important to discuss sources of methane (the principal component of biogas).

TABLE 13.1

U.S. Methane Emissions by Source ($TgCO_2$ Equivalents)

Source Category	1990	2000	2005	2006	2007	2008	2009
Natural gas systems	189.8	209.3	190.4	217.7	205.2	211.8	221.2
Enteric fermentation	132.1	136.5	136.5	138.8	141.0	140.6	139.8
Landfills	147.4	111.7	112.5	111.7	111.3	115.9	117.5
Coal mining	84.1	60.4	56.9	58.2	57.9	67.1	71.0
Manure management	31.7	42.4	46.6	46.7	50.7	49.4	49.5
Petroleum systems	35.4	31.5	29.4	29.4	30.0	30.2	30.9
Wastewater treatment	**23.5**	**25.2**	**24.3**	**24.5**	**24.4**	**24.5**	**24.5**
Remaining forest land	3.2	14.3	9.8	21.6	20.0	11.9	7.8
Rice cultivation	7.1	7.5	6.8	5.9	6.2	7.2	7.3
Stationary combustion	7.4	6.6	6.6	6.2	6.5	6.5	6.2
Abandoned underground coal mines	6.0	7.4	5.5	5.5	5.6	5.9	5.5
Mobile combustion	4.7	3.4	2.5	2.3	2.2	2.0	2.0
Composting	0.3	1.3	1.6	1.6	1.7	1.7	1.7
Petrochemical production	0.9	1.2	1.1	1.0	1.0	0.9	0.8
Iron and steel production and metallurgical coke production	1.0	0.9	0.7	0.7	0.7	0.6	0.4
Field burning of agricultural residue	0.3	0.3	0.2	0.2	0.2	0.3	0.2
International bunker fuels	0.2	0.1	0.1	0.2	0.2	0.2	0.1
Total for United States	674.9	659.9	631.4	672.1	664.6	676.7	686.3

Source: USEPA, *Methane*, U.S. Environmental Protection Agency, Washington, DC, 2011, www.epa.gov/methane/sources.html.

13.4.1 The 411 on Methane

As mentioned, methane (CH_4) is emitted from a variety of both human-related (anthropogenic) and natural sources. In the United States, the largest methane emissions come from the decomposition of wastes in landfills, ruminant digestion and manure management associated with domestic livestock, natural gas and oil systems, and coal mining. Table 13.1 shows the level of emissions in teragrams of CO_2 equivalents (1 teragram equals 1×10^{12} grams or 1 million metric tons) from individual sources for the years 1990, 2000, and 2005 to 2009, with wastewater treatment emissions highlighted.

Primarily known as a fuel for interior heating systems, methane or biogas can also be used as a replacement for natural gas—a fossil fuel for electricity generation and for cooking and heating—and as an alternative fuel to gasoline. Methane is a natural gas produced by the breakdown of organic material in the absence of oxygen in termite mounds, wetlands, and by some animals. Humans are also responsible for the release of methane through biomass burning, rice production, cattle, and release from gas exploration. Methane can also be

DID YOU KNOW?

Methanogens are microorganisms that produce methane as a metabolic byproduct in anoxic conditions. Methanogens are anaerobic, and most are rapidly killed by the presence of oxygen. There are over 50 described species of methanogens.

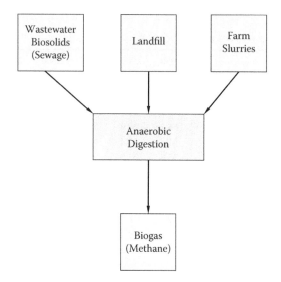

FIGURE 13.1
Production of biogas (methane, CH_4).

obtained directly from the earth; however, other methods of production have been developed, most notably the fermentation or composting of plant and animal waste. The reasons for considering biogas (methane) as a possible biofuel include the following:

- It is viable because of its potential use as an alternative fuel source.
- It is a viable alternative fuel to use to improve air quality.
- It can be produced locally, reducing the need to use imported natural gas.

Methane is produced under anaerobic (no oxygen) conditions where organic material is biodegraded or broken down by a group of microorganisms. The three main sources of feedstock material for anaerobic digestion are given in Figure 13.1 and are described in the following text.

13.5 Wastewater Treatment Plant Biogas

Wastewater treatment plant biogas is produced from the anaerobic digestion of domestic/industrial wastewater biosolids (sludge). During the wastewater treatment process, solids from primary and secondary treatment are collected and further processed, via digestion, to stabilize and reduce the volume of the biosolids. The digestion is performed either aerobically (in the presence of oxygen) or anaerobically (without oxygen) to produce biogas. Anaerobic digestion and wastewater treatment take place in a closed or covered tank to exclude air or oxygen from the waste. Anaerobic treatment has been historically used to biologically stabilize high-strength wastes at a low cost. In many cases, the biogas has not been used as an energy resource but has been burned in a flare and discharged to the atmosphere. Biogas is also generated from other anaerobic wastewater treatment processes, including anaerobic lagoons and facultative lagoons (USEPA, 2007b).

DID YOU KNOW?

Wastewater treatment biogas consists of approximately 55 to 65% methane, 30% CO_2, and other inert gases such as nitrogen. This composition results in a heating value of approximately 550 to 650 Btu/scf.

13.5.1 Anaerobic Digestion

Anaerobic digestion is the traditional method of managing waste, sludge stabilization, and releasing energy. It involves using bacteria that thrive in the absence of oxygen and is slower than aerobic digestion, but it has the advantage that only a small portion of the wastes is converted into new bacterial cells. Instead, most of the organics are converted into carbon dioxide and methane gas.

Cautionary Note: Allowing air to enter an anaerobic digester should be prevented because the mixture of air and gas produced in the digester can be explosive.

13.5.1.1 Stages of Anaerobic Digestion

Anaerobic digestion (see Figure 13.2) has four key biological and chemical stages (Spellman, 2009; USEPA, 1979, 2006c):

1. *Hydrolysis*—Proteins, cellulose, lipids, and other complex organics are broken down into smaller molecules and become soluble by utilizing water to split the chemical bonds of the substances.
2. *Acidogenesis*—The products of hydrolysis are converted into organic acids (where monomers are converted to fatty acids).
3. *Acetogenesis*—The fatty acids are converted to acetic acid, carbon dioxide, and hydrogen.
4. *Methanogenesis*—Organic acids produced during the fermentation step are converted to methane and carbon dioxide.

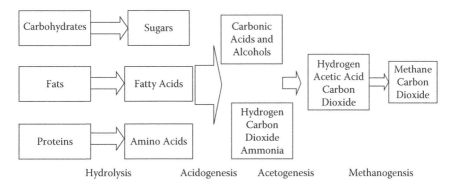

FIGURE 13.2
Key stages of anaerobic digestion.

TABLE 13.2

Typical Contents of Biogas

Matter	Percentage (%)
Methane (CH_4)	50–75
Carbon dioxide (CO_2)	25–50
Nitrogen (N_2)	0–10
Hydrogen (H_2)	0–1
Hydrogen sulfide (H_2S)	0–3
Oxygen (O_2)	0–2

The efficiency of each phase is influenced by the temperature and the amount of time the process is allowed to react. For example, the organisms that perform hydrolysis and volatile acid fermentation (often called *acidogenic bacteria*) are fast-growing microorganisms that prefer a slightly acidic environment and higher temperatures than the organisms that perform the methane formation step (*methanogenic bacteria*).

A simplified generic chemical equation for the overall processes outlined above is as follows:

$$C_6H_{12}O_6 \rightarrow 3CO_2 + 3CH_4$$

Biogas is the ultimate waste product of the bacteria feeding off the input biodegradable feedstock and is composed primarily of methane and carbon dioxide, with a small amount of hydrogen and trace hydrogen sulfide (see Table 13.2). Keep in mind that the ultimate output from a wastewater digester is water; biogas (methane) is more of an off-gas that can be used as an energy source. Wastewater digestion and the production of biogas are discussed in the next section.

13.5.1.2 Anaerobic Digestion of Sewage Biosolids (Sludge)

Equipment used in anaerobic digestion typically includes a sealed digestion tank with either a fixed or a floating cover or an inflatable gas bubble, heating and mixing equipment, gas storage tanks, solids and supernatant withdrawal equipment, and safety equipment (e.g., vacuum relief, pressure relief, flame traps, explosion-proof electrical equipment).

Caution: Biosolids are inherently dangerous as possible sources of explosive gases, and biosolids sites should never be entered without following OSHA's confined space entry permit requirements; only fully trained personnel should enter permit-required confined spaces.

In operation, process residual (thickened or unthickened biosolids/sludge) is pumped into the sealed digester. The organic matter digests anaerobically by a two-stage process. Sugars, starches, and carbohydrates are converted to volatile acids, carbon dioxide, and hydrogen sulfide. The volatile acids are then converted to methane gas. This operation can occur in a single tank (one stage) or in two tanks (two stages). In a single-stage system, supernatant and digested solids must be removed whenever flow is added. In a two-stage operation, solids and liquids from the first stage flow into the second stage each time fresh solids are added. Supernatant is withdrawn from the second stage to provide additional treatment space. Periodically, solids are withdrawn for dewatering or disposal. The methane gas produced in the process may be used for many plant activities.

TABLE 13.3

Sludge Parameters for Anaerobic Digesters

Raw Biosolids (Sludge) Solids	Impact
<4% solids	Loss of alkalinity
	Decreased sludge retention time
	Increased heating requirements
	Decreased volatile acid/alkalinity ratio
4–8% solids	Normal operation
>8% solids	Poor mixing
	Organic overloading
	Decreased volatile acid/alkalinity ratio

Various performance factors affect the operation of the anaerobic digester. The percent volatile matter in raw sludge, digester temperature, mixing, volatile acids/alkalinity ratio, feed rate, percent solids in raw biosolids, and pH are all important operational parameters that the operator must monitor (see Table 13.3). Along with being able to recognize normal and abnormal anaerobic digester performance parameters, digester operators must also know and understand normal operating procedures. Normal operating procedures include biosolids additions, supernatant withdrawal, sludge withdrawal, pH control, temperature control, mixing, and safety requirements.

Sludge must be pumped (in small amounts) several times each day to achieve the desired organic loading and optimum performance, and supernatant withdrawal must be controlled for maximum sludge retention time. All drawoff points are sampled, and the level with the best quality is selected. Digested sludge is withdrawn only when necessary; at least 25% seed remains. A pH of 6.8 to 7.2 is maintained by adjusting the feed rate, sludge withdrawal, or alkalinity additions. Anaerobic digesters must be continuously monitored and tested to ensure proper operation. Testing is performed to determine supernatant pH, volatile acids, alkalinity, biochemical oxygen demand (BOD) or chemical oxygen demand (COD), total solids, and temperature. Sludge (in and out) is routinely tested for percent solids and percent volatile matter. Normal operating parameters are listed in Table 13.4.

Note: The buffer capacity of an anaerobic digester is indicated by the volatile acid/alkalinity relationship. Decreases in alkalinity cause a corresponding increase in ratio.

The temperature in a heated digester must be kept in a normal temperature range of 90 to 95°F. The temperature is never adjusted by more than 1°F per day. In digesters equipped with mixers, the mixing process ensures that organisms are exposed to food materials. Again, anaerobic digesters are inherently dangerous, and several catastrophic failures have been recorded. To prevent such failures, safety equipment, such as pressure relief and vacuum relief valves, flame traps, condensate traps, and gas collection safety devices, is installed. It is important that these critical safety devices be checked and maintained for proper operation.

Note: Because of the inherent danger involved with working inside anaerobic digesters, they are automatically classified as permit-required confined spaces. All operations involving internal entry must be made in accordance with OSHA's confined space entry standard. Questions concerning safe entry into confined spaces of any type should be addressed by a Certified Safety Professional (CSP), Certified Industrial Hygienist (CIH), or Professional Engineer (PE).

TABLE 13.4

Anaerobic Digester: Normal Operating Ranges

Parameter	Normal Range
Sludge retention time	
Heated	30–60 days
Unheated	180+ days
Volatile solids loading	0.04–0.1 lb/day/ft^3
Operating temperature	
Heated	90–95°F
Unheated	Varies with season
Mixing	
Heated—primary	Yes
Unheated—secondary	No
Methane in gas	60–72%
Carbon dioxide in gas	28–40%
pH	6.8–7.2
Volatile acids/alkalinity ratio	≤0.1
Volatile solids reduction	40–60%
Moisture reduction	40–60%

13.6 Cogeneration Using Landfill Biogas

Landfills can be a source of energy. Some wastewater treatment plants with anaerobic digesters located close to landfills harness the methane from the landfill and combine it with methane from their anaerobic digesters to provide additional power for their plant site. Landfills produce methane as organic waste decomposes in the same anaerobic digestion process used to convert wastewater and farm waste slurries into biogas. Most landfill gas results from the degradation of cellulose contained in municipal and industrial solid waste. Unlike animal manure digesters, which control the anaerobic digestion process, the digestion occurring in landfills is an uncontrolled process of biomass decay. To be technically feasible, a landfill must be at least 40 feet deep and have at least a million tons of waste in place for landfill gas collection.

The efficiency of the process depends on the waste composition and moisture content of the landfill, cover material, temperature, and other factors. The biogas released from landfills, commonly called *landfill gas*, is typically 50% methane, 45% carbon dioxide, and 5% other gases. The energy content of landfill gas is 400 to 550 Btu per cubic foot.

Figure 13.3 shows a landfill energy system. Such a system consists of a series of wells drilled into the landfill. A piping system connects the wells and collects the gas. Dryers remove moisture from the gas, and filters remove impurities. The gas typically fuels an engine–generator set or gas turbine to produce electricity. The gas can also fuel a boiler to produce heat or steam. Because waste-generated biogas is considered to be a dirty gas, as compared to natural gas, further gas cleanup is required to improve biogas to pipeline quality, the equivalent of natural gas. Reforming the gas to hydrogen would make possible the production of electricity using fuel cell technology.

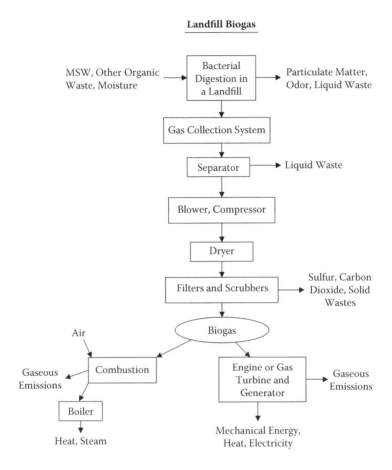

FIGURE 13.3
Landfill biogas system flow diagram.

13.7 Biodiesel

The diesel engine is the workhorse of heavy transportation and industrial processes; it is widely used to power trains, tractors, ships, pumps, and generators. Powering the diesel engine is conventional diesel fuel or biodiesel fuel. Biodiesel is a rather viscous liquid fuel traditionally made up of fatty acid alkyl esters, fatty acid methyl esters (FAMEs), or long-chain mono-alkyl esters. Biodiesel is produced from renewable sources such as new and used vegetable oils and animal fats and is a cleaner burning replacement for petroleum-based diesel fuel. It is nontoxic and biodegradable. This fuel is designed to be used in compression ignition (diesel) engines similar or identical to those that burn petroleum diesel. Biodiesel has physical properties (see Table 13.5) similar to those of petroleum diesel.

In the United States, most biodiesel is made from soybean oil or recycled cooking oils. Animal fats, other vegetable oils, and other recycled oils can also be used to produce biodiesel, depending on the costs and availability. In the future, blends of all kinds of fats and oils may be used to produce biodiesel. At present, municipal sewage biosolids are gaining

TABLE 13.5

Biodiesel Physical Characteristics

Characteristic	Range of Values
Specific gravity	0.87 to 0.89
Kinematic viscosity at 40°C	3.7 to 5.8
Cetane number	46 to 70
Higher heating value (Btu/lb)	16,928 to 17,996
Sulfur (wt%)	0.0 to 0.0024
Cloud point (°C)	–11 to 16
Pour point (°C)	–15 to 13
Iodine number	60 to 135
Lower heating value (lb/lb)	15,700 to 16,735

traction in the United States and around the world as a lipid feedstock for biodiesel production. Biosolids are plentiful and consist of significant concentrations of lipid that can make production of biodiesel from biosolids profitable. In 2010, it was estimated that the cost of production of biodiesel from wastewater biosolids was $3.11 per gallon of biodiesel. To be competitive, this 2010 cost would have to have been reduced to a level that was at or below the petro diesel costs of $3.00 per gallon at that time (Kargbo, 2010).

Before providing a basic description of the process of converting recycled vegetable oil and grease to biodiesel, it is important to define and review a few of the key technological terms associated with the conversion process:

- *Acid esterification*—Oil feedstocks containing more than 4% free fatty acids go through an acid esterification process to increase the yield of biodiesel. These feedstocks are filtered and preprocessed to remove water and contaminants and are then fed to the acid esterification process. The catalyst, sulfuric acid, is dissolved in methanol and then mixed with the pretreated oil. The mixture is heated and stirred, and the free fatty acids are converted to biodiesel. Once the reaction is complete, it is dewatered and then fed to the transesterification process.

- *Transesterification*—Oil feedstocks containing less than 4% free fatty acids are filtered and preprocessed to remove water and contaminants and then fed directly to the transesterification process along with any products of the acid esterification process. The catalyst, potassium hydroxide, is dissolved in methanol and then mixed with the pretreated oil. If an acid esterification process is used, then extra base catalyst must be added to neutralize the acid added in that step. Once the reaction is complete, the major co-products, biodiesel and glycerin, are separated into two layers.

- *Methanol recovery*—The methanol is typically removed after the biodiesel and glycerin have been separated, to prevent the reaction from reversing itself. The methanol is cleaned and recycled back to the beginning of the process.

DID YOU KNOW?

Post-consumer resources such as fats, oil, and grease (FOG) that are usually either dumped in landfills or flushed down drains, clogging pipes and causing costly sewer overflow spills, can be converted to biodiesel.

- *Biodiesel refining*—Once separated from the glycerin, the biodiesel goes through a clean-up or purification process to remove excess alcohol, residual catalyst, and soaps. This consists of one or more washings with clear water. It is then dried and sent to storage. Sometimes the biodiesel goes through an additional distillation step to produce a colorless, odorless, zero-sulfur biodiesel.

- *Glycerin refining*—The glycerin byproduct contains unreacted catalyst and soaps that are neutralized with an acid. Water and alcohol are removed to produce 50 to 80% crude glycerin. The remaining contaminants include unreacted fats and oils. In large biodiesel plants, the glycerin can be further purified, to 99% or higher purity, for sale to the pharmaceutical and cosmetic industries.

The most popular biodiesel production process is transesterification (production of the ester) of vegetable oils or animal fats, using alcohol in the presence of a chemical catalyst. About 3.4 kg of oil or fat are required for each gallon of biodiesel produced (Baize, 2006). The transesterification of degummed soybean oil produces ester and glycerin. The reaction requires heat and a strong base catalyst such as sodium hydroxide or potassium hydroxide. The simplified transesterification reaction is shown below:

Triglycerides + Free fatty acids (<4%) + Alcohol → Alkyl esters + Glycerin

Some feedstocks must be pretreated before they can go through the transesterification process. Feedstocks with less than 4% free fatty acids, which include vegetable oils and some food-grade animal fats, do not require pretreatment. Feedstocks with more than 4% free fatty acids, which include inedible animal fats and recycled greases, must be pretreated in an acid esterification process. In this step, the feedstock is reacted with an alcohol (like methanol) in the presence of a strong acid catalyst (sulfuric acid), converting the free fatty acids into biodiesel. The remaining triglycerides are converted to biodiesel in the transesterification reaction:

Triglycerides + Free fatty acids (>4%) + Alcohol → Alkyl esters + Triglycerides

Figure 13.4 illustrates the basic technology for processing vegetable oils (such as soybeans) and recycled greases (used cooking oil and animal fat). When the feedstock is vegetable oil, the extracted oil is processed to remove all traces of water, dirt, and other contaminants. Free fatty acids are also removed. A combination of methyl alcohol and a catalyst, usually sodium hydroxide or potassium hydroxide, breaks the oil molecules apart in the esterification process. The resulting esters are then refined into usable biodiesel.

DID YOU KNOW?

It is important that the reader has a fundamental understanding of the difference between conventional gasoline and diesel fuel energy output as compared to nonconventional renewable products. Typically, this comparison is made utilizing a standard engineering parameter known as the *gasoline gallon equivalent* (GGE), which is the ratio of the number of British thermal units (Btu) available in 1 U.S. gallon (1 gal) of gasoline to the number of British thermal units available in 1 gal of the alternative substance in question. NIST (2007) defines a gasoline gallon equivalent as 5660 pounds of natural gas. The GGE parameter allows consumers to compare the energy content of competing fuels against a commonly known fuel—gasoline.

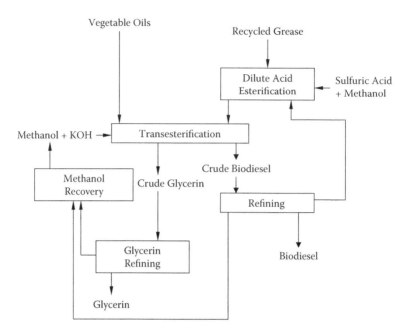

FIGURE 13.4
Biodiesel process description.

When the feedstock is used-up cooking oil and animal fats refined to produce biodiesel, the process is similar to the way biodiesel is derived from vegetable oil, except there is an additional step involved (Figure 13.4). Methyl alcohol and sulfur are used in a process called *dilute acid esterification* to obtain a substance resembling fresh vegetable oil, which is then processed in the same way as vegetable oil to obtain the final product.

References and Recommended Reading

Allen, H.L., Fox, T.R., and Campbell, R.G. (2005). What's ahead for intensive pine plantation silviculture? *Southern Journal of Applied Forestry*, 29, 62–69.

Anon. (2008). Algae eyed as biofuel alternative, *Taipei Times*, January 12, p. 2 (http://www.taipeitimes.com/News/taiwan/archives/2008/01/12/2003396760).

Ashford, R.D. (2001). *Ashford's Dictionary of Industrial Chemicals*, 2nd ed., Wavelength Publications, London.

Aylott, M.J. (2008). Yield and spatial supply of bioenergy poplar and willow short-rotation coppice in the U.K., *New Phytologist*, 178(2), 358–370.

Baize, J. (2006). *Bioenergy and Biofuels*, Agricultural Outlook Forum, February 17, Washington, DC.

Baker, J.B. and Broadfoot, W.M. (1979). *A Practical Field Method of Site Evaluation for Commercially Important Southern Hardwoods*, General Technical Report SO-26, U.S. Department of Agriculture, Forest Service, South Forest Experiment Station, New Orleans, LA.

Bender, M. (1999). Economic feasibility review for community-scale farmer cooperatives for biodiesel, *Bioresource Technology*, 70, 81–87.

Christi, Y. (2007). Biodiesel from microalgae, *Biotechnology Advances*, 25, 294–306.

Coyle, W. (2007). The future of biofuels: a global perspective, *Amber Waves*, November (http://www.thebioenergysite.com/articles/9/the-future-of-biofuels-a-global-perspective).

Dickman, D. (2006). Silviculture and biology of short-rotation wood crops in temperate regions: then and now, *Biomass and Bioenergy*, 30, 696–705.

EEA. (2006). *Combined Heat and Power Database*, Energy and Environmental Analysis, Inc. (ICF International), Fairfax, VA (www.eea-inc.com/chpdata/index.html).

EERE. (2008). *Biomass Program*, U.S. Department of Energy, Energy Efficiency and Renewable Energy, Washington, DC (http://www1.eere.energy.gov/biomass/feedstocks_types.html).

EIA. (2006). *Annual Energy Outlook 2007 with Projections to 2030*, DOE/EIA-0383(2007), U.S. Department of Energy, Energy Information Administration, Office of Integrated Analysis and Forecasting, Washington, DC.

Elliott, D.C., Fitzpatrick, S.W., Bozell, J.J. et al. (1999). Production of levulinic acid and use as a platform chemical for derived products, in Overend, R.P. and Chornet, E., Eds., *Biomass: A Growth Opportunity in Green Energy and Value-Added Products, Proceedings of the Fourth Biomass Conference of the Americas*, Elsevier Science, Oxford, UK, pp. 595–600.

EPRI. (1997). *Renewable Energy Technology Characterizations*, TR-109496, U.S. Department of Energy, Electric Power Research Institute, Washington, DC.

Fahey, J. (2001). Shucking petroleum, *Forbes Magazine*, 168(13), 206–208.

Farrell, A.E. and Gopal, A.R. (2008). Bioenergy research needs for heat, electricity, and liquid fuels, *MRS Bulletin*, 33, 373–387.

Friedman, T. (2010). The fat lady has sung, *The New York Times*, December 12, p. WK8 (http://www.nytimes.com/2010/02/21/opinion/21friedman.html).

Gentille, S.B. (1996). *Reinventing Energy: Making the Right Choices*, American Petroleum Institute, Washington, DC.

Goldstein, L., Hedman, B., Knowles, D. et al. (2003). *Gas-Fired Distributed Energy Resource Technology Characterizations*, NREL/TP-620-34783, Gas Research Institute and National Renewable Energy Laboratory, Washington, DC (www.nrel.gov/docs/fy04osti/34783.pdf).

Graham, R., Nelson, R., Sheehan J., Perlack, R., and Wright L. (2007). Current and potential U.S. corn stover supplies, *Agronomy Journal*, 99, 1–11.

Haas, M., McAloon, A., Yee, W., and Foglia, T. (2006). A process model to estimate biodiesel production costs, *Bioresource Technology*, 97, 671–678.

Harrar, E.S. and Harrar, J.G. (1962). *Guide to Southern Trees*, 2nd ed., Dover, New York.

Hileman, B. (2003). Clashes over agbiotech, *Chemical & Engineering News*, 81, 25–33.

Hoffman, J. (2001). BDO outlook remains healthy, *Chemical Market Reporter*, 259(14), 5.

Hughes, E. (2000). Biomass cofiring: economics, policy and opportunities, *Biomass and Bioenergy*, 19, 457–465.

ILSR. (2002). *Accelerating the Shift to a Carbohydrate Economy: The Federal Role*, Executive Summary of the Minority Report of the Biomass Research and Development Technical Advisory Committee, Institute for Local Self-Reliance, Washington, DC.

Kantor, S.L., Lipton, K., Manchester, A., and Oliveira, V. (1997). Estimating and addressing America's food losses, *Food Review*, 20(1), 3–11.

Kargbo, D.M. (2010). Biodiesel production from municipal sewage sludges, *Energy Fuels*, 24, 2791–2794.

Klass, D.I. (1998). *Biomass for Renewable Energy, Fuels, and Chemicals*, Academic Press, San Diego, CA.

Lee, S. (1996). *Alternative Fuels*, Taylor & Francis, Boca Raton, FL.

Lewis, L. (2005). Seaweed to breathe new life into fight against global warming, *London Times*, May 14.

Little, A. (2001). *Aggressive Growth in the Use of Bioderived Energy and Products in the United States by 2010: Final Report*, U.S. Department of Energy, Washington, DC.

Markarian, J. (2003). New additives and basestocks smooth way for lubricants, *Chemical Market Reporter*, April 28.

McDill, S. (2009). Can algae save the world—again? *Reuters*, February 20 (http://www.reuters.com/article/2009/02/10/us-biofuels-algae-idUSTRE5196HB20090210?pageNumber=2&virtualBrandChannel=0).

McGraw, L. (1999). Three new crops for the future, *Agricultural Research Magazine*, 47(2), 17.

McKeever, D.B. (1998). Wood residual quantities in the United States, *BioCycle: Journal of Composting and Recycling*, 39(1), 65–68; as cited in Antares Group, Inc., *Assessment of Power Production at Rural Utilities Using Forest Thinnings and Commercially Available Biomass Power Technologies*, prepared for the U.S. Department of Agriculture, U.S. Department of Energy, and National Renewable Energy Laboratory, Washington, DC, 2003.

McKendry, P. (2002). Energy production from biomass. Part I. Overview of biomass, *Bioresource Technology*, 83, 37–46.

McLaughlin, S.B. and Kzos, L.A. (2005). Development of switchgrass (*Pancium virgatum*) as a bioenergy feedstock in the United States, *Biomass and Bioenergy*, 28, 515–535.

Miles, T.R., Miles, Jr., T.R., Baxter, L.L., Jenkins, B.M., and Oden, L.L. (1993). Alkali slagging problems with biomass fuels, in *First Biomass Conference of the Americas: Energy, Environment, Agriculture, and Industry*, Vol. 1, National Renewable Energy Laboratory, Burlington, VT, pp. 406–421.

Morey, R.V., Tiffany, D.G., and Hartfield D.L. (2006). Biomass for electricity and process heat at ethanol plants, *Applied Engineering in Agriculture*, 22, 723–728.

Morris, D. (2002). *Accelerating the Shift to a Carbohydrate Economy: The Federal Role*, Institute for Local Self-Reliance, Washington, DC.

Nilles, D. (2005). The Northland's choice, *Biodiesel Magazine*, August 1 (http://www.biodieselmagazine.com/articles/378/the-northlands-choice).

NIST. (2007). Appendix D: definitions, in *Specifications, Tolerances, and Other Technical Requirements for Weighing and Measuring Devices*, NIST Handbook 44, U.S. Department of Commerce, National Institute of Standards and Technology (ts.nist.gov/WeightsAndMeasures/upload/HB44_07_FullDoc_Rev1_LC.pdf).

Parrish, D.J. Fike, J.H. Bransby, D.I., and Samson, R. (2008). Establishing and managing switchgrass as an energy crop, *Forage and Grazinglands*, February (http://www.plantmanagementnetwork.org/fg/element/sum2.aspx?id=6903).

Perlin, J. (2005). *A Forest Journey: The Story of Wood and Civilization*, Countryman Press, Woodstock, VT.

Polagye, B., Hodgson, K., and Malte, P. (2007). An economic analysis of bioenergy options using thinning from overstocked forests, *Biomass and Bioenergy*, 31, 105–125.

Rinehart, L. (2006). *Switchgrass as a Bioenergy Crop*, National Center for Appropriate Technology, Butte, MT.

Rossell, J.B. and Pritchard, J.L.R., Eds. (1991). *Analysis of Oilseeds, Fats and Fatty Foods*. Elsevier, London.

Samson, R. et al. (2008). Developing energy crops for thermal applications: optimizing fuel quality, energy security and GHG mitigation, in Pimental, D., Ed., *Biofuels, Solar and Wind as Renewable Energy Systems: Benefits and Risks*, Springer Science, Berlin.

Sauer, P. (2000). Domestic spearmint oil producers face flat pricing, *Chemical Market Reporter*, 258(18), 14.

Schenk, P., Thomas-Hall, S., Stephens, R., Marx U., Mussgnug, J., Posten, C., Kruse, O., and Hankamer, B. (2008). Second generation biofuels: high-efficiency microalgae for industrial production, *BioEnergy Research*, 1(1), 20–43.

Sedjo, R. (1997). The economics of forest-based biomass supply, *Energy Policy*, 25(6), 559–566.

Silva, B. (1998). Meadowfoam as an alternative crop, *AgVentures*, 2(4), 28.

Sioru, B. (1999). Process converts trash into oil, *Waste Age*, 30(11), 20.

Spellman, F.R. (2009). *Handbook of Water and Wastewater Treatment Plant Operations*, 2nd ed., CRC Press, Boca Raton, FL.

Spellman, F.R. and Whiting, N. (2007). *Concentrated Animal Feeding Operations (CAFOs)*, CRC Press, Boca Raton, FL.

Tillman, D. (2000). Biomass cofiring: the technology, the experience, the combustion consequences, *Biomass and Bioenergy*, 19, 365–384.

Uhlig, H. (1998). *Industrial Enzymes and Their Applications*, John Wiley & Sons, New York.

USDA. (2012). *World Agricultural Supply and Demand Estimates*, U.S. Department of Agriculture, Washington, DC (http://www.usda.gov/oce/commodity/wasde/).

USDOE. (2004). *Industrial Bioproducts: Today and Tomorrow*, U.S. Department of Energy, Washington, DC.

USDOE. (2006a). *The Bioproducts Industry: Today and Tomorrow*, U.S. Department of Energy, Washington, DC.

USDOE. (2006b). *Breaking the Biological Barriers to Cellulosic Ethanol: A Joint Research Agenda*, Report from the December 2005 Workshop, DOE-SC-0095, U.S. Department of Energy, Office of Science, Washington, DC.

USDOE. (2007). *Understanding Biomass: Plant Cell Walls*, U.S. Department of Energy, Washington, DC.

USEPA. (1979). *Process Design Manual: Sludge Treatment and Disposal*, EPA/625/625/1-79-011, U.S. Environmental Protection Agency, Office of Research and Development, Washington, DC.

USEPA. (2004). Overview of biogas technology, in Roos, K.F., Martin, Jr., J.B., and Moser, M.A., Eds., *AgSTAR Handbook*, U.S. Environmental Protection Agency, Washington, DC (http://www.epa.gov/agstar/documents/chapter1.pdf).

USEPA. (2006a). *Auxiliary and Supplemental Power Fact Sheet: Microturbines*, EPA 832-F-05-014, U.S. Environmental Protection Agency, Washington, DC.

USEPA. (2006b). *Auxiliary and Supplemental Power Fact Sheet: Viable Sources*, EPA 832-F-05-009, U.S. Environmental Protection Agency, Washington, DC.

USEPA. (2006c). *Biosolids Technology Fact Sheet: Multi-Stage Anaerobic Digestion*, EPA/832-F-06-031, U.S. Environmental Protection Agency, Office of Water, Washington, DC.

USEPA. (2007a). *Fuel Economy Impact Analysis of RFG*, U.S. Environmental Protection Agency, Washington, DC (http://www.epa.gov/oms/rfgecon.htm).

USEPA. (2007b). *Biomass Combined Heat and Power Catalog of Technologies*, U.S. Environmental Protection Agency, Washington, DC.

USEPA. (2008). *Catalog of CHP Technologies*, U.S. Environmental Protection Agency, Washington, DC (www.epa.gov/chp/documents/catalog_chptech_full.pdf).

Valigra, L. (2000). Tough as soybeans, *The Christian Science Monitor*, January 20.

Wilhelm, W.W., Johnson, J.M.F., Karlen, D.L., and Lightle, D.T. (2002). Corn stover to sustain soil organic carbon further constrains biomass supply, *Agronomy Journal*, 99, 1665–1667.

Wiltsee, G. (1998). *Urban Wood Waste Resource Assessment*, NREL/SR-570-25918, National Renewable Energy Laboratory, Golden, CO.

Wood, M. (2002). Desert shrub may help preserve wood, *Agricultural Research Magazine*, 50(4), 10–11.

Section V

Sustainability Using Renewable Energy

That's human nature. Nobody does anything until it's too late.

—Michael Crichton (*Prey*)

14

Macro- and Microhydropower

03/07/12—The U.S. House will vote today on a bill that would give developers of small hydroelectric projects that use Bureau of Reclamation canals and ditches the ability to bypass current environmental reviews.

—HydroWorld.com

When we speak of water and its many manifestations, we are speaking of that endless quintessential cycle that predates all other cycles. Water is our most precious natural resource; we can't survive without it. There is no more water today than there was yesterday—that is, no more this calendar year than 100 million years ago. The water present today is the same water used by all the animals that ever lived, by cave dwellers, Caesar, Cleopatra, Christ, da Vinci, John Snow, Teddy Roosevelt, and the rest of us—again, there is not one drop more or one drop less of water than there has always been. This life-giving cycle, a unique blend of thermal and mechanical aspects, is dependent on solar energy and gravity for its existence. Nothing on Earth is truly infinite in supply, but the energy available from water sources, in practical terms, comes closest to that ideal. But keep in mind, still waters run no turbines.

— F.R. Spellman

14.1 Introduction

This chapter is about making electricity from water power. More specifically, this chapter describes renewable, sustainable, and efficient energy production at the micro level in conduit hydropower systems. *Conduit hydropower* refers to the generation of hydroelectric power in existing tunnels, canals, aqueducts, pipelines, flumes, ditches, or similar human-made water conveyances that have been equipped with electricity-generating equipment. Before we discuss microhydropower applications in detail, you first need to know something about how water behaves, water hydraulics (computations), and hydropower applications on the macro and micro scales.

14.2 Hydropower

When you get right down to it, there is nothing new about using water power to assist humans in the day-to-day struggle to survive. Almost all human settlements began near some major water body. Not only was it important to have easy access to freshwater for drinking and cooking, but it was also important to be near water for transportation needs and to have a close proximity to moving water to generate power to operate various mechanical devices such as grist mills and later for the generation of electricity. This is somewhat ironic, because when we look at rushing waterfalls and rivers we may

TABLE 14.1

History of Hydropower

Date	Hydropower Event
BC	Hydropower was used by the Greeks to turn water wheels for grinding wheat into flour more than 2000 years ago.
Mid-1770s	French hydraulic and military engineer Bernard Forest de Belidor wrote a four-volume work describing vertical- and horizontal-axis machines.
1775	The U.S. Army Corps of Engineers was founded, with establishment of Chief Engineer for the Continental Army.
1880	Michigan's Grand Rapids Electric Light and Power Company, generating electricity by a dynamo belted to a water turbine at the Wolverine Chair Factory, lit up 16 brush-arc lamps.
1881	Niagara Falls city street lamps were powered by hydropower.
1882	The world's first hydroelectric power plant began operation on the Fox River in Appleton, Wisconsin.
1886	There were about 45 water-powered electric plants in the United States and Canada.
1887	San Bernardino, California, opened the first hydroelectric plant in the West.
1889	Two hundred electric plants in the United States used water power for some or all generation.
1901	First Federal Water Power Act was passed.
1902	Bureau of Reclamation was established.
1907	Hydropower provided 15% of U.S. electrical generation.
1920	Hydropower provided 25% of U.S. electrical generation. The Federal Power Act established the Federal Power Commission authority to issue licenses for hydro development on public lands.
1933	The Tennessee Valley Authority was established.
1935	Federal Power Commission authority was extended to all hydroelectric projects built by utilities engaged in interstate commerce.
1937	Bonneville Dam, the first federal dam, began operation on the Columbia River, and the Bonneville Power Administration was established.
1940	Hydropower provided 40% of electrical generation. Conventional capacity had tripled in the United States since 1920.
1980	Conventional capacity had nearly tripled in the United States since 1900.
2003	About 10% of U.S. electricity comes from hydropower. Today, there are about 80,000 MW of conventional capacity and 18,000 MW of pumped storage.

Source: EERE, *History of Hydropower*, Energy Efficiency & Renewable Energy, U.S. Department of Energy, Washington, DC, 2011.

not immediately think of electricity, but hydroelectric (water-powered) power plants are responsible for lighting many of our homes and neighborhoods. *Hydropower* is the harnessing of water to perform work. The power from falling water has been used in industry for thousands of years (see Table 14.1). The Greeks used water wheels to grind wheat into flour more than 2000 years ago. Besides grinding flour, the power from water was used to saw wood, to power textile mills, and for manufacturing plants.

The technology for using falling water to create hydroelectricity has existed for more than a century. The evolution of the modern hydropower turbine began in the mid-1700s when a French hydraulic and military engineer, Bernard Forest de Belidor, wrote a four-volume work describing using a vertical-axis vs. a horizontal-axis machine. Water turbine development continued during the 1700s and 1800s. In 1880, a brush arc light dynamo driven by a water turbine was used to provide theater and storefront lighting in Grand Rapids, Michigan, and in 1881 a brush dynamo connected to a turbine in a flour mill provided street lighting at Niagara Falls, New York. These two projects used direct-current (DC) technology.

TABLE 14.2

U.S. Energy Consumption by Energy Source, 2008

Energy Source	Energy Consumption (quadrillion Btu)
Total	99.438
Fossil fuels (coal, natural gas, petroleum)	83.532
Electricity net imports	0.113
Nuclear electric power	8.427
Renewable	7.367
Biomass (total)	3.852
Biofuels	1.372
Waste	0.436
Wood and wood-derived fuels	2.044
Geothermal energy	0.360
Hydroelectric conventional	**2.512**
Solar thermal/photovoltaic energy	0.097
Wind energy	0.546

Source: EIA, *Renewable Energy Annual 2008*, Energy Information Administration, U.S. Department of Energy, Washington, DC, 2010.

Alternating current (AC) is used today. That breakthrough came when the electric generator was coupled to the turbine in 1882, which resulted in the world's first hydroelectric plant, which was located on the Fox River in Appleton, Wisconsin. The U.S. Library of Congress (LOC, 2009) lists the Appleton hydroelectric power plant as one of the major accomplishments of the Gilded Age (1878–1889). Soon, people across the United States were enjoying electricity in homes, schools, and offices, reading by electric lamp instead of candlelight or kerosene. Today, we take electricity for granted and cannot imagine life without it. Table 14.2 shows the energy consumption by energy source for 2008; in that year, hydropower accounted for 2.512 quadrillion Btu. Ranging in size from small systems (100 kW to 30 MW) for a home or village to large projects (capacity greater than 30 MW) producing electricity for utilities, hydropower plants are of three types: *impoundment*, *diversion*, and *pumped storage*. Some hydropower plants use dams and some do not. Many dams were built for other purposes and hydropower was added later. In the United States, there are about 80,000 dams, of which only 2400 produce power. The other dams are for recreation, stock/farm ponds, flood control, water supply, and irrigation. The types of hydropower plants are described below.

14.2.1 Impoundment

The most common type of hydroelectric power plant is an impoundment facility. An impoundment facility, typically a large hydropower system, uses a dam to store river water in a reservoir. This type of facility works best in mountainous or hilly terrain where high dams can be built and deep reservoirs can be maintained. Potential energy available in a reservoir depends on the mass of water contained in it, as well as on overall depth of the water. Water released from the reservoir flows through a turbine, spinning it, which in turn activates a generator to produce electricity. The water may be released either to meet changing electricity needs or to maintain a constant reservoir level.

14.2.2 Diversion

A diversion, sometimes called *run-of-river*, facility channels all or a portion of the flow of a river from its natural course through a canal or penstock, and the current through this medium is used to drive turbines. It may not require the use of a dam. This type of system is best suited for locations where a river drops considerably per unit of horizontal distance. The ideal location is near a natural waterfall or rapids. The chief advantage of a diversion system is the fact that, lacking a dam, it has far less impact on the environment than an impoundment facility (Gibilisco, 2007).

14.2.3 Pumped Storage

When the demand for electricity is low, a pumped storage facility stores energy by pumping water from a lower reservoir to an upper reservoir. During periods of high electrical demand, the water is released back to the lower reservoir to generate electricity. Table 14.2 highlights hydropower energy's ranking among current renewable energy sources in 2008. The figure for solar/photovoltaic power is expected to steadily increase to a higher level, which should be reflected in the 2009–2010 figures when they are released.

14.3 Hydropower Basic Concepts[*]

Air pressure (at sea level) = 14.7 pounds per square inch (psi)

The relationship shown above is important because our study of hydropower basics begins with air. A blanket of air, many miles thick, surrounds the Earth. The weight of this blanket on a given square inch of the Earth's surface will vary according to the thickness of the atmospheric blanket above that point. As shown above, at sea level, the pressure exerted is 14.7 pounds per square inch (psi). On a mountain top, air pressure decreases because the blanket is not as thick.

$$1 \text{ ft}^3 \text{ H}_2\text{O} = 62.4 \text{ lb}$$

This relationship is also important; note that both cubic feet and pounds are used to describe a volume of water. A defined relationship exists between these two methods of measurement. The specific weight of water is defined relative to a cubic foot. One cubic foot of water weighs 62.4 lb. This relationship is true only at a temperature of 4°C and at a pressure of 1 atmosphere, conditions that are referred to as standard temperature and pressure (STP). One atmosphere = 14.7 psi at sea level and 1 ft^3 of water contains 7.48 gal.

The weight varies so little that, for practical purposes, this weight is used for temperatures ranging from 0 to 100°C. One cubic inch of water weighs 0.0362 lb. Water 1 ft deep will exert a pressure of 0.43 psi on the bottom area (12 in. × 0.0362 lb/in.^{3}). A column of water 2 ft high exerts 0.86 psi (2 ft × 0.43 psi/ft); one 10 ft high is 4.3 psi (10 ft × 0.43 psi/ ft); and one 55 ft high exerts 23.65 psi (55 ft × 0.43 psi/ft). A column of water 2.31 ft high will exert 1.0 psi (2.31 ft × 0.43 psi/ft). To produce a pressure of 50 psi requires a 115.5-ft water column (50 psi × 2.31 ft/psi).

---

[*] Adapted from Spellman, F.R., *Handbook of Water and Wastewater Treatment Plant Operations*, 2nd ed., CRC Press, Boca Raton, FL, 2009.

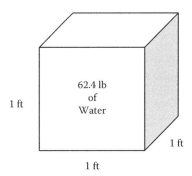

FIGURE 14.1
One cubic foot of water weights 62.4 lb.

Two important points are being made here:

1. 1 ft³ H₂O = 62.4 lb (see Figure 14.1).
2. A column of water 2.31 ft high will exert 1.0 psi.

Another relationship is also important. At standard temperature and pressure, 1 ft³ of water contains 7.48 gal. With these two relationships, we can determine the weight of a gallon of water:

$$\text{Weight of 1 gal water} = 62.4 \text{ lb} \div 7.48 \text{ gal} = 8.34 \text{ lb/gal}$$

Thus,

$$1 \text{ gal water} = 8.34 \text{ lb}$$

Further, this information allows cubic feet to be converted to gallons by simply multiplying the number of cubic feet by 7.48 gal/ft³.

■ *Example 14.1*

Problem: Find the number of gallons in a reservoir that has a volume of 855.5 ft³.

Solution:

$$855.5 \text{ ft}^3 \times 7.48 \text{ gal/ft}^3 = 6399 \text{ gal (rounded)}$$

Note: The term *head* is discussed later but for now it is important to point out that it is used to designate water pressure in terms of the height of a column of water in feet; for example, a 10-ft column of water exerts 4.3 psi. This can be called 4.3-psi pressure or 10 ft of head.

14.3.1 Stevin's Law

Stevin's law deals with water at rest. Specifically, it states: "The pressure at any point in a fluid at rest depends on the distance measured vertically to the free surface and the density of the fluid." Stated as a formula, this becomes

$$p = w \times h \tag{14.1}$$

where

p = Pressure in pounds per square foot (psf)

w = Density in pounds per cubic foot (lb/ft³)

h = Vertical distance in feet

■ Example 14.2

Problem: What is the pressure at a point 18 ft below the surface of a reservoir?

Solution: To calculate this, we must know that the density of the water (w) is 62.4 lb/ft³.

$$p = w \times h = 62.4 \text{ lb/ft}^3 \times 18 \text{ ft} = 1123 \text{ lb/ft}^2 \text{ (psf)}$$

Water practitioners generally measure pressure in pounds per square inch rather than pounds per square foot; to convert, divide by 144 in.²/ft² (12 in. × 12 in. = 144 in.²):

$$p = \frac{1123 \text{ psf}}{144 \text{ in.}^2/\text{ft}^2}$$

14.3.2 Density and Specific Gravity

Table 14.3 shows the relationships among temperature, specific weight, and density of water. When we say that iron is heavier than aluminum, we say that iron has a greater density than aluminum. In practice, what we are really saying is that a given volume of iron is heavier than the same volume of aluminum. A cubic foot of water, for example, weighs 62.4 lb, and a cubic foot of aluminum weighs 178 lb; thus, aluminum is 2.7 times heavier than water.

Note: What is density? Density is the mass per unit volume of a substance.

TABLE 14.3

Water Properties (Temperature, Specific Weight, and Density)

Temperature (°F)	Specific Weight (lb/ft³)	Density (slugs/ft³)	Temperature (°F)	Specific Weight (lb/ft³)	Density (slugs/ft³)
32	62.4	1.94	130	61.5	1.91
40	62.4	1.94	140	61.4	1.91
50	62.4	1.94	150	61.2	1.90
60	62.4	1.94	160	61.0	1.90
70	62.3	1.94	170	60.8	1.89
80	62.2	1.93	180	60.6	1.88
90	62.1	1.93	190	60.4	1.88
100	62.0	1.93	200	60.1	1.87
110	61.9	1.92	210	59.8	1.86
120	61.7	1.92			

Suppose you had a tub of lard and a large box of cold cereal, each having a mass of 600 g. The density of the cereal would be much less than the density of the lard because the cereal occupies a much larger volume than the lard occupies. The density of an object can be calculated by using the formula:

$$\text{Density} = \frac{\text{Mass}}{\text{Volume}} \tag{14.2}$$

Perhaps the most common measures of density are pounds per cubic foot (lb/ft³) and pounds per gallon (lb/gal):

- 1 ft³ of water weighs 62.4 lb; density = 62.4 lb/ft³.
- 1 gal of water weighs 8.34 lb; density = 8.34 lb/gal.

The density of a dry material, such as cereal, lime, soda, and sand, is usually expressed in pounds per cubic foot. The density of a liquid, such as liquid alum, liquid chlorine, or water, can be expressed either as pounds per cubic foot or as pounds per gallon. The density of a gas, such as chlorine gas, methane, carbon dioxide, or air, is usually expressed in pounds per cubic foot. As shown in Table 14.3, the density of a substance like water changes slightly as the temperature of the substance changes. This occurs because substances usually increase in volume (size) by expanding as they become warmer. Because of this expansion with warming, the same weight is spread over a larger volume, so the density is lower when a substance is warm than when it is cold.

Note: What is specific gravity? Specific gravity is the weight (or density) of a substance compared to the weight (or density) of an equal volume of water. The specific gravity of water is 1.

This relationship is easily seen when a cubic foot of water, which weighs 62.4 lb, is compared to a cubic foot of aluminum, which weighs 178 lb. Aluminum is 2.7 times as heavy as water. It is not that difficult to find the specific gravity of a piece of metal. All you have to do is to weigh the metal in air, then weigh it under water. Its loss of weight is the weight of an equal volume of water. To find the specific gravity, divide the weight of the metal by its loss of weight in water.

$$\text{Specific gravity} = \frac{\text{Weight of substance}}{\text{Weight of equal volume of water}} \tag{14.3}$$

■ *Example 14.3*

Problem: Suppose a piece of metal weighs 150 lb in air and 85 lb under water. What is the specific gravity?
Solution:

$$150 \text{ lb} - 85 \text{ lb} = 65 \text{ lb loss of weight in water}$$

$$\text{Specific gravity} = \frac{150}{65} = 2.3$$

As stated earlier, the specific gravity of water is 1, which is the standard, the reference to which all other liquid or solid substances are compared. Specifically, any object that has a specific gravity greater than 1 will sink in water (e.g., rocks, steel, iron, grit, floc, sludge). Substances with a specific gravity of less than 1 will float (e.g., wood, scum, gasoline). Considering the total weight and volume of a ship, its specific gravity is less than 1; therefore, it can float.

The most common use of specific gravity in water operations is in gallon-to-pound conversions. In many cases, the liquids being handled have a specific gravity of 1 or very nearly 1 (between 0.98 and 1.02), so 1 may be used in the calculations without introducing significant error. For calculations involving a liquid with a specific gravity of less than 0.98 or greater than 1.02, the conversions from gallons to pounds must consider specific gravity. The technique is illustrated in the following example.

■ *Example 14.4*

Problem: A basin contains 1455 gal of a certain liquid. If the specific gravity of the liquid is 0.94, how many pounds of liquid are in the basin?

Solution: Normally, for a conversion from gallons to pounds, we would use the factor 8.34 lb/gal (the density of water) if the specific gravity of the substance is between 0.98 and 1.02. In this instance, however, the substance has a specific gravity outside this range, so the 8.34 factor must be adjusted by multiplying 8.34 lb/gal by the specific gravity to obtain the adjusted factor:

$$8.34 \text{ lb/gal} \times 0.94 = 7.84 \text{ lb/gal (rounded)}$$

Then convert 1455 gal to pounds using the corrected factor:

$$1455 \text{ gal} \times 7.84 \text{ lb/gal} = 11{,}407 \text{ lb (rounded)}$$

14.3.3 Force and Pressure

Water exerts force and pressure against the walls of its container, whether it is stored in a tank or flowing in a pipeline. There is a difference between force and pressure, although they are closely related. *Force* is the push or pull influence that causes motion. In the English system, force and weight are often used in the same way. The weight of 1 ft³ of water is 62.4 lb. The force exerted on the bottom of a 1-foot cube is 62.4 lb (see Figure 14.1). If we stack two cubes on top of one another, the force on the bottom will be 124.8 lb. *Pressure* is the force per unit of area. In equation form, this can be expressed as

$$P = \frac{F}{A} \tag{14.4}$$

where

   P = Pressure

   F = Force

   A = Area over which the force is distributed

Pounds per square inch or pounds per square foot are common expressions of pressure. The pressure on the bottom of the cube is 62.4 lb/ft² (see Figure 14.1). It is normal to express pressure in pounds per square inch (psi). This is easily accomplished by determining the weight of 1 in.² of a 1-ft cube. If we have a cube that is 12 in. on each side, the number of square inches on the bottom surface of the cube is 12 × 12 = 144 in.². Dividing the weight by the number of square inches determines the weight on each square inch:

$$\text{psi} = \frac{62.4 \ \text{lb/ft}}{144 \ \text{in.}^2} = 0.433 \ \text{psi/ft}$$

This is the weight of a column of water 1 in. square and 1 ft tall. If the column of water were 2 ft tall, the pressure would be 2 ft × 0.433 psi/ft = 0.866.

Note: 1 foot of water = 0.433 psi.

With this information, feet of head can be converted to psi by multiplying the feet of head by 0.433 psi/ft.

■ *Example 14.5*

Problem: A tank is mounted at a height of 90 ft. Find the pressure at the bottom of the tank.

Solution:

90 ft × 0.433 psi/ft = 39 psi (rounded)

Note: To convert psi to feet, divide the psi by 0.433 psi/ft.

■ *Example 14.6*

Problem: Find the height of water in a tank if the pressure at the bottom of the tank is 22 psi.

Solution:

$$\text{Height in feet} = \frac{22 \ \text{psi}}{0.433 \ \text{psi/ft}} = 51 \ \text{ft (rounded)}$$

14.3.4 Hydrostatic Pressure

Figure 14.2 shows a number of differently shaped, connected, open containers of water. Note that the water level is the same in each container, regardless of the shape or size of the container. This occurs because pressure is developed within a liquid by the weight of the liquid above. If the water level in any one container were to be momentarily higher than that in any of the other containers, then the higher pressure at the bottom of this container would cause some water to flow into the container having the lower liquid level. In addition, the pressure of the water at any level (such as line *T*) is the same in each of the containers. Pressure increases because of the weight of the water. The farther down from the surface, the more pressure is created. This illustrates that the *weight*, not the volume, of water contained in a vessel determines the pressure at the bottom of the vessel. Some very important principles always apply for hydrostatic pressure (Nathanson, 1997):

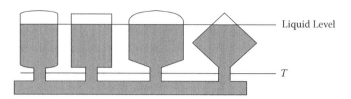

FIGURE 14.2
Hydrostatic pressure.

1. The pressure depends only on the depth of water above the point in question (not on the water surface area).
2. The pressure increases in direct proportion to the depth.
3. The pressure in a continuous volume of water is the same at all points that are at the same depth.
4. The pressure at any point in the water acts in all directions at the same depth.

14.3.5 Head

Head is defined as the vertical distance the water must be lifted from the supply tank to the discharge or as the height a column of water would rise due to the pressure at its base. A perfect vacuum plus atmospheric pressure of 14.7 psi would lift the water 34 ft. If the top of the sealed tube is open to the atmosphere and the reservoir is enclosed, the pressure in the reservoir is increased; the water will rise in the tube. Because atmospheric pressure is essentially universal, we usually ignore the first 14.7 psi of actual pressure measurements and measure only the difference between the water pressure and the atmospheric pressure; we call this *gauge pressure*. Water in an open reservoir is subjected to the 14.7 psi of atmospheric pressure, but subtracting this 14.7 psi leaves a gauge pressure of 0 psi. This indicates that the water would rise 0 feet above the reservoir surface. If the gauge pressure in a water main were 120 psi, the water would rise in a tube connected to the main:

$$120 \text{ psi} \times 2.31 \text{ ft/psi} = 277 \text{ ft (rounded)}$$

14.3.5.1 Total Dynamic (System) Head

Total head includes the vertical distance the liquid must be lifted (*static head*), the loss to friction (*friction head*), and the energy required to maintain the desired velocity (*velocity head*):

$$\text{Total head} = \text{Static head} + \text{Friction head} + \text{Velocity head} \qquad (14.5)$$

14.3.5.2 Static Head

Static head is the actual vertical distance the liquid must be lifted:

$$\text{Static head} = \text{Discharge elevation} - \text{Supply elevation} \qquad (14.6)$$

■ **Example 14.7**

Problem: The supply tank is located at elevation 118 ft. The discharge point is at elevation 215 ft. What is the static head in feet?

Solution:
$$\text{Static head (ft)} = 215 \text{ ft} - 118 \text{ ft} = 97 \text{ ft}$$

14.3.5.3 Friction Head

Friction head is the equivalent distance of the energy that must be supplied to overcome friction. Engineering references include tables showing the equivalent vertical distance for various sizes and types of pipes, fittings, and valves. The total friction head is the sum of the equivalent vertical distances for each component. For friction head,

$$\text{Friction head (ft)} = \text{Energy losses due to friction} \tag{14.7}$$

14.3.5.4 Velocity Head

Velocity head is the equivalent distance of the energy consumed to achieve and maintain the desired velocity in the system:

$$\text{Velocity head (ft)} = \text{Energy losses to maintain velocity} \tag{14.8}$$

14.3.5.5 Pressure and Head

The pressure exerted by water is directly proportional to its depth or head in the pipe, tank, or channel. If the pressure is known, the equivalent head can be calculated:

$$\text{Head (ft)} = \text{Pressure (psi)} \times 2.31 \text{ (ft/psi)} \tag{14.9}$$

■ **Example 14.8**

Problem: The pressure gauge on the discharge line from the influent pump reads 72.3 psi. What is the equivalent head in feet?

Solution:
$$\text{Head (ft)} = 72.3 \times 2.31 \text{ ft/psi} = 167 \text{ ft}$$

14.3.5.6 Head and Pressure

If the head is known, the equivalent pressure can be calculated by

$$\text{Pressure (psi)} = \frac{\text{Head (ft)}}{2.31 \text{ ft/psi}} \tag{14.10}$$

■ **Example 14.9**

Problem: A tank is 22 ft deep. What is the pressure in psi at the bottom of the tank when it is filled with water?

Solution:
$$\text{Pressure (psi)} = \frac{22 \text{ ft}}{2.31 \text{ ft/psi}} = 9.52 \text{ psi (rounded)}$$

14.3.6 Flow and Discharge Rates: Water in Motion

The study of fluid flow is much more complicated than that of fluids at rest, but it is important to have an understanding of these principles because the water used in hydropower applications to propel turbine blades is nearly always in motion. *Discharge* (or flow) is the quantity of water passing a given point in a pipe or channel during a given period. Stated another way for open channels, the flow rate through an open channel is directly related to the velocity of the liquid and the cross-sectional area of the liquid in the channel:

$$Q = A \times V \tag{14.11}$$

where

 Q = Flow, or discharge in cubic feet per second (cfs)

 A = Cross-sectional area of the pipe or channel in square feet (ft²)

 V = Water velocity in feet per second (fps or ft/sec)

■ *Example 14.10*

Problem: A channel is 6 ft wide and the water depth is 3 ft. The velocity in the channel is 4 fps. What is the discharge or flow rate in cubic feet per second?

Solution:

$$\text{Flow (cfs)} = 6 \text{ ft} \times 3 \text{ ft} \times 4 \text{ ft/sec} = 72 \text{ cfs}$$

Discharge or flow can be recorded as gal/day (gpd), gal/min (gpm), or cubic feet per second (cfs). Flows treated by many hydropower systems are large and are often referred to in million gallons per day (MGD). The discharge or flow rate can be converted from cfs to other units such as gpm or MGD by using appropriate conversion factors.

■ *Example 14.11*

Problem: A 12-in.-diameter pipe has water flowing through it at 10 fps. What is the discharge in (a) cfs, (b) gpm, and (c) MGD?

Solution: Before we can use the basic formula, we must determine the area (A) of the pipe. The formula for the area of a circle is

$$\text{Area } (A) = \pi \times \frac{D^2}{4} = \pi \times r^2 \tag{14.12}$$

where

 π = Constant value 3.14159, or simply 3.14

 D = Diameter of the circle in feet

 r = Radius of the circle in feet

Therefore, the area of the pipe is

$$A = \pi \times \frac{D^2}{4} = 3.14 \times \frac{1 \text{ ft}^2}{4} = 0.785 \text{ ft}^2$$

(a) Now we can determine the discharge in cfs:

$$Q = V \times A$$

$$Q = 10 \text{ ft/sec} \times 0.785 \text{ ft}^2 = 7.85 \text{ft}^3/\text{sec (cfs)}$$

(b) We know that 1 cfs is 449 gpm, so 7.85 cfs × 449 gpm/cfs = 3525 gpm (rounded).

(c) 1 million gallons per day is 1.55 cfs, so

$$\frac{7.85 \text{ cfs}}{1.55 \text{ cfs/MGD}} = 5.06 \text{ MGD}$$

14.3.7 Area and Velocity

The *law of continuity* states that the discharge at each point in a pipe or channel is the same as the discharge at any other point (if water does not leave or enter the pipe or channel). That is, under the assumption of steady-state flow, the flow that enters the pipe or channel is the same flow that exits the pipe or channel. In equation form, this can be represented as

$$Q_1 = Q_2 \quad \text{or} \quad A_1 V_1 = A_2 V_2 \tag{14.13}$$

■ *Example 14.12*

Problem: A pipe 12 in. in diameter is connected to a 6-in.-diameter pipe. The velocity of the water in the 12-in. pipe is 3 fps. What is the velocity in the 6-in. pipe?

Solution: Using the equation $A_1 V_1 = A_2 V_2$, we need to determine the area of each pipe:

$$A = \pi \times \frac{D^2}{4}$$

$$A \text{ (12-in. pipe)} = 3.14 \times \frac{(1 \text{ ft})^2}{4} = 0.785 \text{ ft}^2$$

$$A \text{ (6-in. pipe)} = 3.14 \times \frac{(0.5 \text{ ft})^2}{4} = 0.196 \text{ ft}^2$$

The continuity equation now becomes

$$0.785 \text{ ft}^2 \times 3 \text{ ft/sec} = 0.196 \text{ ft}^2 \times V_2$$

Solving for V_2,

$$V_2 = \frac{0.785 \text{ ft}^2 \times 3 \text{ ft/sec}}{0.196 \text{ ft}^2} = 12 \text{ ft/sec (fps)}$$

14.3.8 Pressure and Velocity

In a closed pipe flowing full (under pressure), the pressure is indirectly related to the velocity of the liquid:

$$\text{Velocity}_1 \times \text{Pressure}_1 = \text{Velocity}_2 \times \text{Pressure}_2 \qquad (14.14)$$

or

$$V_1 P_1 = V_2 P_2$$

14.3.9 Conservation of Energy

Many of the principles of physics are important to the study of hydraulics. When applied to problems involving the flow of water, few of the principles of physical science are more important and useful to us than the *law of conservation of energy*. Simply, the law of conservation of energy states that energy can be neither created nor destroyed, but it can be converted from one form to another. In a given closed system, the total energy is constant.

14.3.10 Energy Head

Hydraulic systems have three forms of mechanical energy: potential energy due to elevation, potential energy due to pressure, and kinetic energy due to velocity. Energy has the units of foot-pounds (ft-lb). It is convenient to express hydraulic energy in terms of *energy head* in feet of water. This is equivalent to foot-pounds per pound of water (ft-lb/lb = ft).

14.3.11 Energy Available

Energy available is directly proportional to flow rate and to the hydraulic head, and the head is equivalent to stored potential energy. This is shown as

$$\text{Head} = m \times g \times h$$

where

    m = Mass of water

    g = Acceleration due to gravity (taken as 10 ms^{-2} in most applications)

    h = Head difference

The diameter of the pipe must be large enough to handle the volume of water flowing. Friction in the pipes will reduce the effective head of water and larger diameters are used, although cost then has a bearing. Ideally, the pipes should narrow as one proceeds downhill; however, friction losses are highest where the velocity is highest, so there is usually little change in pipe diameter. Friction losses in piping are classified as either major head loss or minor head loss (Tovey, 2005).

14.3.12 Major Head Loss

Major head loss consists of pressure decreases along the length of pipe caused by friction created as water encounters the surfaces of the pipe. It typically accounts for most of the pressure drop in a pressurized or dynamic water system. The components that contribute to major head loss are roughness, length, diameter, and velocity:

- *Roughness*—Even when new, the interior surfaces of pipes are rough. The roughness varies, of course, depending on pipe material, corrosion (tuberculation and pitting), and age. Because normal flow in a water pipe is turbulent, the turbulence increases with pipe roughness, which, in turn, causes pressure to drop over the length of the pipe.
- *Pipe length*—With every foot of pipe length, friction losses occur. The longer the pipe, the more head loss. Friction loss because of pipe length must be factored into head loss calculations.
- *Pipe diameter*—Generally, small-diameter pipes have more head loss than large-diameter pipes. In large-diameter pipes, less of the water actually touches the interior surfaces of the pipe (encountering less friction) than in a small-diameter pipe.
- *Water velocity*—Turbulence in a water pipe is directly proportional to the speed (or velocity) of the flow; thus, the velocity head also contributes to head loss.

14.3.12.1 Calculating Major Head Loss

Darcy, Weisbach, and others developed the first practical equation used to determine pipe friction in about 1850. The equation or formula now known as the *Darcy–Weisbach* equation for circular pipes is

$$h_f = f \frac{LV^2}{D^2 g} \tag{14.15}$$

In terms of the flow rate Q, the equation becomes:

$$h_f = \frac{8 f L Q^2}{\pi^2 g D^5} \tag{14.16}$$

where

h_f = Head loss (ft)
f = Coefficient of friction
L = Length of pipe (ft)
V = Mean velocity (ft/sec)
D = Diameter of pipe (ft)
g = Acceleration due to gravity (32.2 ft/sec^2)
Q = Flow rate (ft^3/sec)

The Darcy–Weisbach formula as such was meant to apply to the flow of any fluid, and into this friction factor was incorporated the degree of roughness and an element known as the *Reynold's number*, which is based on the viscosity of the fluid and the degree of turbulence of flow. The Darcy–Weisbach formula is used primarily for determining head loss calculations in pipes. For making this determination in open channels, the *Manning equation* was developed during the latter part of the 19th century. Later, this equation was used for both open channels and closed conduits. In the early 1900s, a more practical equation, the *Hazen–Williams equation*, was developed for use in making calculations related to water pipes and wastewater force mains:

$$Q = 0.435 \times CD^{2.63} \times S^{0.54} \tag{14.17}$$

where

Q = Flow rate (ft^3/sec)

C = Coefficient of roughness (which decreases with roughness)

D = Hydraulic radius R (ft)

S = Slope of energy grade line (ft/ft)

14.3.12.2 C Factor

The C factor, as used in the Hazen–Williams formula, designates the coefficient of roughness. C does not vary appreciably with velocity, and by comparing pipe types and ages it includes only the concept of roughness, ignoring fluid viscosity and Reynold's number. Based on experience (experimentation), accepted tables of C factors have been established for pipe (see Table 14.4). Generally, the C factor decreases by one with each year of pipe age. Flow for a newly designed system is often calculated with a C factor of 100, based on averaging it over the life of the pipe system.

TABLE 14.4

C Factors

Type of Pipe	C Factor
Asbestos cement	140
Brass	140
Brick sewer	100
Cast iron	
10 years old	110
20 years old	90
Ductile iron (cement-lined)	140
Concrete or concrete-lined	
Smooth, steel forms	140
Wooden forms	120
Rough	110
Copper	140
Fire hose (rubber-lined)	135
Galvanized iron	120
Glass	140
Lead	130
Masonry conduit	130
Plastic	150
Steel	
Coal-tar-enamel-lined	150
New unlined	140
Riveted	110
Tin	130
Vitrified	120
Wood stave	120

Source: Adapted from Lindeburg, M.R., *Civil Engineering Reference Manual*, 4th ed., Professional Publications, San Carlos, CA, 1986.

DID YOU KNOW?

In practice, minor head loss less than 5% of the total head loss is usually ignored.

14.3.12.3 Slope

Slope is defined as the head loss per foot. In open channels, where the water flows by gravity, slope is the amount of incline of the pipe and is calculated as feet of drop per foot of pipe length (ft/ft). Slope is designed to be just enough to overcome frictional losses, so the velocity remains constant, the water keeps flowing, and solids will not settle in the conduit. In piped systems, where pressure loss for every foot of pipe is experienced, slope is not provided by slanting the pipe but instead by adding pressure to overcome friction.

14.3.13 Minor Head Loss

In addition to the head loss caused by friction between the fluid and the pipe wall, losses also are caused by turbulence created by obstructions (i.e., valves and fittings of all types) in the line, changes in direction, and changes in flow area.

14.4 Reservoir Stored Energy[*]

A major component of a hydroelectric dam is the area behind the dam, its reservoir. The water temporarily stored there provides *gravitational potential energy*. The water is in a stored position above the rest of the dam facility so as to allow gravity to carry the water down to the turbines. Because this higher altitude is different than where the water would naturally be, the water is considered to be at an altered equilibrium. The result is stored energy of position—that is, gravitational potential energy. The water has the potential to do work because of the position it is in. As shown in Figure 14.3, gravity will force the water to fall to a lower position through the intake and the control gate. When the control gate installed within the dam is opened, the water from the reservoir goes through the intake and becomes translational kinetic energy as it falls through the next main part of the system, the penstock. Translational kinetic energy is the energy due to motion from one location to another. The falling water is moving from the reservoir toward the turbines through the penstock. The water is carried through the penstock's long shaft toward the turbines, where the kinetic energy becomes mechanical energy. The force of the water is used to turn the turbines that turn the generator shaft. The generators convert the energy of water into electricity, and step-up transformers then increase the voltage produced to higher voltage levels. Because the potential energy stored in the reservoir is converted into kinetic energy at the inlet to the water turbine, we can equate:

$$m \times g \times h = 1/2mV^2 \tag{14.18}$$

---

[*] Adapted from Spellman, F.R. and Bieber, R.M., *Environmental Health and Science Desk Reference*, Government Institutes Press, Lanham, MD, 2012.

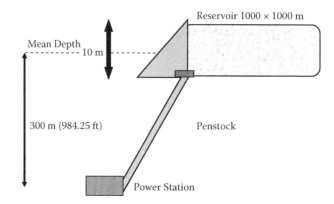

FIGURE 14.3
Schematic representation of a hydro scheme. (Adapted from Tovey, N.K., *ENV-2E02 Energy Resources (2004–2005)* [lecture], University of East Anglia, Norwich, UK, 2005.)

where

    m = Mass of water

    g = Acceleration due to gravity (10 mm^{-2} in most applications)

    h = Head difference

    V = Velocity of water at the inlet

■ *Example 14.13*[*]

Problem: A reservoir has an area of 1 km², and the difference between the crest of the dam and the inlet to a hydro station is 10 m. The station runs at an overall efficiency of 80% and is situated 305 m below the crest of the dam. The rainfall is 1000 mm per annum, the catchment area of the reservoir is 10 times the area of the reservoir, and the run is 50%. What should the rated output of the turbine be if its maximum output is designed to be 5 times the mean output at the site? What is the maximum time the station could operate at full power during a sustained drought?

Solution: Mean head between max and min levels = 305 − 10/2 = 300 m. Average annual flow into the reservoir is equal to 50% of 10 times the area multiplied by the rainfall:

$$\text{Average annual flow} = 0.5 \times 10 \times 1000 \times 1000 \times 1 = 5{,}000{,}000 \text{ m}^3$$

Mean energy generated per annum at 80% efficiency = $mgh \times 8$

$$= 5{,}000{,}000 \times 1000 \times 10 \times 300 \times 0.8$$
$$= 12{,}000{,}000 \text{ MJ}$$

$$\text{Rated output (i.e., mean power)} = 12{,}000{,}000/(60 \times 60 \times 24 \times 365) = 0.381 \text{ MW}$$

So, max power out = 5 × 0.381 = 190 MW, and time at max power assuming the reservoir falls by 10 m is

---

[*] Adapted from Tovey, N.K., *ENV-2E02 Energy Resources (2004–2005)* [lecture], University of East Anglia, Norwich, UK, 2005.

$$\text{Days} = \frac{\text{Area} \times \text{Depth} \times \text{Density} \times g \times h \times 0.8}{\text{Maximum power}}$$

$$= \frac{1000 \times 1000 \times 10 \times 1000 \times 300 \times 0.8}{1,900,000 \times 60 \times 60 \times 24}$$

$$= 146.2 \text{ days}$$

14.5 Hydroturbines

The two main types of hydroturbines are *impulse* and *reaction* (EERE, 2008). The type of hydropower turbine selected for a project is based on the height of standing water—as mentioned, this is referred to as *head*—and the flow, or volume of water, at the site. Cost, efficiency desired, and how deep the turbine must be set are other deciding factors.

14.5.1 Impulse Turbine

The impulse turbine uses the velocity of the water to move the runner and discharges to atmospheric pressure. The water stream hits each bucket on the runner. There is no suction on the down side of the turbine, and water flows out the bottom of the turbine housing after hitting the runner. An impulse turbine is generally suitable for high-head, low-flow applications.

14.5.2 Reaction Turbine

A reaction turbine develops power from the combined action of pressure and moving water. The runner is placed directly in the water stream flowing over the blades rather than striking each individually. Reaction turbines are generally used for sites with lower head and higher flows than compared with impulse turbines.

14.6 Advanced Hydropower Technology

The U.S. Department of Energy (USDOE) and its associated technical activities support the development of technologies that will enable existing hydropower power to generate more electricity with less environmental impact. This will be done by (1) developing new turbine systems that have improved overall performances; (2) developing new methods to optimize hydropower operations at the unit, plant, and reservoir system levels; and (3) conducting research to improve the effectiveness of the environmental mitigation practices required at hydropower projects.

The main objective of research into advanced hydropower technology is to develop new system designs and operation modes that will enable both improved environmental performance and competitive generation of electricity. The products of USDOE research will allow hydropower projects to generate cleaner electricity. USDOE-sponsored projects will develop and demonstrate new equipment and operational techniques that will optimize water-use efficiency, increase generation, and improve environmental performance and mitigation practices in existing plants. Ongoing research efforts contributing to the success of these objectives will enable up to a 10% increase in the hydropower generation at existing dams; these objectives include

- Test a new generation of large turbines in the field to demonstrate that these turbines are commercially viable, compatible with today's environmental standards, and capable of balancing environmental, technical, operational, and cost considerations.
- Develop new tools to improve water use efficiency and operations optimization within hydropower units, plants, and river systems with multiple hydropower facilities.
- Identify improved practices that can be applied at hydropower plants to mitigate for environmental effects of hydro development and operation.

14.7 Hydropower Generation: Dissolved Oxygen Concerns

With regard to the benefits derived from the use of hydropower, it is a clean fuel source; it is a fuel source that is domestically supplied; it relies on the water cycle and thus is a renewable power source; it is generally available as needed; it creates reservoirs that offer a variety of recreational opportunities, notably fishing, swimming, and boating; and it supplies water where needed and assists in flood control—many of these benefits are well known and often taken for granted.

Coins are two sided, of course; that is, for the good side of anything there generally is an accompanying bad side. Many view this to be the case with hydropower. The bad side or disadvantages of hydropower include the impact on fish populations, such as salmon, when the fish cannot migrate upstream past impoundment dams to spawning grounds or if they cannot migrate downstream to the ocean. Hydropower can also be impacted by drought, because when water is not available the plant cannot produce electricity. Hydropower plants also compete with other uses for the land.

Other lesser known negatives of hydropower plants concern their impact on water flow and quality; hydropower plants can cause low water levels that impact riparian habitats. Water quality is also affected by hydropower plants. The low dissolved oxygen (DO) levels in the water, a problem that is harmful to riparian (riverbank) habitats, can result when reservoirs stratify, or develop layers of water of different temperatures (Figure 14.4). Stratification could affect the water temperature with resultant effects on dissolved oxygen levels, nutrient levels, productivity, and the bioavailability of heavy metals. During the summer, stratification, a natural process, can divide the reservoir into distinct vertical strata, such as a warm, well-mixed upper layer (epilimnion) overlying a cooler, relativity stagnant lower layer (hypolimnion). Plant and animal respiration, bacterial decomposition of organic matter, and chemical oxidation can all act to progressively remove DO from hypolimnetic waters. This decrease in hypolimnetic DO is not generally offset by the renewal mechanisms of atmospheric diffusion, circulation, and photosynthesis that operate in the epilmnion (Spellman, 2008). In temperate regions, the decline in hypolimnetic

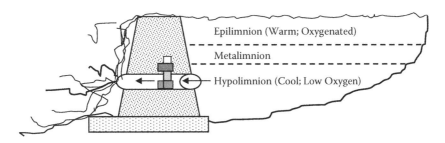

FIGURE 14.4
Thermal stratification of a hydropower reservoir.

DO concentrations begins at the onset of stratification (spring or summer) and continues until either anaerobic conditions predominate or reoxygenation occurs during the fall turnover of the water body.

Numerous structural, operational, and regulatory techniques are available that a hydropower operator can use to resolve a low DO issue. Levels of DO can be increased through modifications in dam operations. These include such techniques as fluctuating the timing and duration of flow releases, spilling or sluicing water, increasing minimum flows, flow mixing, turbine aeration, and, at some sites, injection of air or oxygen in weir aeration. The most effective strategy for addressing a DO problem is dependent on the particular situation.

14.8 Bottom Line on Macrohydropower

Hydropower offers advantages over the other energy sources but faces unique environmental challenges. As mentioned, the advantages of using hydropower begin with the fact that hydropower does not pollute the air like power plants that burn fossil fuels, such as coal and natural gas. Moreover, hydropower does not have to be imported into the United States as foreign oil does; it is produced in the United States. Because hydropower relies on the water cycle, driven by the sun, it is a renewable resource that will be around for at least as long as humans. Hydropower is controllable; that is, engineers can control the flow of water through the turbines to produce electricity on demand. Finally, hydropower impoundment dams create huge lake areas for recreation, irrigation of farm lands, reliable supplies of potable water, and flood control.

Hydropower also has some disadvantages. Fish populations can be impacted if fish cannot migrate upstream past impoundment dams to spawning grounds or if they cannot migrate downstream to the ocean. Many dams have installed fish ladders or elevators to aid upstream fish passage. Downstream fish passage is aided by diverting fish from turbine intakes using screens or racks or even underwater lights and sounds, and by maintaining a minimum spill flow past the turbine. Hydropower can also impact water quality and flow. Hydropower plants can cause low dissolved oxygen levels in the water, a problem that is harmful to riparian habitats and is addressed using various aeration techniques to oxygenate the water. Maintaining minimum flows of water downstream of a hydropower installation is also critical for the survival of riparian habitats. Hydropower is also susceptible to drought. When water is not available, the hydropower plants cannot produce electricity. Finally, construction of new hydropower facilities impacts investors and others by competing with other uses of the land. Preserving local flora and fauna and historical or cultural sites is often more highly valued than electricity generation.

It cannot be denied that generating electricity from water power, hydroelectricity, represents the largest source of renewable energy in the world today. This electrical power is primarily supplied by macrohydropower systems such as the Grand Coulee, Chief Joseph, Wells, Rocky Reach, Rock Island, Wanapum, Priest Rapids, McNary, John Day, The Dalles, Bonneville, and several other macro-sized hydroelectric dams. The electrical power derived from any of these macro dams can easily provide all the electricity required to power a water and wastewater treatment plant with several kilowatt-hours of electricity left over for other applications.

14.9 Microhydropower Concepts

This book, of course, is about energy sustainability and efficiency in water and wastewater treatment plants. Although many of the country's water and wastewater treatment plants receive their electrical power generated from coal-fired, nuclear, and macrohydropower dams and energy sustainability and efficiency can be enhanced via good engineering and management practices, it is also possible to maximize energy sustainability and efficiency by designing, engineering, and installing microhydropower systems in plant sites where appropriate. Micro systems can accommodate a variety of site applications, including providing the electricity for lighting within a plant building, site telemetry systems, alarm systems, and electric fences, as well as providing a few hundred kilowatts for operating small unit process equipment. Experience has demonstrated that microhydropower is one of the most cost-effective and reliable energy technologies to be considered for providing generation of clean electricity. The key advantages of micro systems include the following:

- High capacity factor (i.e., the ratio of the actual output of a power plant over a period of time, typically >50%)
- High efficiency (70–90%; by any calculation the best of all energy technologies)
- High level of predictability, varying with the plant flow rate and annual rainfall patterns
- Output power that varies only gradually from day to day (not minute to minute)
- Good correlation with demand
- Long stay time (can be engineered to last for 50+ years)
- Environmentally benign (little or no water is stored for run-of-the-river micro systems, thus lessening their environmental impact)

14.9.1 Microhydropower Key Terms

As shown in Figure 14.5, small run-of-the-river hydropower systems consist of the following basic elements (USDOE, 2012b):

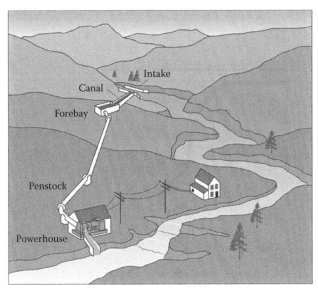

In this microhydropower system, water is diverted into the penstock. Some generators can be placed directly into the stream.

FIGURE 14.5
A type of microhydropower system.

- *Water conveyance*—Pipeline, channel, or pressurized pipeline (penstock) that delivers the water
- *Waterwheel or turbine*—Transforms the energy of flowing water into rotational energy
- *Generator or alternator*—Transforms the rotational energy into electricity
- *Regulator*—Controls the generator
- *Wiring*—Delivers the electricity
- *Inverter*—Converts low-voltage direct-current (DC) electricity produced by the system into 120 or 240 volts of alternating current (AC).
- *Batteries*—Used to store the electricity generated by the system

Before water enters the waterwheel or turbine, it is first channeled or funneled through a series of components that control its flow and filter out debris (e.g., logs, paper, rocks, branches). These components include the following (see Figure 14.5):

- *Headrace*—Waterway running parallel to the water source; often required for microhydropower systems when insufficient head is provided
- *Forebay*—Functions as a settling pond for large debris that would otherwise damage downstream equipment
- *Water conveyance*—Channel, pipeline, or penstock that funnels water directly to the waterwheel or turbine
- *Waterwheel*—Available but rarely used in microhydropower systems because they are too bulky and slow
- *Turbine*—Used more frequently because they are more compact than waterwheels

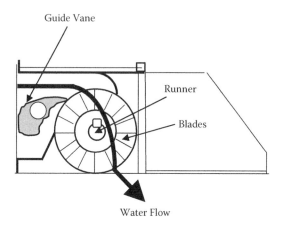

FIGURE 14.6
Basic cross-flow turbine.

As noted earlier, there are two general classes of turbines: impulse and reaction. Impulse turbines are commonly used for high-head microhydropower systems. Figure 14.6 shows a cross-flow impulse turbine, which is commonly used in microhydropower systems. Reaction turbines are primarily used in macrohydropower systems and are seldom used in microhydropower systems because of their high cost and complexity.

14.9.2 Potential Microhydropower Sites

To build a plant-site microhydropower system, access to naturally flowing water (streams and rivers), channels, ditches, conduit conveyance systems, and treated wastewater outfalls is required (USDOE, 2012a). A sufficient quantity of falling water must be available, which usually, but not always, means that hilly or mountainous sites are best. Other considerations for a potential microhydropower site include its power output, economics, permits, and water rights. To make a microhydropower system more feasible, it is good practice to make every effort to reduce plant electricity usage.

To determine if a microhydropower system would be feasible, it is necessary to determine the amount of power that can be obtained from flowing water on the plant site. This involves determining two important factors:

- *Head*—The vertical distance the water falls (see Section 14.9.3)
- *Flow*—The quantity of water falling (see Section 14.9.4)

Once the head and flow have been calculated, Equation 14.19 can be used to estimate the power output for a system with 50 to 60% (approximate) efficiency, which is representative of most microhydropower systems. As shown in Equation 14.19, to determine the system's output in watts (W) simply multiply net head (the vertical distance available after subtracting losses from pipe or conveyance friction) by flow (use U.S. gallons per minute) and divide by 10:

$$[\text{Net head (ft)} = \text{Flow (gpm)}] \div 10 = \text{Watts} \tag{14.19}$$

14.9.3 Head at Potential Microhydropower Site

As mentioned, *head* is the vertical distance that water falls. When evaluating a water or wastewater treatment plant site, head is usually measured in feet, meters, or units of pressure (USDOE, 2012a). Head also is a function of the characteristics of the channel or pipe through which it flows. To determine the site's potential power output, use the site's head and flow calculations.

Most microhydropower sites are categorized as low or high head. The higher the head, the less water is needed to produce a given amount of power; also, smaller, less expensive equipment can be used. Low head refers to a change in elevation of less than 10 ft (3 m). A vertical drop of less than 2 ft (0.6 m) will probably make a small-scale hydroelectric system unfeasible. For extremely small power generation amounts, a flowing stream with as little as 13 inches of water can support a submersible-type turbine. This type of turbine was originally used to power scientific instruments towed behind oil exploration and oceanic survey ships.

When determining head, both *gross head* and *net head* must be considered. Gross (or *static*) head is the vertical distance between the penstock (the pipe that takes water from the stream) and where the water leaves the turbine. Net (or *dynamic*) head is the pressure at the bottom of the pipeline when water is actually flowing to the turbine; that is, net head is the gross head minus pressure losses due to friction and turbulence.

The most accurate way to determine gross head is to have a professional hydrologist or licensed engineer survey the plant site. To get a rough estimate, U.S. Geological Survey maps for the area can be consulted or plant personnel can use the *hose–tube* method.

14.9.3.1 Hose–Tube Method

The hose–tube method involves taking stream-depth measurements across the width of the stream intended for use in the system—from the point at which the penstock is to be placed to the point at which the turbine is to be placed. The following will be needed:

- An assistant
- A 20- to 30-ft (6- to 9-m) length of small-diameter garden hose or other flexible tubing
- A funnel
- A yardstick or measuring tape

Instructions:

1. Stretch the hose or tubing down the stream channel from the point that is the most practical elevation for the penstock intake.
2. Have your assistant hold the upstream end of the hose, with the funnel in it, underwater as near the surface as possible.

DID YOU KNOW?

Net head can be increased by minimizing the length of, and turns in, the pipeline, which can prevent some losses to pressure.

3. Meanwhile, lift the downstream end until water stops flowing from it.

4. Measure the vertical distance between your end of the tube and the surface of the water. This is the gross head for that section of stream.

5. Have your assistant move to where you are and place the funnel at the same point where you took your measurement.

6. Then walk downstream and repeat the procedure. Continue taking measurements until you reach the point where you plan to site the turbine.

The sum of the measurements will give you a rough approximation of the gross head for your site.

Note: Due to the water's force into the upstream end of the hose, water may continue to move through the hose after both ends of the hose are actually level. You may wish to subtract an inch or two (2 to 5 cm) from each measurement to account for this. It is best to be conservative in these preliminary head measurements.

If preliminary estimates look favorable, you will want to acquire more accurate measurements. As mentioned earlier, the most accurate way to determine head is to have a professional survey your site.

14.9.4 Flow at Potential Microhydropower Site

The quantity of water falling from a potential microhydropower system site is called *flow* (USDOE, 2012a). It is measured in gallons per minute, cubic feet per second, or liters per second. The site's flow calculation along with its head calculation can be used to determine the site's potential power output. The easiest way to determine a stream's flow is to obtain data from the following local offices:

- U.S. Geological Survey
- U.S. Army Corps of Engineers
- U.S. Department of Agriculture
- County engineers
- Local water supply or flood control authorities

If existing data are unobtainable, plant site personnel can conduct their own flow measurements. Flow can be measured using the *bucket* or *weighted-float* method.

14.9.4.1 Bucket Method

The bucket method involves damming the stream with logs or boards to divert its flow into a bucket or container. The rate at which the container fills is the flow rate; for example, when a 5-gallon bucket fills in 1 minute that means that the stream's water is flowing at 5 gallons per minute.

14.9.4.2 Weighted-Float Method

As long as the water is not fast flowing and the depth is not above calf height, another way to measure flow involves measuring stream depths across the width of the stream and releasing a weighted float upstream from the measurements. The following is needed:

- An assistant
- A tape measure
- A yardstick or measuring rod
- A weighted float, such as a plastic bottle filled halfway with water
- A stopwatch
- Some graph paper

With this equipment it is possible to calculate flow for a cross-section of the streambed at its lowest water level.

Instructions:

1. First, select a stretch of stream with the straightest channel and the most uniform depth and width possible.
2. At the narrowest point, measure the width of the stream.
3. Holding the yardstick vertically, walk across the stream and measure the water depth at 1-ft increments. To help with the process, stretch across the stream a string or rope upon which the increments are marked.
4. Plot the depths on graph paper to give yourself a cross-sectional profile of the stream.
5. Determine the area of each section by calculating the areas of the rectangles (area = length × width) and right triangles (area = 1/2 base × height) in each section.
6. From the same point where you measured the stream's width, mark a point at least 20 ft upstream.
7. Release the weighted float in the middle of the stream and record the time it takes for the float to travel to the original point downstream. Don't let the float drag along the bottom of the streambed; if it does, use a smaller float.
8. Divide the distance between the two points by the float time in seconds to get flow velocity in feet per second. The more times this procedure is repeated, the more accurate the flow velocity measurement will be.

DID YOU KNOW?

Stream flows can be quite variable over a year, so the season during which you take flow measurements is important. Unless you are considering building a storage reservoir, you can use the lowest average flow of the year as the basis for your system's design. However, if you're legally restricted on the amount of water you can divert from your stream at certain times of the year, use the average flow during the period of the highest expected electricity demand.

9. Multiply the average velocity by the cross-sectional area of the stream.

10. Multiply your result by a factor that accounts for the roughness of the stream channel (0.8 for a sandy streambed, 0.7 for a bed with small to medium sized stones, and 0.6 for a bed with many large stones). The result will give you the flow rate in cubic feet or meters per second.

14.9.5 Economics

If it is determined that the estimated power output that a microhydropower system would provide is feasible, then the next step is to determine whether it economically makes sense. To do this we simply add up all the estimated costs of developing and maintaining the site over the expected life of the microhydropower system equipment, and divide the amount by the system's capacity in watts. This will determine how much the system will cost in dollars per watt. Then compare that to the cost of utility-provided power or other alternative power sources. Whatever the upfront costs, a hydroelectric system will typically last a long time and, in many cases, maintenance is not expensive or time consuming. In addition, sometimes there are a variety of financial incentives available on the state, utility, and federal level for investments in renewable energy systems. They include income tax credits, property tax exemptions, state sales tax exemptions, load programs, and special grant programs, among others.

14.10 Permits and Water Rights

When deciding whether to install a microhydropower system on a particular plant site, local permit requirements and water rights must be checked for legality and compliance reasons. Whether the proposed system will be a grid-connected or stand-alone system will affect what requirements must be followed. If the proposed microhydropower system will have minimal impact on the environment, and there is no plan to sell power to a utility, there is a good chance that the process involved in obtaining a permit will not be too complex. Locally, the first point of contact should be the county engineer. The state energy office may be able to provide advice and assistance as well. In addition, the Federal Energy Regulatory Commission and U.S. Army Corps of Engineers must be contacted. Moreover, a determination of how much water can be diverted from the plant stream channel must be made. Each state controls water rights; the plant may need a separate water right to produce power, even if the plant already has a water right for another use.

References and Recommended Reading

EERE. (2011). *History of Hydropower*, Energy Efficiency & Renewable Energy, U.S. Department of Energy, Washington, DC (http://www1.eere.energy.gov/water/hydro_history.html).

EERE. (2012). *SunShot Initiative*, Energy Efficiency & Renewable Energy, U.S. Department of Energy, Washington, DC (http://www1.eere.Energy.gov/solar/printable_versions/about.html).

EIA. (2010). *Renewable Energy Annual 2008*, Energy Information Administration, U.S. Department of Energy, Washington, DC (ftp://ftp.eia.doe.gov/renewables/060308.pdf).

Enger, E., Kormelink, J.R., Smith, B.F., and Smith, R.J. (1989). *Environmental Science: The Study of Interrelationships*, William C. Brown, Dubuque, IA.

Gibilisco, S. (2007). *Alternative Energy Demystified*, McGraw-Hill, New York.

Liebig, J. (1840). *Chemistry and Its Application to Agriculture and Physiology*, Taylor & Walton, London.

Lindeburg, M.R. (1986). *Civil Engineering Reference Manual*, 4th ed., Professional Publications, San Carlos, CA.

LOC. (2009). *The World's First Hydroelectric Power Plant*, Library of Congress, Washington, DC (http://www.americaslibrary.gov/jb/gilded/jb_gilded_hydro_1.html).

March, P.A., Brice, T.A., Mobley, M.H., and Cybularz, J.M. (1992). Turbines for solving the DO dilemma, *Hydro Review*, 11(1), 30–36.

Masters, G.M. (1991). *Introduction to Environmental Engineering and Science*, Prentice Hall, Englewood Cliffs, NJ.

Miller, G.T. (1988). *Environmental Science: An Introduction*, Wadsworth, Belmont, CA.

Nathanson, J.A. (1997). *Basic Environmental Technology: Water Supply, Waste Management, and Pollution Control*, 2nd ed., Prentice Hall, Upper Saddle River, NJ, pp. 21–22.

NREL. (2012). *Concentrating Solar Power*, National Renewable Energy Laboratory, U.S. Department of Energy, Washington, DC (http://www.nrel.gov/learning/re_csp.html).

Searchinger, T. et al. (2008). Use of U.S. croplands for biofuels increases greenhouse gases through emissions from land-use change, *Science*, 319(5867), 1238–1240.

Simon, J.L. (1980). Resources, population, environment: an oversupply of false bad news, *Science*, 208, 1431–1437.

Spellman, F.R. (2008). *The Science of Water*, 2nd ed., CRC Press, Boca Raton, FL.

Spellman, F.R. (2009). *Handbook of Water and Wastewater Treatment Plant Operations*, 2nd ed., CRC Press, Boca Raton, FL.

Spellman, F.R. and Bieber, R.M. (2012). *Environmental Health and Science Desk Reference*, Government Institutes Press, Lanham, MD.

Tovey, N.K. (2005). *ENV-2E02 Energy Resources (2004–2005)* [lecture], University of East Anglia, Norwich, UK (http://www.uea.ac.uk/~e680/energy/Old_modules/env2e02/env2e02.htm).

Turchin, P. (2001). Does population ecology have general laws? *Oikos*, 94, 17–26.

UCS. (2008). *Land Use Changes and Biofuels*, Union of Concerned Scientists, Cambridge, MA (http://www.ucsusa.org/clean_vehicles/smart-transportation-solutions/cleaner_fuels/ethanol-and-other-biofuels/Land-Use-Changes-and-Biofuels.html).

USDOE. (2009). *Fossil Fuels*, U.S. Department of Energy, Washington, DC (http://www.energy.gov/energysources/fossilfuels.htm).

USDOE. (2012a). *Planning a Microhydropower System*, U.S. Department of Energy, Washington, DC (http://energy.gov/energysaver/articles/planning-microhydropower-system).

USDOE. (2012b). *Microhydropower Systems*, U.S. Department of Energy, Washington, DC (http://energy.gov/energysaver/articles/microhydropower-systems).

USEPA. (2008). *Catalog of CHP Technologies*, U.S. Environmental Protection Agency, Washington, DC (www.epa.gov/chp/documents/catalog_chptech_full.pdf).

15

Solar Power

One generation passes away, and another generation cometh: but the Earth abides for-ever. The sun also rises.

—Ecclesiastes 1:4–5

Busy old fool, unruly Sun,
Why dost thou thus,
Through windows and through curtains, call on us?

—John Dunne, "The Sun Rising"

The Sun, with all the planets revolving around it and dependent on it, can still ripen a bunch of grapes as though it had nothing else in the universe to do.

—Galileo Galilei

15.1 Introduction[*]

Solar power is one of the most promising renewable energy sources today. Solar cells, also known as photovoltaic (PV) cells, convert sunlight directly into electricity. They can gen-erate electricity with no moving parts, they can be operated quietly with no emissions, they require little maintenance, and they are ideal for remote locations. If these advan-tages outweigh a few disadvantages associated with the use of solar cells—good weather and location are essential, although they require little maintenance they are difficult to repair, and their initial cost is high—they may be considered a good fit as auxiliary and supplemental power sources (ASPSs) for water treatment plants (WTPs) and wastewater treatment plants (WWTPs).

Solar energy (a term used interchangeably with solar power) uses various technologies to take advantage of the power of the sun to produce energy. Solar energy is one of the best renewable energy sources available because it is one the cleanest sources of energy. Direct solar radiation absorbed in solar collectors can provide space heating and hot water. Passive solar can be used to enhance the solar energy used in buildings for space heating and lighting requirements. Solar energy can also be used to produce electricity, and this is the renewable energy area that is the focus of our attention in this section.

---

[*] Adapted from Spellman, F.R. and Bieber, R., *The Science of Renewable Energy*, CRC Press, Boca Raton, FL, 2011.

TABLE 15.1

U.S. Energy Consumption by Energy Source, 2008

Energy Source	Energy Consumption (quadrillion Btu)
Total	99.438
Fossil fuels (coal, natural gas, petroleum)	83.532
Electricity net imports	0.113
Nuclear electric power	8.427
Renewable	7.367
Biomass (total)	3.852
Biofuels	1.372
Waste	0.436
Wood and wood-derived fuels	2.044
Geothermal energy	0.360
Hydroelectric conventional	2.512
Solar thermal/photovoltaic energy	**0.097**
Wind energy	0.546

Source: EIA, *Renewable Energy Annual 2008*, Energy Information Administration, U.S. Department of Energy, Washington, DC, 2010.

Radiant energy from the sun, in the form of photons, strikes the surface of the Earth with the average equivalent of about 168 kWh of energy (equal to 575,000 BTUs thermal energy) per square foot per year; it varies, of course, with location, cloud cover, and orientation with the surface (Hanson, 2004). The question, then, becomes one of how much of this energy can be used by consumers in the United States. Table 15.1 shows the energy consumption by energy source for 2008; in that year, solar energy accounted for 0.097 quadrillion Btu, a figure that is expected to steadily increase, which should be reflected in later figures when they are released.

In addition to the advantages offered by solar power already mentioned, it is important to point out that solar technologies diversify the energy supply and reduce the country's dependence on imported fuel. Moreover, solar energy technologies have a beneficial environmental impact in that they help to improve air quality and offset greenhouse gas emissions. An additional (and significant) benefit of a growing solar technology industry is that it stimulates our economy by creating jobs in solar manufacturing and installation. The two solar electric technologies with the greatest potential (based on cost-effectiveness) are concentrating solar power (CSP) and photovoltaics (PV).

Harnessing the sun's energy using solar technologies to ensure sustainability and efficiency in water and wastewater treatment plant operations is not only a reliable concept but also quite practicable. This is especially the case in the Sun Belt states. The energy in sunlight striking the Earth for 42 minutes is equivalent to global energy consumption for a year. At least 250,000 square miles of land in the Southwest alone are suitable for constructing solar power plants.

The U.S. Department of Energy lists three types of solar technologies: concentrating solar power, photovoltaics, and solar heat. Before discussing current applications of solar power at water and wastewater treatment plants, a discussion of the fundamental concepts of each of these technologies is provided.

15.2 Concentrating Solar Power[*]

Concentrating solar power (CSP) offers a utility-scale, reliable, firm, dispatchable, renewable energy option that can help meet a nation's demand for electricity. Nine trough plants producing more than 400 megawatts (MW) of electricity have been operating reliably in the California Mojave Desert since the 1980s (USDOE, 2008). CSP plants produce power by first using mirrors to concentrate and focus sunlight onto a thermal receiver, similar to a boiler tube. The receiver absorbs and converts sunlight into heat. Ultimately, this high-temperature fluid is used to spin a turbine or power an engine that drives a generator that produces electricity. Concentrating solar power systems can be classified by how they collect solar energy: linear concentrators, dish/engines, or power tower systems (NREL, 2012c). All three are typically engineered with tracking devices for following the sun, both seasonally and throughout the day, to maximize the electrical output of the system (Chiras, 2002).

15.2.1 Linear Concentrators

Linear CSP collectors capture or collect the sun's energy using long, rectangular, curved (U-shaped) mirrors. The mirrors, tilted toward the sun, focus sunlight on tubes (or receivers) that run the length of the mirrors. The reflected sunlight heats a fluid flowing through the tubes. The hot fluid is then used to boil water to create superheated steam that spins a conventional steam-turbine generator to produce electricity. Alternatively, steam can be generated directly in the solar field, eliminating the need for costly heat exchangers. Linear concentrating collector fields consist of a large number of collectors in parallel rows that are typically aligned in a north–south orientation to maximize both annual and summertime energy collection. A single-axis sun-tracking system allows the mirrors to track the sun from east to west during the day, ensuring that the sun reflects continuously onto the receiver tubes (USDOE, 2009a). The two major types of linear concentrator systems are *parabolic trough systems*, where receiver tubes are positioned along the focal line of each parabolic mirror, and *linear Fresnel reflector systems*, where one receiver tube is positioned above several mirrors to allow the mirrors greater mobility in tracking the sun.

15.2.1.1 Parabolic Trough Systems

Parabolic trough systems have been developed in the United States, Spain, and Japan (Chiras, 2002); in the United States today, the predominant CSP systems currently in operation are linear concentrators using parabolic trough collectors (USDOE, 2009a). Parabolic trough systems have long, curved arrays of mirrors (usually coated silver or polished aluminum). All rays of light that enter parallel to the axis of a parabolic-shaped mirror will be reflected to one point, the focus; hence, it is possible to concentrate virtually all of the radiation incident upon the mirror in a relatively small area in the focus (Perez-Blanco, 2009). The receiver tube (Dewar tube) runs its length and is positioned along the focal line of each parabola-shaped reflector (Duffie and Beckman, 1991; Patel, 1999). The tube is fixed to the

---

[*] Adapted from EERE, *Concentrating Solar Power: Energy from Mirrors*, DOE/GO-102001-1147, Energy Efficiency and Renewable Energy, U.S. Department of Energy, Washington, DC, 2001; Spellman, F.R. and Bieber, R., *The Science of Renewable Energy*, CRC Press, Boca Raton, FL, 2011.

mirror structure, and the heated fluid—either a heat-transfer fluid (usually oil) or water or steam—flows through and out of the field of solar mirrors to where it will be used to create steam; in the case of a water/steam receiver, the flow is sent directly to the turbine (i.e., the turbine is the prime mover of the electrical generator that produces electrical current flow in an accompanying electrical distribution system). Temperatures in these systems range from 150 to 750°F (80 to 400°C). Currently, the largest individual trough systems can generate 80 MW of electricity; however, individual systems being developed will generate 250 MW. In addition, individual systems can be collocated in power parks. The potential capacity of such power parks would be constrained only by transmission capacity and the availability of contiguous land area. Trough designs can incorporate thermal storage (discussed in more detail later). In thermal storage systems, the collector field is oversized to heat a storage system during the day that can be used in the evening or during cloudy weather to generate additional steam to produce electricity.

Parabolic trough plants can be designed as hybrids, meaning that they use fossil fuel to supplement the solar output during periods of low solar radiation. In such a design, a natural-gas-fired heater or gas–steam boiler/reheater is used. In the future, troughs may be integrated with existing or new combined-cycle, natural gas, and coal-fired plants (USDOE, 2009a).

15.2.1.2 Fresnel Reflector Systems

Another type of linear concentrator technology is the linear Fresnel reflector system. Flat or slightly curved mirrors mounted on trackers on the ground are configured to reflect sunlight onto a water-filled receiver tube fixed in space above these mirrors. A small parabolic mirror is sometimes added atop the receiver to further focus the sunlight. The key advantage of Fresnel reflectors systems is their simplicity as compared to other systems.

15.2.2 Dish/Engine Systems

Dish/engine systems use a mirrored, dish-shaped, parabolic mirror similar to a very large satellite dish. The dish-shaped surface directs and concentrates sunlight onto a thermal receiver (the interface between the dish and the engine or generator), which absorbs and collects the heat and transfers it to the engine generator. Solar dish/engine systems convert the energy from the sun into electricity at a very high efficiency. The power conversion unit includes the thermal receiver and the engine or generator. It absorbs the concentrated beams of solar energy, converts them to heat, and transfers the heat to the engine or generator. A thermal receiver can be a bank of tubes with a cooling fluid—usually hydrogen or helium—that typically is the heat-transfer medium and also the working fluid for an engine. Alternative thermal receivers are heat pipes, where the process of boiling and condensing an intermediate fluid transfers the heat to the engine. The most common type of heat engine used today in dish/engine systems is the Stirling engine (conceived in 1816). This system uses the fluid heated by the receiver to move pistons and create mechanical power. The mechanical power is then used to run a generator or alternator to produce electricity (USDOE, 1998). Disk/engine systems use dual-axis collectors to track the sun. As mentioned, the ideal concentrator shape is parabolic, created with a single reflective surface or multiple reflectors, or facets. Many options exist for receiver and engine type, including the Stirling cycle, microturbine, and concentration photovoltaic modules. Because of the high concentration ratios achievable with parabolic dishes and the small size of the receiver, solar dishes are efficient (as high as 30%) at collecting solar energy at very high temperatures (900 to 2700°F, or 500 to 1500°C). Solar dish/engine systems have environmental, operational, and potential economical advantages over more conventional power generation options because (USDOE, 1998):

- They produce zero emissions when operating on solar energy.
- They operate more quietly than diesel or gasoline engines.
- They are easier to operate and maintain than conventional engines.
- They start up and shut down automatically.
- They operate for long periods with minimal maintenance.

Solar dish/engine systems are well suited and often used for nontraditional power generation because of their size and durability. Individual units range in size from 10 to 25 kW. They can operate independently of power grids in remote sunny locations for uses such as pumping water and providing power to people living in isolated locations.

The high temperatures produced by dish/engine systems can be used to produce steam that can be used for either electricity production or various high-temperature industrial processes (Chiras, 2002). Dish/engine systems also can be linked together to provide utility-scale power to a transmission grid. Such systems could be located near consumers, substantially reducing the need for building or upgrading transmission capacity. Largely because of their high efficiency, the cost of these systems is expected to be lower than that of other solar systems for these applications (USDOE, 1998).

15.2.3 Power Tower System

The power tower or the central receiver system consists of a large field of flat, sun-tracking mirrors known as *heliostats* (silver-laminated acrylic membranes) focused to concentrate sunlight onto a receiver on the top of a tower. A heat-transfer fluid heated in the receiver is used to generate steam, which, in turn, is used in a conventional turbine generator to produce electricity. Some power towers use water or steam as the heat-transfer fluid. Other advanced designs are experimenting with molten (liquefied) nitrate salt because of its superior heat-transfer and energy-storage (heat-retention) capabilities, allowing continued electricity production for several consecutive cloudy days (NREL, 2010). The energy-storage capability, or thermal storage, allows the system to continue to dispatch electricity during cloudy weather or at night. A power tower system is composed of five main components: heliostats, receiver, heat transport and exchange, thermal storage, and controls (Figure 15.1). Individual commercial plants can be sized to produce up to 200 MW of electricity.

DID YOU KNOW?

Smaller CSP systems can be located directly where the power is needed; for example, a single dish/engine system can produce 3 to 25 kW of power and is well suited for such distributed applications. Larger, utility-scale CSP applications provide hundreds of megawatts of electricity for the power grid. Both linear concentrator and power tower systems can be easily integrated with thermal storage, helping to generate electricity during cloudy periods or at night. Alternatively, these systems can be combined with natural gas, and the resulting hybrid power plants can provide high-value, dispatchable power throughout the day.

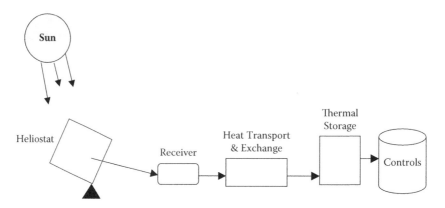

FIGURE 15.1
Solar power tower main components.

15.2.4 Thermal Energy Storage

Thermal energy storage (TES) has become a critical aspect of any concentrating solar power system deployed today (USDOE, 2009b). One challenge facing the widespread use of solar energy is the reduced or curtailed energy production when the sun sets or is blocked by clouds. Thermal energy storage provides a workable solution to this challenge. In a CSP system, the sun's rays are reflected onto a receiver, creating heat that is then used to generate electricity. If the receiver contains oil or molten salt as the heat-transfer medium, then the thermal energy can be stored for later use. This allows CSP systems to be a cost-competitive option for providing clean, renewable energy. Current steam-based receivers cannot store thermal energy for later use.

An important criterion in selecting a storage medium, such as the oil or molten salt referred to above, is the *specific heat* of the substance. Specific heat (c) is the measure of the heat energy required to increase the temperature of a unit quantity of a substance by a unit of temperature. For example, at a temperature of 15°C, the heat energy required to raise the temperature of water 1 Kelvin (equal to 1°C) is 4186 joules per kilogram (J/kg). Different materials absorb different amounts of heat when undergoing the same temperature increase. The relationship between a temperature change (ΔT) and the amount of heat (Q) added (or subtracted) is given by

$$Q = mc\Delta T \qquad (15.1)$$

where c is the specific heat and m the mass of the substance. Table 15.2 lists some specific heats for different materials per pound and per cubic foot. Another important parameter

DID YOU KNOW?

Concentrated solar power plants use water not only for lens cleaning but also for steam cycle production. Water consumption is an issue because these plants require water but they are most effective in locations where the sun is most intense, which in turn often corresponds to places like the Mohave Desert where there is little water (USDOE, 2008).

TABLE 15.2

Thermal Energy Storage Materials

Material	Specific Heat (Btu/lb °F)	Density (lb/ft³)	(kg/m³)	Heat Capacity (Btu/ft³ °F)	(kJ/m³ °C)
Water	1.00	62	1000	62	4186
Iron	0.12	490	7860	59	3521
Copper	0.09	555	8920	50	3420
Aluminum	0.22	170	2700	37	2430
Concrete	0.23	140	2250	32	2160
Stone	0.21	170	2700	36	2270
White pine	0.67	2	435	18	1220
Sand	0.19	95	1530	18	1540
Air	0.24	0.075	1.29	0.02	1.3

related to specific heat and thermal energy storage materials is *heat capacity*, which is defined as the ratio of the heat energy absorbed by a substance to the substance's increase in temperature. It is given by

$$\text{Heat capacity} = \text{Specific heat} \times \text{Density} \qquad (15.2)$$

Other media for thermal storage are *phase-change materials*, which are classified as latent heat storage units. Specifically, a phase-change material is a substance with a high heat of fusion, which is the amount of thermal energy that must be absorbed or evolved for 1 mole of a substance to change states from a solid to a liquid or vice versa without a temperature change. A phase-change material melts and solidifies at a certain temperature and is capable of storing and releasing large amounts of energy. Heat is absorbed or released when the material changes from solid to liquid and vice versa. One common group of substances being used as phase-change materials for solar applications (and elsewhere) are eutectic salts, such as such as sodium sulfate decahydrate, also known as Glauber's salt. This salt melts at 91°F with the addition of 108 Btu/lb. Conversely, when the temperature drops below 91°F, 108 Btu/lb of heat energy is released as the salt solidifies (Hinrichs and Kleinbach, 2006). Several TES technologies have been tested and implemented since 1985. These include the two-tank direct system, two-tank indirect system, and single-tank thermocline system.

15.2.4.1 Two-Tank Direct System

Solar thermal energy in this system is stored in the same fluid used to collect it. The fluid is stored in two tanks—one at high temperature and the other at low temperature. Fluid from the low-temperature tank flows through the solar collector or receiver, where solar energy heats it to the high temperature; it then flows back to the high-temperature tank for storage. Fluid from the high-temperature tank flows through a heat exchanger, where it generates steam for electricity production. The fluid exits the heat exchanger at the low temperature and returns to the low-temperature tank. Two-tank direct storage was used in early parabolic trough power plants and at the Solar Two power tower in California. The trough plants used mineral oil as the heat-transfer and storage fluid, and the Solar Two power tower used molten salt.

15.2.4.2 Two-Tank Indirect System

The two-tank indirect system functions in the same way as the two-tank direct system, except that different fluids are used for the heat-transfer fluid and for the storage fluid. This system is used in plants where the heat-transfer fluid is too expensive or not suited for use as the storage fluid. The storage fluid from the low-temperature tank flows through an extra heat exchanger, where it is heated by the high-temperature heat-transfer fluid. The high-temperature storage fluid then flows back to the high-temperature storage tank. The fluid exits this heat exchanger at a low temperature and returns to the solar collector or receiver, where it is heated back to the high temperature. Storage fluid from the high-temperature tank is used to generate steam in the same manner as the two-tank direct system. The indirect system requires an extra heat exchanger, which adds cost to the system. This system will be used in many of the parabolic power plants in Spain and has also been proposed for several U.S. parabolic plants. The plants will use organic oil as the heat-transfer fluid and molten salt as the storage fluid.

15.2.4.3 Single-Tank Thermocline System

This system stores thermal energy in a solid medium—most commonly silica sand—located in a single tank. At any time during operation, a portion of the medium is at high temperature and a portion is at lower temperature. The hot- and cold-temperature regions are separated by a temperature gradient, or *thermocline* (see Figure 15.2). High-temperature heat-transfer fluid flows into the top of the thermocline and exits the bottom at low temperature. This process moves the thermocline downward and adds thermal energy to the system for storage. Reversing the flow moves the thermocline upward and removes thermal energy from the system to generate steam and electricity. Buoyancy effects create thermal stratification of the fluid within the tank, which helps to stabilize and maintain the thermocline.

Note: Using a solid storage medium and only requiring one tank can reduce the cost of the single-tank thermocline system relative to two-tank systems.

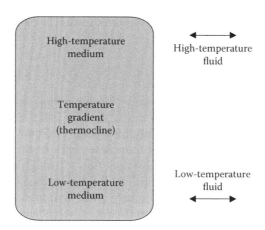

FIGURE 15.2
Single-tank thermocline thermal energy storage system.

15.3 Photovoltaics (PV)

Photovoltaics (PV) has been and will continue to be one of the more glamorous technologies in the energy field (Hinrichs and Kleinbach, 2006). Photovoltaic (*photo* from the Greek word for "light" and *volt* for electricity pioneer Alessandro Volta) technology makes use of the abundant energy in the sun, and it has little impact on our environment. Photovoltaics is the direct conversion of light (photons) into electricity (voltage) at the atomic level. Some materials exhibit a property known as the *photoelectric effect* (discovered and described by Becquerel in 1839) that causes them to absorb photons of light and release electrons. When these free electrons are captured, an electric current (flow of free electrons) results, which can be used as electricity. The first photovoltaic module (billed as a solar battery) was built by Bell laboratories in 1954. In the 1960s, the space program began to make the first serious use of the technology to provide power aboard spacecraft. Use of this technology in the space program produced giant advancements in its reliability and helped to lower costs; however, it was the oil embargo of the 1970s (the so-called energy crisis) that focused attention on using photovoltaic technology for applications other than the space program. Photovoltaics can be used in a wide range of products, from small consumer items to large commercial solar electric systems.

Figure 15.3 illustrates the photoelectric effect when light shines on a negative plate; electrons are emitted with an amount of kinetic energy inversely proportional to the wavelength of the incident light. Figure 15.4 illustrates the operation of a basic *photovoltaic cell*, also called a *solar cell*. Solar cells are made of silicon and other semiconductor materials such as germanium, gallium arsenide, and silicon carbide that are used in the microelectronics industry. For solar cells, a thin semiconductor wafer is specially treated to form an electric field, positive on one side and negative on the other. When light energy strikes the solar cell, electrons are jarred loose from the atoms in the semiconductor material (see Figures 15.3 and 15.4). If electrical conductors are attached to the positive and negative sides, forming an electrical circuit, the electrons can be captured in the form of an electrical current (recall that electron flow is electricity). This electricity can then be used to power a load, such as for a light, tool, or toaster, for example.

A *photovoltaic module* is comprised of a number of solar cells electrically connected to each other and mounted on a support panel or frame (see Figure 15.5). Solar panels used to power homes and businesses are typically made from solar cells combined into modules that hold about 40 cells. Modules are designed to supply electricity at a certain voltage (e.g., 12 volts). The current produced is directly dependent on how much light strikes the module.

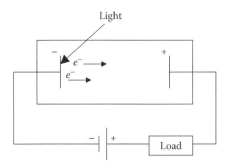

FIGURE 15.3
Photoelectric effect.

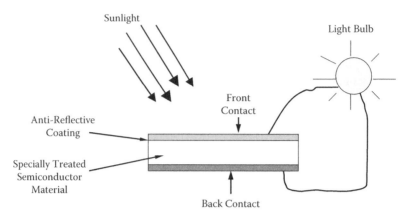

FIGURE 15.4
Operation of basic photovoltaic cell.

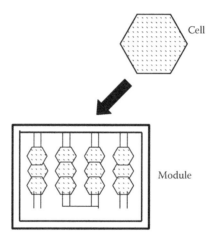

FIGURE 15.5
Single solar cell and solar cell module.

Multiple modules can be wired together to form an array. In general, the larger the area of a module or array, the more electricity that will be produced. Photovoltaic modules and arrays produce direct-current (DC) electricity. They can be connected in both series and parallel electrical arrangements to produce any required voltage and current combination.

15.4 Solar Power Applications

When thinking about a potential renewable energy source, we first want to know if it is reliable, clean, and affordable. This is especially the case for industry owners contemplating solar applications for industrial use. Until recently, solar energy has been used for both domestic and industrial applications, but only on a somewhat limited basis. This trend is changing, especially in those locations where sunlight is prevalent throughout the year. With the soaring costs of fossil fuel supplies and their pending decreased availability, solar power is beginning to receive more attention. Solar technologies for domestic use include photovoltaics, passive heating, window and structure daylighting, and water heating. Industrial, commercial, and municipal treatment facilities may use the same solar technologies that are used for residential buildings, in addition to solar energy technologies that would be impractical for a home, such as ventilation air preheating, solar process heating, and solar cooling.

15.4.1 Solar Hot Water

Just as the sun heats the surface layers of exposed bodies of water—ponds, lakes, streams, and oceans—it can also heat water used in buildings and swimming pools. Harnessing the ability of the sun to heat water requires a solar collector and a storage tank. The most common collector is the *flat-plate collector*. The flat-plate collector system is the most commonly used collector and is ideal for applications that require water temperatures under 140°F (60°C). Mounted on the roof, the system consists of a thin, flat, rectangular, insulated box with a transparent glass cover that faces the sun. Small copper tubes run through the box and carry the fluid (water or other fluid such as an antifreeze solution) to be heated. The tubes, typically arranged in series (i.e., water flows in one end and out the other end), are attached to an absorber plate, which, along with the tubes, is painted black to absorb the heat. As heat builds up in the collector, it heats the fluid (water or propylene glycol) passing through the tubes. The motive force for passing fluid through the tubes can be accomplished by either active or passive means. In an active system, the most common type of system, water heaters rely on an electric pump and controller to circulate water or other heat-transfer fluids through the collectors. Passive solar water-heater systems rely on gravity and the tendency for water to naturally circulate as it is heated. Because passive systems have no moving parts, they are more reliable, are easier to maintain, and often have a longer life than active systems. The storage tank holds the hot liquid (see Figure 15.6). It is usually a large well-insulated tank, but modified water

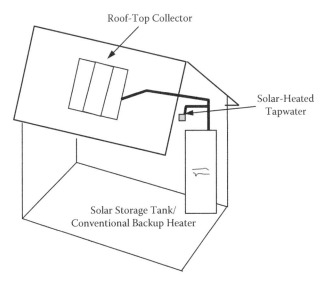

FIGURE 15.6
Simplified representation of a home or industrial solar water-heating system.

heaters can also be used to store the hot liquid. When the fluid used is other than water, the water is heated by passing it through a coil of tubing in the tank, which is full of hot fluid. Active solar systems usually include a storage tank along with a conventional water heater. In two-tank systems, the solar water heater preheats water before it enters the conventional water heater.

15.4.2　Solar Process Heat

Industrial, commercial, and municipal treatment facilities may use the same solar technologies that are used for residential buildings (photovoltaics, passive heating, daylighting, and water heating), in addition to solar energy technologies that would be impractical for a home, such as ventilation air preheating, solar process heating, and solar cooling.

15.4.2.1　Space Heating

In order to maintain indoor air quality (IAQ), many large buildings require ventilated air. Although all that fresh air can be great for indoor air quality, heating this air in cold climates can use large amounts of energy and can be very expensive. An elegantly simple solar ventilation system can preheat the air, though, saving both energy and money. This type of system typically uses a transpired air collector. Transpired air collector systems essentially consist of a dark-colored, perforated facade installed on the south-facing wall of a building. A fan or the existing ventilation system draws ventilation air into the building through the perforated absorber plate on the facade and up the plenum (the air space between the absorber and the south wall). Solar energy absorbed by the dark absorber and transferred to the air flowing through it can preheat the intake air by as much as 40°F (22°C). Reduced heating costs will pay for the systems in 3 to 12 years (USDOE, 2006).

In addition to meeting a portion of a building's heating load with clean free solar energy, the transpired collector helps save energy and money in other ways. It recaptures heat loss through a building's south-facing wall, as heat that escapes through the south wall is captured in the air space between the structural wall and the transpired collector and returned to the interior. Also, by introducing make-up air though ceiling-mounted ducts, the system eliminates the wasteful air stratification that often plagues high-ceiling buildings (USDOE, 2006).

To illustrate an active, real-world example of transpired air collectors used as solar preheaters for outdoor ventilation air, it is probably best to use an actual example of a working system in an industrial application. The following case study is such an example; it has been monitored extensively by experts.

> **CASE STUDY 15.1. GENERAL MOTORS BATTERY PLANT**[*]
>
> The General Motors battery plant in Oshawa, Canada, is a 100,000-ft^2 facility in which automotive batteries are manufactured. The plant was built in the 1970s and consists of an open shop floor and a 28-foot-high ceiling. GM operates two full-time production shifts within the plant and conducts maintenance activities at night and on weekends, so the building is continuously occupied.

---

[*] Adapted from USDOE, *Transpired Collectors (Solar Preheaters for Outdoor Ventilation Air)*, Federal Technology Alert, U.S. Department of Energy, Washington, DC, 1998.

Until the early 1990s, GM relied solely on a steam-operated fan coil system for space heating, but the system was incapable of providing the necessary quantities of heated outdoor air. As a result, the plant was not being adequately ventilated. In 1991, plant management installed a transpired collector to correct the ventilation problems. Over the next two years the transpired collector system was modified slightly to further improve airflow; the original fans and motors were replaced with vane axial fans and high-efficiency motors, and the original ducting was replaced with upgraded fabric ducting.

The GM plant collector is comprised of 4520 ft^2 of absorber sheeting. The lower 21 feet of the transpired collector is black, perforated, aluminum wall cladding with 1.6-mm holes totaling 2% porosity. The average depth of the plenum between the transpired collector and the plant's structural wall is 6 inches. The canopy at the top of the wall acts as both a manifold for air flow and a solar heat collection device. The canopy face is made of perforated plate with 1% porosity. The transpired collector covers about 50% of the total area of the plant's south-facing wall; the remainder of the south facade has shipping doors and other obstructions that make it unsuitable for mounting collector cladding.

The GM transpired collector has two fan/distribution systems, each consisting of a constant-speed fan, a recirculation damper system, and a fabric distribution duct. The total airflow delivered by the system's fans is 40,000 cfm. Both recirculated air and air drawn through the solar collectors make up this flow; the percentages of each depend on the temperature of the air coming from the collector.

The GM battery plant's transpired collector has been monitored extensively since it was installed. The data in this case study reflect the performance of the system during the 1993–94 heating season. An in-depth report on the monitoring program is available (Enermodal Engineering, Ltd., 1995). The data show that the annual energy savings for the 4520-ft^2 collector were 940 million Btu/year: 678 MBtu resulted from the thermal energy gained directly from the outside air as it passed through the absorber, and 262 MBtu resulted from heat loss recaptured by the wall from inside the building. Other possible energy-saving mechanisms—such as destratification and heat recapture—likely contributed to improved system performance; however, these effects are highly structure specific and have not been incorporated into the savings reported here.

The cost of the transpired collector system at the GM plant was $66,530, or $14.72/ ft^2 of installed collector. The cost per square foot is higher than typical installations for two reasons: (1) this system was installed soon after the technology was introduced, before design and installation procedures had been streamlined; and (2) the cost includes the fan and ducting modifications that were implemented during the first 2 years of operation.

It is important to point out that the GM transpired collector experienced a number of operational problems. After the system was initially installed, employees complained about fan noise and feeling cold drafts, and employees occasionally disabled the system. The fan and duct upgrades described previously eliminated the problems on one of the fan systems; the other fan continues to generate noise, and employees still disable it when working in the immediate vicinity. The manufacturer has addressed these complaints by specifying smaller, but more numerous, fans in subsequent installations.

Also, both bypass dampers and a recirculation damper required additional maintenance. The recirculation damper became stuck in the full recirculation mode, and a new modulating motor was installed to fix the problem. The bypass dampers occasionally became bound, which led to unacceptably high leakage rates. These dampers were kept closed manually through the 1993–1994 heating season.

15.4.2.2 Water Heating

Solar water-heating systems are designed to provide large quantities of hot water for non-residential buildings. A typical system includes solar collectors that work along with a pump, heat exchanger, and one or more large storage tanks. The two main types of solar collectors used for nonresidential buildings—evacuated-tube collector and linear concentrator—can operate at high temperatures with high efficiency. An evacuated-tube collector is a set of many double-walled glass tubes and reflectors to heat the fluid inside the tubes. A vacuum between the two walls insulates the inner tube, retaining the heat. Linear concentrators use long, rectangular, curved (U-shaped) mirrors tilted to focus sunlight on tubes that run along the length of the mirrors. The concentrated sunlight heats the fluid within the tubes.

15.4.2.3 Space Cooling

Space cooling can be accomplished using a thermally activated cooling system (TACS) driven by solar energy. Because of a high initial cost, the TACS is not widespread. The two systems currently in operation are solar absorption systems and solar desiccant systems. Solar absorption systems use thermal energy to evaporate a refrigerant fluid to cool the air. In contrast, solar desiccant systems use thermal energy to regenerate desiccants that dry the air, thereby cooling the air. These systems also work well with evaporative coolers in more humid climates.

15.5 Structure Daylighting

Another beneficial practice that water and wastewater treatments plants can incorporate into their energy sustainability and efficiency strategies and methodologies is the incorporation of daylighting into plant buildings and structures. Structure daylighting is the practice of placing windows or other openings and reflective surfaces so that during the day natural light provides effective internal lighting. When properly designed and effectively integrated with the electric lighting system, daylighting can offer significant energy savings by offsetting a portion of the electric lighting load. A related benefit is the reduction in cooling capacity and use by lowering a significant component of internal gains. In addition to energy savings, daylighting generally improves occupant satisfaction and comfort. Windows also provide visual relief, a contact with nature, time orientation, the possibility of ventilation, and emergency egress.

15.5.1 Daylight Zone

High daylight potential is found particularly in those spaces that are predominantly occupied in the daytime. Site solar analysis should assess the access to daylight by considering what can be viewed from the various potential window orientations. What proportion of the sky is seen from typical task locations in the room? What are the exterior obstructions and glare sources? Is the building design going to shade a neighboring building or landscape feature that is dependent on daylight or solar access? It is important to establish which spaces will most benefit from daylight and which spaces have little or no need for daylight. Within the spaces that can use daylight, place the most critical visual tasks in positions near the window. Try to group tasks by similar lighting requirements and occupancy patterns. Avoid placing the windows in the direct line of sight of the occupant as this can cause extreme contrast and glare. It is best to orient the occupant at 90° from the window. Where privacy is not a major concern, consider interior glazing (known as *relights* or *borrow lights*), which allows light from one space to be shared with another. This can be achieved with transom lights, vision glass, or translucent panels if privacy is required. The floor plan configuration should maximize the perimeter daylight zone. This may result in a building with a higher skin-to-volume ratio than a typical compact building design. A standard window can produce useful illumination to a depth of about 1.5 times the height of the window. With light shelves or other reflector systems this can be increased to 2.0 times or more. As a general rule of thumb, the higher the window is placed on the wall, the deeper the daylight penetration.

15.5.2 Window Design Considerations

The daylight that arrives at a work surface comes from three sources:

1. *Exterior reflected component*—This includes ground surfaces, pavement, adjacent buildings, wide windowsills, and objects. Remember that excessive rough reflectance will result in glare.
2. *Direct sun/sky component*—Typically, the direct sun component is blocked from occupied spaces because of heat gain, glare, and ultraviolet degradation issues. The sky dome then becomes an important contribution to daylighting the place.
3. *Internal reflected component*—Once the daylight enters the room, the surrounding wall, ceiling, and floor surfaces are important light reflectors. Using high-reflectance surfaces will better bounce the daylight around the room, and it will reduce extreme brightness contrast. Window-frame materials should be light colored to reduce contrast with the view and should have a nonspecular finish to eliminate glare spots. The window jambs and sills can be beneficial light reflectors. Deep jambs should be splayed (angled toward the interior) to reduce the contrast around the perimeter of the window.

Major room furnishings such as office cubicles or partitions can have a significant impact on reflected light so select light-colored materials. Suggested reflectance values for various room surfaces are

- Ceilings, >80%
- Walls, 50 to 70%
- Floors, 20 to 40%
- Furnishings, 25 to 45%

Because light essentially has no scale for architectural purposes, the proportions of the room are more important than the dimensions. A room that has a higher ceiling compared to the room depth will have deeper penetration of daylight, whether from side lighting (windows) or top lighting (skylights and clerestories). Raising the window to head height will also result in deeper penetration and more even illumination in the room. Punched

DID YOU KNOW?

The most important interior light-reflecting surface is the ceiling. High reflectance paints and ceiling tiles are now available with .90 or higher reflectance values. Tilting the ceiling plane toward the daylight source increases the daylight that is reflected from this surface. The rear wall of a small room should also have a highly reflective finish. Side walls, followed by the floor, have less impact on the reflected daylight in the space.

window openings, such as small, square windows separated by wall area, result in uneven illumination and harsh contrast between the window and adjacent wall surfaces. A more even distribution is achieved with horizontal strip windows.

15.5.3 Effective Aperture (EA)

One method of assessing the relationship between visible light and the size of the window is the effective aperture method. The *effective aperture* (EA) is defined as the product of the visible transmittance and the *window-to-wall ratio* (WWR). The window-to-wall ratio is the proportion of window area compared to the total wall area where the window is located; for example, if a window covers 25 square feet in a 100-square-foot wall then the WWR is 25/100, or 0.25. A good starting target for EA is in the range of 0.20 to 0.30. For a given EA number, a higher WWR (larger window) results in a lower visible transmittance. For example, a WWR = 0.5 (half the wall is glazing) has a VT = 0.6 and EA = 0.3. For WWR = 0.75, VT = 0.4 for the same EA of 0.3.

15.5.4 Light Shelves

A light shelf is an effective horizontal light-reflecting overhangs placed above eye level with a transom window placed above it. Light shelves enhance the lighting from windows of the equator-facing side of a building. Exterior shelves are more effective shading devices than interior shelves. A combination of exterior and interior will work best in providing an even illumination gradient.

15.5.5 Toplighting Strategies

Large single-level floor areas and the top floors of multiple-story buildings can benefit from toplighting. Types of toplighting include skylights, clerestories, roof monitors, and saw-tooth roofs. Horizontal skylights can be an energy problem because they tend to receive maximum solar gain at the peak of the day; the daylight contribution peaks at midday and falls off severely in the morning and afternoon. Some high-performance skylight designs

DID YOU KNOW?

There is no direct sunlight on the polar-side wall of a structure from the autumnal equinox to the spring equinox in parts of the globe north of the Tropic of Cancer and in parts of the globe south of the Tropic of Capricorn.

incorporate reflectors or prismatic lenses that reduce the peak daylight and heat gain while increasing early and late afternoon daylight contributions. Another option is light pipes, high-reflectance ducts that channel light from a skylight down to a diffusing lens in the room. These may be advantageous in deep roof constructions.

Clerestory windows are vertical glazings located high on an interior wall. South-facing clerestories can be effectively shaded from direct sunlight by a properly designed horizontal overhang. In this design, the interior north wall can be sloped to better reflect the light down into the room. Use light-colored overhangs and adjacent roof surfaces to improve the reflected component. If exterior shading is not possible, consider interior vertical baffles to better diffuse the light. A south-facing clerestory will produce higher daylight illumination than a northern-facing clerestory. East- and west-facing clerestories present the same problems as east and west windows: difficult shading and potentially high heat gains.

A roof monitor consists of a flat roof section raised above the adjacent roof with vertical glazing on all sides. This design often results in excessive glazing area, which results in higher heat losses and gains than a clerestory design. The multiple orientations of the glazing can also create shading problems. The sawtooth roof is an old design often seen in industrial buildings. Typically, one sloped surface is opaque and the other is glazed. A contemporary sawtooth roof may have solar collectors or photovoltaic cells on the south-facing slope and daylight glazing on the north-facing slope. Unprotected glazing on the south-facing sawtooth surface may result in high heat gains. In these applications, an insulated diffusing panel may be a good choice.

15.6 Water and Wastewater Treatment Plant Applications

Several water and wastewater treatment plants have installed solar cells to generate electricity for process controls and to increase their energy self-reliance. Many plant sites also have retrofitted or included various daylighting techniques in order to reduce energy needs and use at their plant sites. Oroville, a town in Northern California, operates a 6.5-MGD WWTP that services 15,000 households and many industrial users. In 2002, amidst an energy crisis that saw the price of utilities rise 41%, the Oroville Sewage Commission (SC-OR) decided to pursue solar power as a way to reduce costs and increase energy self-reliance. That same year, the utility installed a 520-kW ground-mounted solar array capable of being manually adjusted seasonally to maximize the solar harvest. The solar array consists of 5184 solar panels covering 3 acres of land adjacent to the WWTP. The total cost of the solar system, which is the fifth-largest solar energy system in the United States, was $4.83 million, with a rebate to the utility of $2.34 million from the Self-Generation Incentive Program of Pacific Gas and Electric (PG&E) and managed through the California Public Utilities Commission (CPUC). SC-OR was able to save an 80% reduction in power costs. The SC-OR solar array is designed to produce more power than the utility requires during peak hours, and because the system is connected to the local energy grid it can feed all of the excess energy back to the power utility so SC-OR can receive credit on its power bill. This credit goes toward paying for the off-peak power that the treatment plant uses at night. SC-OR saved $58,000 in the first year and expects the solar array to pay for itself in 9 years (USEPA, 2007).

In addition to using wind power, the 40-MGD Atlantic County Utilities Authority (ACUA) Wastewater Treatment Facility in Atlantic City, New Jersey, installed five solar array totaling 500 kW for the facility (NREL, 2007). The five solar arrays were placed at different locations

DID YOU KNOW?

The benefits of using solar-powered aerators include not only energy savings but also reductions in odor, greenhouse gas emissions, and biosolids volume at the bottom of a pond or basin that would otherwise have to be dredged and disposed.

throughout the facility and include two ground-mount arrays, two roof-mount arrays, and a canopy array. The roof-mount arrays are mounted such that they can withstand hurricane force winds. The solar array can generate electricity at rates lower that 5¢ per kWh for the next 20 years. This is the second largest solar array in the state, producing over 660,000 kWh of electricity annually or about 3% of the facility's 20-GWh annual electricity needs. This equivalent amount of energy displaces 388 barrels of oil and over 400,000 pounds of carbon dioxide. Energy rebates of $1.9 million were obtained from the New Jersey Board of Public Utilities, and an additional anticipated savings of over $35,000 is expected each year. The total cost of the project was about $3.9 million (ACUA, 2012).

In 2001, SolarBee, Inc., developed the SolarBee®, which is a floating solar-powered circulator that is capable of moving up to 10,000 gallons of water per minute for long distances. The SolarBee® possesses battery storage for up to 24-hour operation, which is beneficial during low sunlight conditions. A single SolarBee® unit can effectively aerate a 35-acre lake or treat a 25-million-gallon drinking water reservoir or tank. Since its creation, over 1000 units have been installed in many treatment applications, including wastewater lagoons. Use of the SolarBee® circulator can effectively improve biochemical oxygen demand (BOD) and biosolids reductions, control odor, and reduce total solids and ammonia concentrations in the effluent (USEPA, 2007).

In 2005, the City of Myrtle Beach in South Carolina installed five SolarBee® units in the first three cells of the city's 50-acre wastewater lagoon. Improvements in dissolved oxygen (DO) and H_2S levels within the lagoon after a few months prompted the city to budget for five additional SolarBee® units for the following year. The electrical savings due to installing these units were expected to average $100,000 per year (Shelly, 2007).

In southwest Arizona, the City of Somerton replaced a 40-hp wastewater lagoon aeration motor with four solar-power aerators. The project cost was about $100,000, with an expected electric energy cost savings of $25,000. Other applications of solar-powered aerators include the wastewater treatment facility in Bennett, Colorado, and the Town of Discovery Bay Community Services District (TDBCSD) wastewater treatment plant in California (USEPA, 2007).

References and Recommended Reading

ACUA. (2012). *Solar Array Project Fact Sheet*, Atlantic County Utilities Authority, Atlantic City, NJ.

Baker, M.S. (1990). Modeling complex daylighting with DOE 2.1-C, *DOE-2 User News*, 11(1).

Baylon, D. and Storm, P. (2008). Comparison of commercial LEED buildings and non-LEED buildings within the 2002–2004 Pacific Northwest commercial building stock, in *Proceedings of ACEEE Summer Study on Energy Efficiency in Buildings*, American Council for an Energy-Efficient Economy, Washington, DC.

Bierman, A. (2007). Photosensors: dimming and switching systems for daylight harvesting, *NLPIP Specifier Reports*, 11(1), 1–54.

Birt, B. and Newsham, G.R. (2009). Post-occupancy evaluation of energy and indoor environment quality in green building: a review, in *Proceedings of 3rd International Conference on Smart and Sustainable Built Environments*, Delft, The Netherlands, June 15–19.

Brookfield, H.C. (1989). *Sensitivity to Global Change: A New Task for Old/New Geographers*, Norma Wilkinson Memorial Lecture, University of Reading, Reading, UK.

Brundtland, G.H. (1987). *Our Common Future*, Oxford University Press, New York.

Brunger, A. and Hollands, K. (1996). Back-of-Plate Heat Transfer in Unglazed Perforated Collectors Operated under Non-Uniform Air Flow conditions, paper presented at the 22nd Annual Conference of the Solar Energy Society of Canada, Inc., Orillia, Ontario, Canada, June 9–10.

Calder, W.A. (1996). *Size, Function and Life History*, Dover, Mineola, NY.

Chiras, D.D. (2002). *The Solar House: Passive Heating and Cooling*, Chelsea Green Publishing Co., White River Junction, VT.

Christensen, C., Hancock, E., Barker, G., Kutscher, C., 1990. Cost and Performance Predictions for Advanced Active Solar Concepts. *Proceedings of the American Solar Energy Society Annual Meeting*, Austin, Texas.

Debres, K. (2005). Burgers for Britain: a cultural geography of McDonald's UK, *Journal of Cultural Geography*, 22(2), 115–139.

Dengler, J. and Wittwear, V. (1994). Glazings with granular aerogels, *SPIE*, 255, 718–727.

Duffie, J. and Beckman, W. (1991). *Solar Engineering of Thermal Processes*, John Wiley & Sons, New York.

Dumortier, D. (1997). Evaluation of luminous efficacy models according to sky types and atmospheric conditions, in *Proceedings of Lux Europa '97*, Ecole Nationale des Travaux Publics de l'Etat, Vaulx-en-Velin, France.

EERE. (2006). *Solar Energy Technologies Program: Overview and Highlights*, Energy Efficiency & Renewable Energy, U.S. Department of Energy, Washington, DC.

EERE. (2011). *History of Hydropower*, Energy Efficiency & Renewable Energy, U.S. Department of Energy, Washington, DC (http://www1.eere.energy.gov/water/hydro_history.html).

EERE. (2012). *SunShot Initiative*, Energy Efficiency & Renewable Energy, U.S. Department of Energy, Washington, DC (http://www1.eere.Energy.gov/solar/printable_versions/about.html).

Enermodal Engineering, Ltd. (1995). *Performance of the Perforated-Plate/Canopy Solarwall at GM Canada, Oshawa*, Energy Technology Branch, Department of Natural Resources, Ottawa, ON.

Fenchel, T. (1974). Intrinsic rate of natural increase: the relationship with body size, *Oecologia*, 14, 317–326.

Fontoynont, M., Place, W., and Bauman, P. (1984). Impact of electric lighting efficiency on the energy saving potential of daylighting from roof monitors, *Energy and Buildings*, 6(4), 375–386.

Galasiu, A.D., Atif, M.R., MacDonald, R.A. (2004). Impact of window blinds on daylight-linked dimming and automatic on/off lighting controls, *Solar Energy*, 76(5), 523–544.

Galasiu, A.D., Newsham, G.R., Suvagau, C., and Sander, D.M. (2007). Energy saving lighting control system for open-plan offices: a field study, *Leukos*, 4(1), 7–29.

Gunnewick, L. (1994). An Investigation of the Flow Distribution Through Unglazed Transpired-Plate Solar Air Heaters, master's thesis, Department of Mechanical Engineering, University of Waterloo, Canada.

Hanson, B.J. (2004). *Energy Power Shift: Benefiting from Today's New Technologies*, Lakota Scientific Press, Maple, WI.

Hardin, G. (1986). Cultural carrying capacity: a biological approach to human problems, *BioScience*, 36, 599–606.

Heschong, L. and McHugh, J. (2003). A report on integrated design of commercial ceiling, *PIER*.

Heschong, L. and McHugh, J. (2006). Skylights: calculating illumination levels and energy impacts, *Journal of Illuminating Engineering Society*, 29(1), 90–100.

Hinrichs, R.A. and Kleinbach, M.H. (2006). *Energy: Its Use and the Environment*, 4th ed., Thomson Learning, New York.

Holdren, J.P. (1991). Population and the energy problem, *Population and the Environment*, 12, 231–255.

Howlett, O. et al. (2006). Sidelighting photocontrols field study, in *Proceedings of ACEEE Summer Study on Energy Efficiency in Buildings*, Pacific Grove, CA, August 13–18, pp. 3-148–3-159.

Hubbell, S.P. and Johnson, L.K. (1977). Competition and nest spacing in a tropical stingless bee community, *Ecology*, 58, 949–963.

James J. Hirsch & Assoc. (2004). *DOE 2.2 Building Energy Use and Cost Analysis Program*. Vol. 3. *Topics*, Lawrence Berkeley National Laboratory, University of California, Berkeley.

James J. Hirsch & Assoc. (2009). *eQuest Introductory Tutorial, Version 3.63*, James J. Hirsch & Assoc., Camarillo, CA.

Janak, M. (1997). Coupling building energy and lighting simulation, in *Proceedings of the Fifth International IBPSA Conference*, Prague, September 8–10.

Janak, M. and Macdonald, I. (1999). Current state-of-the-art of integrated thermal and lighting simulation and future issues, in *Proceedings of the Sixth International IBPSA Conference*, Kyoto, Japan, September 13–15.

Lederer, E.M. (2008). UN says half the world's population will live in urban areas by end of 2008, *Associated Press*, February 26 (http://www.iht.com/aritcles/ap/2008/02/26/news/UN-GEN-UB-Growing-Cities.php).

Krebs, R.E. (2001). *Scientific Laws, Principles and Theories*, Greenwood Press, Westport, CT.

Kuhlken, R. (2002). Intensive agricultural landscapes of Oceania, *Journal of Cultural Geography*, 19(2), 161–195.

Kutscher, C.F. (1992). An Investigation of Heat Transfer for Air Flow Through Low Porosity Perforated Plates, doctoral thesis, University of Colorado, Boulder.

Kutscher, C.F., Christensen, C., and Barker, G. (1993). Unglazed transpired solar collectors: heat loss theory, *ASME Journal of Solar Energy Engineering*, 115, 182–188.

Lauoadi, A. and Aresnault, C. (2004). *Validation of Skyvision*, IRC-RR-167, National Research Council, Washington, DC.

Lauoadi, A. and Atif, M.R. (1999). Predicting optical and thermal characteristics of transparent single-glazed domed skylights, *ASHRAE Transactions*, 105(2), 325–333.

LOC. (2009). *The World's First Hydroelectric Power Plant*, Library of Congress, Washington, DC (http://www.americaslibrary.gov/jb/gilded/jb_gilded_hydro_1.html).

Lund, J.W. (2007). Characteristics, development and utilization of geothermal resources, *Institute of Technology*, 28(2), 1–9.

Moore, F. (1991). *Concepts and Practice of Architectural Daylighting*, John Wiley & Sons, New York.

Newsham, G.R. (2004). *American National Standard Practice for Office Lighting*, RP-1, ANSI/IESNA, Washington, DC.

NREL. (2010). *Concentrating Solar Power*, National Renewable Energy Laboratory, Washington, DC (http://www.nrel.gov/csp/pdfs/capability_factsheet_overview_of_csp_capabilities.pdf).

NREL. (2012a). *Dynamic Maps, GIS Data, & Analysis Tools: Solar Maps*, National Renewable Energy Laboratory, U.S. Department of Energy, Washington, DC (http://www.nrel.gov/gis/solar.html#csp).

NREL. (2012b). *Learning About Renewable Energy: Concentrating Solar Power Basics*, National Renewable Energy Laboratory, U.S. Department of Energy, Washington, DC (http://www.nrel.gov/learning/re_csp.html).

NREL. (2012c). *Concentrating Solar Power Research*, National Renewable Energy Laboratory, U.S. Department of Energy, Washington, DC (http://www.nrel.gov/csp/).

O'Connor, J, Lee, E., Rubinstein, L.E., and Selkowitz, S. (1997). *Tips for Daylighting with Windows*, Lawrence Berkeley National Laboratory, University of California, Berkeley.

Patel, M. (1999). *Wind and Solar Power Systems*, CRC Press, Boca Raton, FL.

Perez-Blanco, H. (2009). *The Dynamics of Energy: Supply, Conversion, and Utilization*, CRC Press, Boca Raton, FL.

Place, W. et al. (1984). The predicted impact of roof aperture design on the energy performance of office buildings, *Energy and Building*, 6(4), 361–373.

Reid, W. et al. (1988). *Bankrolling Successes*, Environmental Policy Institute and National Wildlife Federation, Washington, DC.

Richardson, J.M., Ed. (1982). *Making It Happen: A Positive Guide to the Future*, U.S. Association for the Club of Rome, Washington, DC.

Rubinstein, F., Ward, G., and Verderber, R. (1989). Improving the performance of photo-electrically controlled lighting systems, *Journal of the Illuminating Engineering Society*, 18(1), 70–94.

Searchinger, T. et al. (2008). Use of U.S. croplands for biofuels increases greenhouse gases through emissions from land-use change, *Science*, 319(5867), 1238–1240.

Shelly, P. (2006). City of Myrtle Beach cleaned up by "bees," *South Carolina Energy Office Newsletter*, December 1 (news.medoraco.com/city-of-myrtle-beach-cleaned-up-by-bees).

Simon, J.L. (1980). Resources, population, environment: an oversupply of false bad news, *Science*, 208, 1431–1437.

UCS. (2008). *Land-Use Changes and Biofuels*, Fact Sheet, Union of Concerned Scientists, Cambridge, MA.

USDOE. (1998). *Solar Buildings: Transpired Air Collectors*, U.S. Department of Energy, Washington, DC (http://www1.eere.energy.gov/femp/pdfs/FTA_trans_coll.pdf).

USDOE. (2006). *Geothermal Technologies Program: A History of Geothermal Energy in the United States*, U.S. Department of Energy, Washington, DC (http://www1.eere.energy.gov/geothermal/history.html).

USDOE. (2008). *Concentrating Solar Power Commercial Application Study: Reducing Water Consumption of Concentrating Solar Power Electricity Generation*, U.S. Department of Energy, Washington, DC (www1.eere.energy.gov/solar/pdfs/csp_water_study.pdf).

USDOE. (2009a). *Linear Concentrators Research and Development*, U.S. Department of Energy, Washington, DC (http://www1.eere.energy.gov/solar/linear_concentrator_rnd.html).

USDOE. (2009b). *Thermal Storage Research and Development*, U.S. Department of Energy, Washington, DC (http://www1.eere.energy.gov/solar/sunshot/index.html).

USDOE. (2010). *Dish/Engine Systems for Concentrating Solar Power*, U.S. Department of Energy, Washington, DC (http://www.eere.energy.gov/basics/renewable_energy/dish_engine.html).

USDOE. (2012). *Energy Sources: Fossil*, U.S. Department of Energy, Washington, DC (http://energy.gov/science-innovation/energy-sources/fossil).

USEPA. (2007). *Auxiliary and Supplemental Power Fact Sheet: Solar Power*, U.S. Environmental Protection Agency, Washington, DC.

USEPA. (2008). *Catalog of CHP Technologies*, U.S. Environmental Protection Agency, Washington, DC (www.epa.gov/chp/documents/catalog_chptech_full.pdf).

USEPA. (2011). *The Science of Ozone Layer Depletion*, U.S. Environmental Protection Agency, Washington, DC (http://www.epa.gov/ozone/science/index.html).

Vitousek, P.M., Ehrlich, A.H., and Matson, P.A. (1986). Human appropriation of the products of photosynthesis, *BioScience*, 36, 368–373.

Volterra, V. (1926). Variazioni e fluttuazioni del numero d'indivudui in specie animali conviventi, *Mem. Accad. Lincei*, 2(Ser. VI), 31–113.

World Commission on Environment and Development. (1987). *Our Common Future*, Oxford University Press, New York.

Wilson, E.O. (2001). *The Future of Life*, Knopf, New York.

Winkelman, F. and Selkowitz, S. (1985). Daylighting simulation in the DOE-2 building energy analysis program, *Energy and Buildings*, 8(4), 271–286.

16

Wind Power

The wind goeth toward the south, and turneth about unto the north; it whirleth about continually, and the wind returneth again.

—**Ecclesiastes 1:6**

The Good, Bad, and Ugly of Wind Energy

Good: As long as Earth exists, the wind will always exist (Figure 16.1). The energy in the winds that blow across the United States each year could produce more than 16 billion GJ of electricity—more than one and one-half times the electricity consumed in the United States in 2000.

Bad: Turbines are expensive. Wind doesn't blow all the time, so they have to be part of a larger plan. Turbines make noise. Turbine blades kill birds.

Ugly: Some look upon giant wind turbine blades cutting through the air as grotesque scars on the landscape, as visible polluters.

The bottom line: Do not expect Don Quixote, mounted in armor on his old nag, Rocinate, with or without Sancho Panza, to lead the charge to build those windmills. Instead, expect—you can count on it, bet on it, and rely on it—that the charge will be made by the rest of us to satisfy our growing, inexorable need for renewable energy. What other choice do we have?

16.1 Introduction[*]

Wind turbines can be used as auxiliary and supplemental power sources (ASPSs) for wastewater treatment plants (WWTPs). The use of wind power to generate electricity at WWTPs offers several advantages. Depending upon the size of the wind generation unit, energy production can be inexpensive when compared to conventional power production methods. The cost to generate the electricity decreases as the size of the generating units (e.g., from a wind farm) increases. Wind turbine power is an infinitely sustainable form of energy that does not require any fuel for operation and generates no harmful air or water pollution; it produces no greenhouse gases or toxic or radioactive waste. In addition, the land below each turbine can still be used for other purposes.

---

[*] Adapted from Spellman, F.R. and Whiting, N.E., *Environmental Science and Technology*, Government Institutes Press, Lanham, MD, 2006; Spellman, F.R. and Bieber, R., *The Science of Renewable Energy*, CRC Press, Boca Raton, FL, 2011.

FIGURE 16.1
Oh, how things have changed! (Artwork by Tracy L. Teeter and commissioned by F.R. Spellman.)

Disadvantages of using wind turbines include the need for more plant site space to support a single turbine or a number of turbines and the difficulty in having a location with enough wind to produce maximum efficiency and power. The placement of turbines in areas of high population density can also result in aesthetic problems. Other drawbacks include the deaths of birds and bats due to collisions with spinning turbine blades and turbine obstruction in their migratory flight paths (Nicholls and Racey, 2007; RReDC, 2010). In cold climates, ice and rime formation on turbine blades can result in turbine failure (Seifert, 2004). Moreover, the potential for ice to be thrown great distances during windy conditions is a potential health hazard. A recommended safety zone should be factored into the design specification to reduce public access and potential risks.

16.2 It's All about the Wind

Obviously, wind energy or power is all about wind. In simple terms, wind is the response of the atmosphere to uneven heating conditions. Earth's atmosphere is constantly in motion. Anyone observing the constant weather changes and cloud movement around them is well aware of this phenomenon. Although its physical manifestations are obvious, the importance of the dynamic state of our atmosphere is much less obvious. The constant motion of Earth's atmosphere (air movement) is both horizontal (*wind*) and vertical (*air currents*). The air movement is the result of thermal energy produced from heating of the surface of the Earth and the air molecules above. Because of differential heating of the surface of the Earth, energy flows from the equator poleward. The energy resources contained in the wind in the United States are well known and mapped in detail (Hanson, 2004). It is clear that air movement plays a critical role in transporting the energy of the lower atmosphere, bringing the warming influences of spring and summer and the cold chill of winter, and wind and air currents are fundamental to how nature functions. Still, though, the effects of air movements on our environment are often overlooked. All life on Earth has evolved or has been sustained with mechanisms dependent on air movement; for example, pollen is carried by the winds for plant reproduction, animals sniff the wind for essential information, and wind power was the motive force during the earliest stages of the industrial revolution. We can also see other effects of winds. Wind causes weathering (erosion) of the Earth's surface, wind influences ocean currents, and the wind carries air pollutants and contaminants such as radioactive particles that impact our environment.

16.3 Air in Motion

With respect to wind and air currents, as in all dynamic situations, forces are necessary to produce motion and changes in motion. The atmosphere, which is made up of various gases, is subject to two primary forces: gravity and pressure differences from temperature variations. *Gravity* (gravitational forces) holds the atmosphere close to the Earth's surface. Newton's law of universal gravitation states that everybody in the universe attracts another body with a force equal to

$$F = G\left(\frac{m_1 m_2}{r^2}\right) \tag{16.1}$$

where F is the magnitude of the gravitational force between the two bodies, m_1 and m_2 are the masses of the two bodies, G is the gravitational constant $\approx 6.67 \times 10^{-11}$ N (m^2/kg^2), and r is the distance between the two bodies. The force of gravity decreases as an inverse square of the distance between the two bodies. Thermal conditions affect density, which in turn affects gravity and thus vertical air motion and planetary air circulation. This, in turn, affects how air pollution is naturally removed from the atmosphere. Although forces in other directions often overrule gravitational force, gravity constantly acts vertically downward, on every gas molecule, which accounts for the greater density of air near the Earth.

Atmospheric air is a mixture of gases, so the gas laws and other physical principles govern its behavior. The pressure of a gas is directly proportional to its temperature. Pressure is equal to the force per unit area (pressure = force/area). Because the pressure of a gas is directly proportional to its temperature, temperature variations in the air give rise to differences in pressure or force. These differences in pressure cause air movement, both large and small scale, from a high-pressure region to a low-pressure region.

Horizontal air movements (*advective winds*) result from temperature gradients, which give rise to density gradients and, subsequently, pressure gradients. The force associated with these pressure variations (*pressure gradient force*) is directed perpendicular to lines of equal pressure (*isobars*) and is directed from high to low pressure. In Figure 16.1, the pressures over a region are mapped by taking barometric readings at different locations. Lines drawn through the points (locations) of equal pressure are the isobars. All points on an isobar are of equal pressure, which means there is no air movement along the isobar. The wind direction is at right angles to the isobar in the direction of the lower pressure. In Figure 16.2, notice that air moves down a pressure gradient toward a lower isobar like a ball rolls down a hill. If the isobars are close together, the pressure gradient force is large, and such areas are characterized by high-wind speeds. If isobars are widely spaced, the winds are light because the pressure gradient is small.

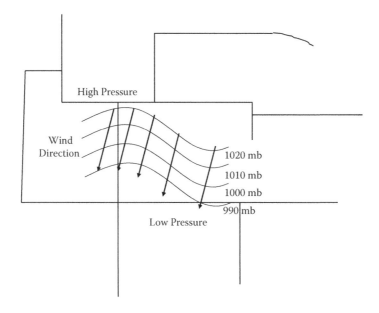

FIGURE 16.2
Isobars drawn through locations having equal atmospheric pressures. The air motion, or wind direction, is at right angles to the isobars and moves from a region of high pressure to a region of low pressure.

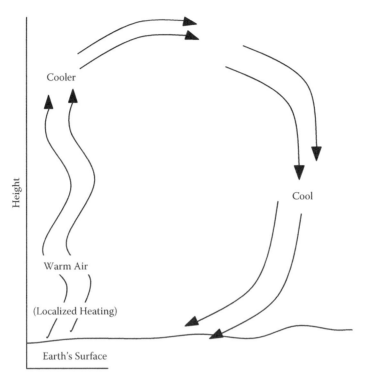

FIGURE 16.3
Thermal circulation of air. Localized heating, which causes air in the region to rise, initiates the circulation. As the warm air rises and cools, cool air near the surface moves horizontally into the region vacated by the rising air. The upper, still cooler, air then descends to occupy the region vacated by the cool air.

Localized air circulation gives rise to *thermal circulation* (a result of the relationship based on a law of physics whereby the pressure and volume of a gas are directly related to its temperature). A change in temperature causes a change in the pressure and volume of a gas. With a change in volume comes a change in density (as density = mass/volume), so regions of the atmosphere with different temperatures may have different air pressures and densities. As a result, localized heating sets up air motion and gives rise to thermal circulation. To gain an understanding of this phenomenon, consider Figure 16.3.

Once the air has been set into motion, secondary (velocity-dependent) forces come into play. These secondary forces are caused by Earth's rotation and contact of the air with the rotating Earth. The *Coriolis force*, named after its discoverer, French mathematician Gaspard Coriolis (1772–1843), is the effect of rotation on the atmosphere and on all objects on the Earth's surface. In the Northern Hemisphere, it causes moving objects and currents to be deflected to the right. The reverse is true in the Southern Hemisphere. Air, in large-scale north or south movements, appears to be deflected from its expected path; that is, air moving poleward in the Northern Hemisphere appears to be deflected toward the east, and air moving southward appears to be deflected toward the west. Figure 16.4 illustrates the Coriolis effect on a propelled particle (analogous to the apparent effect of an air mass flowing from Point A to Point B). The Earth's rotation from east to west acts on the air particle as it travels north over the Earth's surface. The particle will actually reach Point C because even though it is moving in a straight line (deflected) the Earth rotates east to west beneath it.

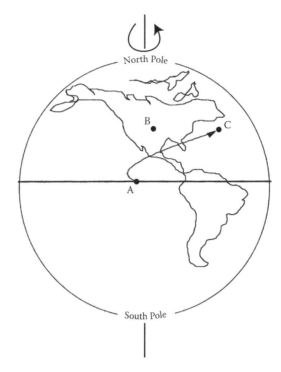

FIGURE 16.4
The effect of the Earth's rotation on the trajectory of a propelled particle.

DID YOU KNOW?

The magnitude of the frictional force along a surface is dependent on the air's magnitude and speed, and the opposing frictional force is in the opposite direction of the air motion.

Friction (drag) can also deflect or decelerate air movement and can be either internal or external. Internal friction is caused by air molecules running into each other; external friction is due to contact of the air with terrestrial surfaces.

16.4 Wind Energy[*]

Wind energy is power produced by the movement of air. Since early recorded history, people have been harnessing the energy of the wind to, for example, mill grain and pump water. Wind energy propelled boats along the Nile River as early as 5000 BC. By 200 BC, simple windmills in China were pumping water, and vertical-axis windmills with woven reed sails were grinding grain in Persia and the Middle East. The use of wind energy

---

[*] Adapted from USEPA, *History of Wind Energy*, U.S. Environmental Protection Agency, Washington, DC, 2011.

spread around the world, and by the 11th century people in the Middle East were using windmills extensively for food production; returning merchants and crusaders carried this idea back to Europe. The Dutch refined the windmill and adapted it for draining lakes and marshes in the Rhine River delta. When settlers took this technology to the New World in the later 19th century, they used windmills to pump water for farms and ranches and, later, to generate electricity for homes and industry.

The first known wind turbine designed to produce electricity was built in 1888 by Charles F. Brush, in Cleveland, Ohio; it was a 12-kW unit that charged batteries in the cellar of a mansion. The first wind turbine used to generate electricity outside of the United States was built in Denmark in 1891 by Poul la Cour, who used electricity from his wind turbines to electrolyze water to make hydrogen for the gas lights at the local schoolhouse. By the 1930s and 1940s, hundreds of thousands of wind turbines were used in rural areas of the United States that were not yet served by the grid. The oil crisis in the 1970s created a renewed interest in wind, until the U.S. government stopped giving tax credits.

Today, several hundred thousand windmills are in operation around the world, many of which are used for pumping water. The use of wind energy as a pollution-free means of generating electricity on a significant scale is attracting the most interest in the subject today. As a matter of fact, due to current and pending shortages and high costs of fossil fuels to generate electricity, as well as the green movement toward the use of cleaner fuels, wind energy is the world's fastest-growing energy source and could power industry, businesses, and homes with clean, renewable electricity for many years to come. In the United States, wind-based electricity generating capacity has increased markedly since the 1970s. Today, though, it still represents only a small fraction of total electric capacity and consumption (see Table 16.1), despite the advent of $4/gal gasoline, increases in the cost of electricity, high heating and cooling costs, and worldwide political unrest or uncertainty in oil-supplying countries. Traveling the wind corridors of the United States (primarily Arizona, New Mexico, Texas, Missouri, and north through the Great Plains to the Pembina

TABLE 16.1

U.S. Energy Consumption by Energy Source, 2008

Energy Source	Energy Consumption (quadrillion Btu)
Total	99.438
Fossil fuels (coal, natural gas, petroleum)	83.532
Electricity net imports	0.113
Nuclear electric power	8.427
Renewable	7.367
Biomass (total)	3.852
Biofuels	1.372
Waste	0.436
Wood and wood-derived fuels	2.044
Geothermal energy	0.360
Hydroelectric conventional	2.512
Solar thermal/PV energy	0.097
Wind energy	**0.546**

Source: EIA, *Renewable Energy Annual 2008*, Energy Information Administration, U.S. Department of Energy, Washington, DC, 2010.

DID YOU KNOW?

We can classify wind energy as a form of solar energy. As mentioned, winds are caused by uneven heating of the atmosphere by the sun, irregularities of the Earth's surface, and rotation of the Earth. As a result, winds are strongly influenced and modified by local terrain, bodies of water, weather patterns, and vegetative cover, among other factors. The wind flow, or motion of energy when harvested by wind turbines, can be used to generate electricity.

Escarpment and Turtle Mountains of North Dakota) gives some indication of the considerable activity and seemingly exponential increase in wind energy development and wind turbine installations.

16.5 Wind Power Basics

The terms *wind energy* and *wind power* reflect the process by which the wind is used to generate mechanical power or electricity. Wind turbines convert the kinetic energy in the wind into mechanical power. This mechanical power can be used for specific tasks (such as grinding grain or pumping water) or a generator can convert this mechanical power into electricity (EERE, 2006a). We have been harnessing the wind's energy for hundreds of years. From old Holland to farms in the United States, windmills have been used for pumping water or grinding grain; today, the modern equivalent of a windmill—a wind turbine—can use the energy of the wind to generate electricity. The blades of a wind turbine spin like aircraft propeller blades. Wind turns the blades, which in turn spin a shaft connected to a generator and produce electricity (Wind Energy Programmatic EIS, 2009). Unlike fans, which use electricity to make wind, wind turbines use wind to make electricity.

16.6 Wind Turbine Types

Whether referred to as a *wind-driven generator, wind generator, wind turbine, wind-turbine generator,* or *wind energy conversion system*, modern wind turbines fall into two basic groups: the horizontal-axis wind turbine (HAWT), like the traditional farm windmills used for water pumping, and the vertical-axis wind turbine (VAWT), like the eggbeater-style Darrieus rotor model, named after its French inventor, which was the only vertical-axis machine with any

DID YOU KNOW?

Whenever wind energy is being considered as a possible source of renewable energy it is important to consider the amount of land area required, accessibility to generators, and aesthetics.

commercial success. Wind hitting the vertical blades, called *aerofoils*, generates lift to create rotation. No yaw (rotation about the vertical axis) control is required to keep them facing into the wind. The heavy machinery in the nacelle (cowling) is located on the ground. Blades are closer to the ground, where wind speeds are lower. Most large modern wind turbines are horizontal-axis turbines; therefore, they are highlighted and described in detail in this text.

16.6.1 Horizontal-Axis Wind Turbines

Wind turbines are available in a variety of sizes and power ratings. Utility-scale turbines range in size from 100 kW to as large as several megawatts. Horizontal-axis wind turbines typically have either two or three blades. Downwind horizontal-axis wind turbines have a turbine with the blades behind (downwind from) the tower. No yaw control is needed because they naturally orient themselves in line with the wind; however, these downwind HAWTs experience a shadowing effect, in that when a blade swings behind the tower the wind it encounters is briefly reduced and the blade flexes. Upwind HAWTs usually have three blades in front (upwind) of the tower. These upwind wind turbines require a somewhat complex yaw control to keep them facing into the wind. They operate more smoothly and deliver more power and thus are the most commonly used modern wind turbines. The largest machine has blades that span more than the length of a football field, stands 20 building stories high and produces enough electricity to power 1400 homes.

16.6.1.1 Inside the HAWT

Basically, a horizontal-axis wind turbine consists of three main parts: a turbine, a nacelle, and a tower. Several other important parts are contained within the tower and nacelle, including anemometer, blades, brake, controller, gear box, generator, high-speed shaft, low-speed shaft, rotor, tower, wind vane, yaw drive, and yaw motor (see Figure 16.5).

16.7 Turbine Features

Anemometer—An anemometer measures the wind speed and transmits wind speed data to the controller.

Blades—Most turbines have either two or three blades. Wind blowing over the blades causes the blades to lift and rotate; they capture the kinetic energy of the wind and help the turbine rotate.

Brake—A disc brake can be applied mechanically, electrically, or hydraulically to stop the rotor in emergencies.

Controller—The controller starts up the machine at wind speeds of about 8 to 16 mph and shuts off the machine at about 55 mph. Turbines do not operate at wind speeds above about 55 mph because they might be damaged by the high winds.

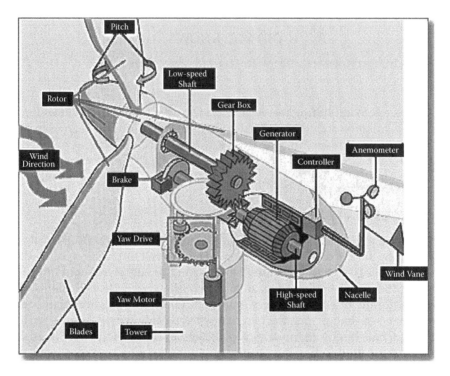

FIGURE 16.5
Horizontal-axis wind turbine components.

Gear box—Gears connect the low-speed shaft to the high-speed shaft and increase the rotational speeds from about 30 to 60 rotations per minute (rpm) to about 1000 to 1800 rpm, the rotational speed required by most generators to produce electricity. The gear box is a costly (and heavy) part of the wind turbine, and engineers are exploring direct-drive generators that operate at lower rotational speeds and do not require gear boxes.

Generator—Usually an off-the-shelf induction generator, it produces 60-hertz AC electricity.

High-speed shaft—This shaft drives the generator.

Low-speed shaft—This shaft turns at about 30 to 60 rotations per minute.

Nacelle—The nacelle unit sits atop the tower and contains the gearbox, generator, low- and high-speed shafts, controller, and brake.

Pitch—Bales are turned, or pitched, out of the wind to control the rotor speed and keep the rotor turning in winds that are too high or too low to produce electricity.

Rotor—The rotor is comprised of the blades and the hub.

Tower—The tower is made from tubular steel, concrete, or steel lattice. Because wind speed increases with height, taller towers enable turbines to capture more energy and generate more electricity.

Wind direction—An upwind turbine operates facing into the wind; other turbines are designed to run downwind, facing away from the wind.

DID YOU KNOW?

During rotation of the nacelle, it is possible for the cables inside the tower to twist, with the cables becoming more and more twisted if the turbine keeps turning in the same direction. The wind turbine is therefore equipped with a cable twist counter, which notifies the controller that it is time to straighten the cables (Khaligh and Onar, 2010).

Wind vane—The wind vane measures wind direction and communicates with the yaw drive to orient the turbine properly with respect to the wind.

Yaw drive—The yaw drive of upwind turbines is used to keep the entire nacelle and thus the rotor facing into the wind as the wind direction changes. Downwind turbines do not require a yaw drive, as the wind blows the rotor downwind.

Yaw motor—The yaw motor powers the yaw drive.

16.8 Wind Energy and Power Calculations

A wind turbine is a machine that converts the kinetic energy in wind into the mechanical energy of a shaft. Calculating the energy and power available in the wind relies on knowledge of basic physics and geometry. The *kinetic energy* of an object is the extra energy it possesses because of its motion. It is defined as the work necessary to accelerate a body of a given mass from rest to its current velocity. Once in motion, a body maintains it kinetic energy unless its speed changes. The kinetic energy of a body is given by the following equation:

$$\text{Kinetic energy} = 0.5 \times m \times v^2 \tag{16.2}$$

where

m = Mass

v = Velocity

■ *Example 16.1. Determining Power in the Wind*

Let's say we have a large packet of wind of thickness D that passes through the plane of a wind turbine's blades, which sweep out at cross-sectional area A (see Figure 16.6).

Step 1: To determine power in the wind, we must first consider the kinetic energy of the packet of air shown in Figure 16.6, along with its mass (m) and velocity (v), as shown in Equation 16.2. Next, we need to divide by time to get power:

$$\text{Power through area } A = \frac{1}{2}\left(\frac{m \text{ passing through } A}{t}\right)v^2 \tag{16.3}$$

Step 2: The mass flow rate can be expressed as follows (where ρ is air density, the mass per unit volume of Earth's atmosphere):

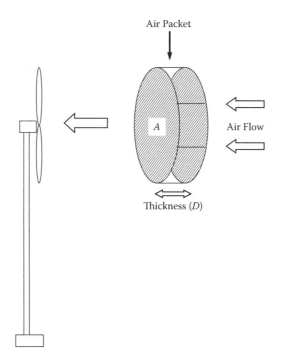

FIGURE 16.6
A packet of air passing through the plane of a wind turbine blade, with the thickness (*D*) passing through the plane over a given time.

$$m = \left(\frac{m \text{ passing through } A}{t} \right) \rho A v \qquad (16.4)$$

Step 3: Combining Equations 16.3 and 16.4, we obtain

$$A = \frac{1}{2} (\rho A v) v^2$$

$$P_W = \frac{1}{2} (\rho A v^3) \qquad (16.5)$$

where

A = Cross-sectional area that wind passes through perpendicular to the wind (m²)

v = Wind speed normal to A (m/s; 1 m/s = 2.237 mph)

ρ = Air density (2.225 kg/m³ at 15°C and 1 atm)

P_W = Power in the wind (watts)

DID YOU KNOW?

Air density decreases with increasing altitude, as does air pressure. It also changes with variations in temperature or humidity.

From Equation 16.5, it is evident that wind power is a function of the cube of the wind speed; that is, doubling the wind speed increases the power by eight. As an example, the energy produced during 1 hour of 20-mph winds is the same as the energy produced during 8 hours of 10-mph winds. When we speak of power (W/m²) and wind speed (mph) we cannot use average wind speed because the relationship is nonlinear. Power in the wind is also proportional to A. For a conventional horizontal-axis wind turbine, $A = (\pi/4)D^2$, so wind power is proportional to the blade diameter squared. Because cost is approximately proportional to blade diameter, larger wind turbines are more cost effective.

16.8.1 Air-Density Correction Factors

Earlier we pointed out that air density is affected by different temperature and pressure. Air-density correction factors can correct air density for temperature and altitude. Correction factors for both temperature and altitude correction can be found in standardized tables. Equation 16.6 is used to determine air density (ρ) for various temperatures and pressures:

$$\rho = \frac{P \times MW \times 10^{-3}}{RT} \tag{16.6}$$

where

P = Absolute gas pressure (atm)

MW = Molecular weigh of air (= 28.97 g/mol)

R = Ideal gas constant (= 8.2056×10^{-5} m³ · atm · K⁻¹ · mol⁻¹)

T = Absolute temperature (degrees Kelvin = 273 + °C)

16.8.2 Elevation and Earth's Roughness

The speed of wind is affected by its elevation above Earth and the roughness of Earth. Because power increases like the cube of wind speed, we can expect a significant economic impact from even a moderate increase in wind speed; thus, in the operation of wind turbines, wind speed is a very important parameter. The surface features of Earth cannot be ignored when deciding where to place wind turbines and calculating their output productivity. Natural obstructions such as mountains and forests and human-made obstructions such as buildings cause friction as winds flow over them. Generally, a lot of friction is encountered in the first few hundred meters above ground (although smooth surfaces such as water offer little resistance). For this reason, taller wind turbine towers are better.

When actual measurements are not possible or available, it is possible to characterize or approximate the impact of rough surfaces and height on wind speed. This is accomplished using Equation 16.7:

$$\frac{v}{v_0} = \left(\frac{H}{H_0} \right)^{\alpha} \tag{16.7}$$

where

α = Friction coefficient, obtained from standardized tables (typical values of α are 0.10 for calm water or smooth, hard ground; 0.15 for open terrain; and 0.40 for a large city with tall buildings)

v = Wind speed at height H

v_0 = Wind speed at height H_0 (H_0 is usually 10 m)

16.8.3 Wind Turbine Rotor Efficiency

Generally, when we think or talk about efficiency we think about input vs. output and know that if we put 100% into something and get 100% output, we have a very efficient machine, operation, or process. In engineering, we can approximate efficiency (input vs. output) by performing mass balance calculations. We know that, according to the laws of conservation, mass cannot disappear or be destroyed; thus, it must continue to exist in one form or another. With regard to this input vs. output concept, we should address a couple views on maximum rotor efficiency that are misinformed and make no sense but are stated here to point out what really does make sense. The two wrong assumptions are that (1) it can be assumed that downwind velocity is zero and the turbine extracts all of the power from the wind, and (2) downwind velocity is the same as the upwind velocity and the turbine has not reduced the wind speed at all. Wrong! Albert Betz set the record straight in 1919 when he proposed that there exists a maximum theoretical efficiency for extracting kinetic energy from the wind. Betz's wind turbine efficiency theory (with regard to power input vs. power output and overall efficiency in general) suggests that the efficiency rule or laws of conservation, as stated above, do not apply. Betz's law states the maximum possible energy that can be derived from a wind turbine. To better understand Betz's law and its derivation, we have provided the following explanation.

16.8.4 Derivation of Betz's Law

To understand Betz's law, we must first understand the constraints on the ability of a wind turbine to convert kinetic energy in the wind into mechanical power. Visualize wind passing through a turbine (see Figure 16.7)—it slows down and the pressure is reduced so it expands. Equation 16.8 is used to determine the power extracted by the blades:

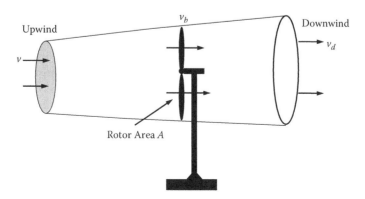

FIGURE 16.7
Wind passing through a turbine.

$$P_b = \frac{1}{2} m \left(v^2 - v_d^2 \right) \tag{16.8}$$

where

m = Mass flow rate of air within stream tube

v = Upwind undisturbed wind speed

v_d = Downwind wind speed

Determining the mass flow rate is the next step. In making the determination, it is easiest to use the cross-sectional area A at the plane of the rotor because we know what this value is; thus, the mass flow rate is

$$m = p \times A \times v_b \tag{16.9}$$

Assume that the velocity through the rotor (v_b) is the average of the upwind velocity (v) and downwind velocity (v_d):

$$v_b = \frac{v + v_d}{2} \quad \rightarrow \quad m = \rho A \left(\frac{v + v_d}{2} \right)$$

Equation 16.8 becomes

$$P_b = \frac{1}{2} \rho A \left(\frac{v + v_d}{2} \right) \left(v^2 - v_d^2 \right) \tag{16.10}$$

Before moving on in the derivation process, it is important that we define the operating speed ratio (λ):

$$\lambda = \frac{v_d}{v} \tag{16.11}$$

We can now rewrite Equation 16.9 as

$$P_b = \frac{1}{2} \rho A \left(\frac{v + \lambda v}{2} \right) \left(v^2 - \lambda^2 v^2 \right) \tag{16.12}$$

$$\downarrow$$

$$\left(\frac{v + \lambda v}{2} \right) \left(v^2 - \lambda^2 v^2 \right) = \frac{v^3}{2} - \frac{\lambda^2 v^3}{2} + \frac{\lambda v^3}{2} - \frac{\lambda^3 v^3}{2} = \frac{v^3}{2} \left[(1 + \lambda) - \lambda^2 (1 + \lambda) \right] = \frac{v^3}{2} \left[(1 + \lambda)(1 - \lambda^2) \right]$$

$$P_b = \left(\frac{1}{2} \rho A v^3 \right) \quad \times \quad \frac{1}{2} \left[(1 + \lambda)(1 - \lambda^2) \right]$$

$$\downarrow \qquad\qquad\qquad \downarrow$$

$$(P_W = \text{Power in the wind}) \quad (C_P = \text{Rotor efficiency})$$

The next step is to find the operating speed ratio (λ) that maximizes the rotor efficiency (C_P):

$$C_P = \frac{1}{2}\left[\left(1+\lambda\right)\left(1-\lambda^2\right)\right] = \frac{1}{2} - \frac{\lambda^2}{2} + \frac{\lambda}{2} - \frac{\lambda^3}{2}$$

Set the derivative of rotor efficiency to zero and solve for λ:

$$\frac{\partial C_P}{\partial \lambda} = -2\lambda + 1 - 3\lambda^2 = 0$$

$$\frac{\partial C_P}{\partial \lambda} = 3\lambda^2 + 2\lambda - 1 = 0$$

$$\frac{\partial C_P}{\partial \lambda} = \left(3\lambda - 1\right)\left(\lambda + 1\right) = 0$$

$$\vdots$$

$$\lambda = 1/3$$

We plug the optimal value for λ back into the equation for C_P to find the maximum rotor efficiency:

$$C_P = \frac{1}{2}\left[\left(1+\frac{1}{3}\right)\left(1-\frac{1}{3}\right)\right] = \frac{16}{27} = 59.3\% \qquad (16.13)$$

The maximum efficiency of 59.3% occurs when air is slowed to 1/3 of its upstream rate. This value is called the *Betz efficiency* (Betz, 1966). Betz's law states that all wind power cannot be captured by the rotor; otherwise, air would be completely still behind the rotor and would not allow more wind to pass through. For illustrative purposes, in Table 16.2, we list wind speed, power of the wind, and power of the wind based on the Betz limit (59.3%).

16.8.5 Tip Speed Ratio (TSR)

Efficiency is a function of how fast the rotor turns. The tip speed ratio (TSR), an extremely important factor in wind turbine design, is the ratio of the speed of the rotating blade tip to the speed of the free stream wind (see Figure 16.8). Stated differently, TSR is the speed of the outer tip of the blade divided by wind speed. There is an optimum angle of attack that creates the highest lift-to-drag ratio. If the rotor of the wind turbine spins too slowly, most of the wind will pass straight through the gap between the blades, thus giving it no power. But, if the rotor spins too fast, the blades will blur and act like a solid wall to the wind. Moreover, rotor blades create turbulence as they spin through the air. If the next blade arrives too quickly, it will hit that turbulent air. Thus, it is actually better to slow down the blades. Because the angle of attack is dependent on wind speed, there is an optimum tip speed ratio:

$$TSR = \Omega R / V \qquad (16.14)$$

TABLE 16.2

Betz Limit for 80-M Rotor Turbines

Wind Speed (mph)	Wind Speed (m/sec)	Wind Power (kW)	Power Based on 59.3% Betz Limit (kW)
5	2.2	36	21
10	4.5	285	169
15	6.7	962	570
20	8.9	2280	1352
25	11.2	4453	2641
28	12.5	6257	3710
30	13.4	7695	4563
35	15.6	12,220	7246
40	17.9	18,241	10,817
45	20.1	25,972	15,401
50	22.4	35,626	21,126
55	24.6	47,419	28,119
56 (cutoff speed)	25.0	50,053	29,681
60	26.8	61,563	36,507

Source: Adapted from Devlin, L., *Wind Turbine Efficiency*, K0LEE.com, 2007, http://k0lee.com/turbineeff.htm.

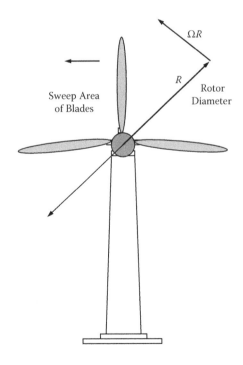

FIGURE 16.8
Tip speed ratio (TSR).

where

Ω = Rotational speed (rad/sec)

R = Rotor radius

V = Wind "free stream" velocity

16.9 Small-Scale Wind Power

To meet the typical domestic user's need for electricity, a small-scale wind machine may be the answer. Such a wind turbine has rotors between 8 and 25 feet in diameter, stands around 30 feet, and can supply the power needs of an all-electric home or small business. Utility-scale turbines range in size from 50 to 750 kW. Single small turbines below 50 kW in size are used for homes, telecommunication dishes, or water pumping

Potential users of small wind turbines should ask many questions before installing such a system (EERE, 2012). Let's ask the obvious question first: "What are the benefits to homeowners from using wind turbines?" Wind energy systems provide a cushion against electricity price increases. Wind energy systems reduce U.S. dependence on fossil fuels, and they don't emit greenhouse gases. If you are building a home in a remote location, a small wind energy system can help you avoid the high cost of extending utility power lines to your site.

Although wind energy systems involve a significant initial investment, they can be competitive with conventional energy sources when you account for a lifetime of reduced or altogether avoided utility costs. The length of payback time—the time before the savings resulting from your system equal the system cost—depends on the system you choose, the wind resources at your site, the electric utility rates in your area, and how you use your wind system.

Another frequently asked question is "Is wind power practical for me?" Small wind energy systems can be used in connection with an electricity transmission and distribution system (*grid-connected systems*) or in stand-alone applications that are not connected to the utility grid. A grid-connected wind turbine can reduce the consumption of utility-supplied electricity for lighting, appliances, and electric heat. If the turbine cannot deliver the amount of energy required, the utility makes up the difference. When the wind system produces more electricity than the household requires, the excess can be sold to the utility. With the interconnections available today, switching takes place automatically. Stand-alone wind energy systems can be appropriate for homes, farms, or even entire communities that are far from the nearest utility lines.

DID YOU KNOW?

The difference between power and energy is that power (kilowatts, kW) is the rate at which electricity is consumed and energy (kilowatt-hours, kWh) is the quantity consumed.

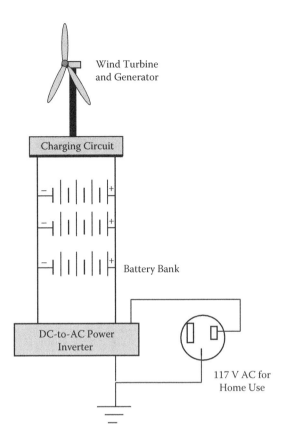

FIGURE 16.9
Stand-alone, small-scale wind power system.

Stand-alone systems (systems not connected to the utility grid) require batteries to store excess power generated for use when the wind is calm. They also must have a charge controller to keep the batteries from overcharging. Deep-cycle batteries, such as those used for golf carts, can discharge and recharge 80% of their capacity hundreds of times, which makes them a good option for remote renewable energy systems. Automotive batteries are shallow-cycle batteries and should not be used in renewable energy systems because of their short life for deep-cycling operations.

Safety Note: For safety, batteries should be isolated from living areas and electronics because they contain corrosive and explosive substances. Lead–acid batteries also require protection from temperature extremes.

Small wind turbines generate direct-current (DC) electricity. In very small systems, DC appliances operate directly off the batteries. To use standard appliances that use conventional household alternating current (AC), an inverter must be installed to convert DC electricity from the batteries to AC (see Figure 16.9). Although the inverter slightly lowers the overall efficiency of the system, it allows the home to be wired for AC, a definite plus with lenders, electrical code officials, and future homebuyers.

In grid-connected (or interactive) systems, the only additional equipment required is a power condition unit (inverter) that makes the turbine output electrically compatible with the utility grid. Usually, batteries are not needed. Either type of system can be practical if the following conditions exist:

- *Conditions for stand-alone systems*

  You live in an area with average annual wind speeds of at least 4.0 mps (9 mph).

  A grid connection is not available or can only be made through an expensive extension. The cost of running a power line to a remote site to connect with the utility grid can be prohibitive, ranging from $15,000 to more than $50,000 per mile, depending on terrain.

  You have an interest in gaining energy independence from the utility.

  You would like to reduce the environmental impact of electricity production.

  You acknowledge the intermittent nature of wind power and have a strategy for using intermittent resources to meet your power needs.

- *Conditions for grid-connected systems*

  You live in an area with average annual wind speeds of at least 4.5 mps (10 mph).

  Utility-supplied electricity is expensive in your area (about 10 to 15¢ per kilowatt-hour).

  The utility's requirements for connecting your system to its grid are not prohibitively expensive.

  Local building codes or covenants allow you to legally erect a wind turbine on your property.

  You are comfortable with long-term investments.

After comparing stand-alone systems and grid-connected systems and determining which is best suited for your particular circumstance, the next question to consider is whether your location is the appropriate site for installing a small-scale wind turbine system. Is it legal to install the system on your property? Are there any environmental or economic issues? Does the wind blow frequently and hard enough to make a small wind turbine system economically worthwhile? This is a key question that is not always easily answered. The wind resource can vary significantly over an area of just a few miles because of local terrain influences on the wind flow; yet, there are steps you can take that will go a long way toward answering the above question.

Wind resource maps like the ones included in the U.S. Department of Energy's *Wind Energy Resource Atlas of the United States* (RReDC, 2010) can be used to estimate the wind resources in your region. The highest average wind speeds in the United States are generally found along seacoasts, on ridge lines, and on the Great Plains; however, many other areas have wind resources strong enough to power a small wind turbine economically. The wind resource estimates provided by the *Wind Energy Resource Atlas* generally apply to terrain features that are well exposed to the wind, such as plains, hilltops, and ridge crests. Local terrain features may cause the wind resources at specific sites to differ considerably from these estimates.

Average wind speed information can be obtained from a nearby airport; however, caution should be used because local terrain and other factors may cause the wind speed to differ from that recorded at an airport. Airport wind data are generally measured at heights about 20 to 33 ft (6 to 10 m) above ground. Also, average wind speeds increase with height and may be 15 to 25% greater at a typical wind turbine hub height of 80 ft (24 m) than those measured at airport anemometer heights. The *Wind Energy Resource Atlas* contains data from airports in the United States and makes wind data summaries available.

Again, it is important to have site-specific data to determine the wind resource at a particular location. If wind speed data for a particular site are not available, it may be necessary to measure wind speeds at that location for a year. A recording anemometer generally costs $500 to $1500. The most accurate readings are taken at hub height (i.e., the elevation at the top of the wind turbine tower). This requires placing the anemometer high enough to avoid turbulence created by trees, buildings, and other obstructions. The standard wind sensor height used to obtain data for USDOE maps is 10 meters (33 feet).

Within the same property it is not unusual to have varied wind resources. Those living in complex terrains should take care when selecting an installation site. A wind turbine installed on the top of a hill or on the windy side of a hill, for example, will have greater access to prevailing winds than if it were located in a gully or on the leeward (sheltered) side of a hill on the same property. Consider existing obstacles and plan for future obstructions, including trees and buildings, that could block the wind. Also recall that the power in the wind is proportional to its speed cubed. This means that the amount of power provided by a generator increases exponentially as the wind speed increases; for example, a site that has an annual average wind speed of about 5.6 mps (12.6 mph) has twice the energy available as a site with an average of 4.5 mps (10 mph) ($12.6/10^3$).

Another useful indirect measurement of the wind resource is observation of an area's vegetation. Trees, especially conifers or evergreens, can be permanently deformed by strong winds. This deformity, known as *flagging*, has been used to estimate the average wind speed for an area (see Figure 16.10).

In addition to ensuring the proper siting of a small wind turbine system, other legal, environmental, and economic issues must be addressed:

- Research potential legal and environmental obstacles.
- Obtain cost and performance information from manufacturers.
- Perform a complete economic analysis that accounts for a multitude of factors.
- Understand the basics of a small wind system.
- Review possibilities for combining the system with other energy sources, backups, and energy efficiency improvements.

With regard to economic issues, because energy efficiency is usually less expensive than energy production, making your house more energy efficient first will likely result in spending less money on a wind turbine, because a smaller one may meet your needs. *A word of caution:* Before investing in a wind turbine, research potential legal and environmental obstacles to installing one. Some jurisdictions, for example, restrict the height of the structures permitted in residentially zoned areas, although variances are often obtainable. Neighbors might object to a wind machine that blocks their view, or they might be concerned about noise. Consider obstacles that might block the wind in the future (large planned developments or saplings, for example). Saplings that will grow into large trees can be a problem in the future. As mentioned, trees can affect wind speed (see Figure 16.10). If you plan to connect the wind generator to your local utility company's grid, find out its requirements for interconnections and buying electricity from small independent power producers. When you are convinced that a small wind turbine is what you want and there are no obstructions restricting its installation, approach buying a wind system as you would any major purchase.

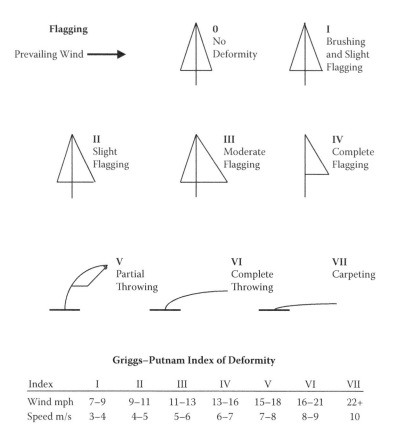

Griggs–Putnam Index of Deformity

Index	I	II	III	IV	V	VI	VII
Wind mph	7–9	9–11	11–13	13–16	15–18	16–21	22+
Speed m/s	3–4	4–5	5–6	6–7	7–8	8–9	10

FIGURE 16.10
A crude method of approximating average annual wind speed from the deformation of trees (and other foliage).

16.10 Wind Power Applications in Water/Wastewater Treatment

Several WWTPs throughout the United States have installed or are considering the installation of wind turbines to temper the rising costs of electricity. The 40-MGD Atlantic County Utilities Authority Wastewater Treatment Facility in Atlantic City, New Jersey, supplements its energy needs using wind turbines (ACUA, 2012). When operating at design wind speeds of over 12 mph, the five 1.5-MW wind turbines at this facility are capable of producing up to 7.5 MW of electrical energy. Because this is much more than the average 2.5 MW of power required each day by this facility, the remaining energy is sold to the local power grid. Power production occurs only when wind speed is greater than 7 mph and shuts down at speeds in excess of 45 mph to protect the machinery inside. On an annual basis, the ACUA wind farm can supply more than 60% of the electricity required by the plant. The remaining electricity can be bought from the local power grid when windmills are not at peak capacity (during calm or gusty weather). The cost of wind-generated electricity is 7.9¢/kWh delivered for the next 20 years, while the current cost delivered by the electrical grid is 12¢/kWh and rising. The estimated cost of the 7.5-MW wind farm was $12.5 million with an expected cost saving of $350,000 (USEPA, 2007).

To encourage the use of renewable energy resources, the town of Browning, Montana, and the Blackfeet Indian Tribe have installed four Bergey Excel 10-kW wind turbines adjacent to the town's sewage treatment plant (Town of Browning, 2012). The turbines provide about one-quarter of the plant's electricity, displacing energy bought from the grid. In the City of Fargo, North Dakota, the installation of a 1.5-MW wind turbine to provide 85% of the annual electricity used by the city's wastewater treatment plant is being considered. The Fargo wind turbine is estimated to cost $2.4 million and could save the plant about $203,000 annually. The Lynn wastewater treatment plant in Massachusetts, which services the counties of Lynn, Saugus, Swampscott, and Nahant, is considering the installation of one or more wind turbine generators to supply a substantial portion of the plant's electricity. As of 2007, information was being collected on possible wind turbine model options to comply with the Federal Aviation Administration height restriction of 254.ft (77.4 m) above ground level, as well as information on each model's estimated energy production, setback requirements, and potential sound impact (USEPA, 2007).

References and Recommended Reading

ACUA. (2012). *Solar Array Project Fact Sheet*, Atlantic County Utilities Authority, Atlantic City, NJ.
Alternative Energy. (2010). *Wind Turbines*, http://www.alternative-energy-news.info/technology/wind-pworer/wind-turbines/.
Archer, C. and Jacobson, M.Z. (2004). *Evaluation of Global Wind Power*, Department of Civil and Environmental Engineering, Stanford University, Stanford, CA.
Betz, A. (1966). *Introduction to the Theory of Flow Machines* (D.G. Randall, trans.), Pergamon Press, Oxford.
Burton, T., Sharpe, D., Jenkins, N., and Bossanyi, E. (2001). *Wind Energy Handbook*, John Wiley & Sons, New York.
EERE. (2006a). *Wind & Water Power Program*, Energy Efficiency & Renewable Energy, U.S. Department of Energy, Washington, DC (http://www1.eere.energy.gov/windandhydro/).
EERE. (2006b). *How Wind Turbines Work*, Energy Efficiency & Renewable Energy, U.S. Department of Energy, Washington, DC (http://www1.eere.energy.gov/Windandhydro/wind_how.html).
EERE. (2007). *Annual Report on U.S. Wind Power Installation, Cost, and Performance Trends: 2006*, DOE/GO-102007-2433, Energy Efficiency & Renewable Energy, U.S. Department of Energy, Washington, DC (www.nrel.gov/docs/fy07osti/41435.pdf).
EERE. (2012). *Wind Program*, Energy Efficiency & Renewable Energy, U.S. Department of Energy, Washington, DC (http://www1.eere.energy.gov/wind/small_wind_system_faqs.html).
Hanson, B.J. (2004). *Energy Power Shift*, Lakota Scientific Press, Maple, WI.
Jespersen, B. (2007). Wind power to fuel waste plant, *Morning Sentinel*, October 25.
Khaligh, A. and Onar, O.C. (2010). *Energy Harvesting*, Taylor & Francis, Boca Raton, FL.
Nicholls, B. and Racey, P.A. (2007). Bats avoid radar installations: could electromagnetic fields deter bats from colliding with wind turbines? *PLoSONE*, 2(3), e297.
RReDC. (2010). *Wind Energy Resource Atlas of the United States*, Renewable Resource Data Center, National Renewable Energy Laboratory, Washington, DC (http://rredc.nrel.gov/wind/pubs/atlas/).
Seifert, H. (2004). Technical requirements for rotor blades operating in cold climates, *DEWI Magazine*, 24.
Serway, R.A. and Jewett, J.W. (2004). *Physics for Scientists and Engineers*, 6th ed., Thomson/Brooks/Cole, New York.
Town of Browning. (2012). *Wind Power for the Wastewater Treatment Plant*, Town of Browning, Montana (http://www.browningmontana.com/wind.html).

Tripler, P. (2004). *Physics for Scientists and Engineers*. Vol. 1. *Mechanics, Oscillations and Waves, Thermodynamics*, 5th ed., W.H. Freeman, New York.

Tripler, P. and Llewellyn, R. (2002). *Modern Physics*, 4th ed., W.H. Freeman, New York.

USEPA. (2007). *Auxiliary and Supplemental Power Fact Sheet: Wind Turbines*, EPA 832-F-05-013, U.S. Environmental Protection Agency, Washington, DC.

USEPA. (2008). *Catalog of CHP Technologies*, U.S. Environmental Protection Agency, Washington, DC (www.epa.gov/chp/documents/catalog_chptech_full.pdf).

USEPA. (2011). *History of Wind Energy*, U.S. Environmental Protection Agency, Washington, DC (http://www1.eere.energy.gov/wind/wind_history.html).

White, F.M. (1988). *Fluid Mechanics*, 2nd ed., McGraw-Hill, Singapore.

Wind Energy Programmatic EIS. (2009). *Wind Energy Basics,* Wind Energy Development Programmatic Environmental Impact Statement, U.S. Department of the Interior Bureau of Land Management and Argonne National Laboratory, Washington, DC (http://windeis.anl.gov/guide/basics/index.cfm).

17

Energy Conservation Measures for Wastewater Treatment*

Providing reliable wastewater services and safe drinking water is a highly energy-intensive activity in the United States. A report prepared for the Electric Power Research Institute (EPRI) in 1996 estimated that by the end of that year, the energy demand for the water and wastewater industry would be approximately 75 billion kilowatt hours (kWh) per year, or about 3 percent of the electricity consumed in the U.S.

—**Burton (1996)**

17.1 Introduction

In this final chapter, it is only fitting to discuss in more detail the energy conservation measures (ECMs) that have been presented in the preceding chapters, focusing particularly on actual applications in the wastewater treatment industry. I have chosen to focus on the wastewater industry in particular because it is a much larger user of energy than the water treatment industry; however, the reader should understand that the ECMs discussed in this chapter can also apply to water treatment operations. The purpose of this chapter is to encourage the implementation of ECMs at publicly owned treatment works (POTWs) by providing accurate performance and cost/benefit information for such projects.

It is important to understand that energy is used throughout the wastewater treatment process; however, pumping and aeration operations are typically the largest energy users. Energy costs in the wastewater industry are rising due to many factors, including (USEPA, 2008):

- Implementation of more stringent effluent requirements, including enhanced removal of nutrients and other emerging contaminants of concern that may, in some cases, lead to the use of more energy-intensive technologies
- Enhanced treatment of biosolids, including drying/pelletizing
- Aging wastewater collection systems that result in additional inflow and infiltration, leading to higher pumping and treatment costs
- Increase in electricity rates

As a consequence of these rising costs, many wastewater facilities have developed energy management strategies and implemented energy conservation measures. Accordingly, this chapter describes ECMs being employed in wastewater treatment plants by

---

* Material in this chapter is adapted from USEPA, *Evaluation of Energy Conservation Measures for Wastewater Treatment Facilities*, EPA 832-R-10-005, U.S. Environmental Protection Agency, Washington, DC, 2010.

- Providing an overview of conventional ECMs related to pumping design, variable-frequency drives (VFDs), and motors
- Providing detailed information on ECMs related to the design of aeration systems and automated aeration control, including conventional control based on dissolved oxygen (DO) measurements and emerging control strategies, in addition to discussing innovative and emerging technologies for automated control of biological nitrogen removal
- Describing innovative ECMs related to blower and diffuser equipment, including a summary of various blower types such as single-stage centrifugal, high-speed turbo, and screw compressors in addition to new diffuser technology
- Providing a discussion of ECMs for advanced technologies (ultraviolet disinfection, membranes, and anoxic zone mixing) and presenting full-scale plant test results where available

17.2 Pumping System Energy Conservation Measures

Pumping operations can be a significant energy draw at wastewater treatment plants (WWTPs), in many cases second only to aeration. Pumps are used for many applications. At the plant headworks, they may be used to provide hydraulic head for the downstream treatment processes. Within the plant, they are used to recycle and convey waste flows, solids, and treated effluent to and from a variety of treatment processes. Pumps are also found in remote locations in the collection system to help convey wastewater to the plant.

The overall efficiency of a pumping system, also called the *wire-to-water efficiency*, is the product of the efficiency of the pump itself, the motor, and the drive system or method of flow control employed. Pumps lose efficiency from turbulence, friction, and recirculation with the pump (Spellman and Drinan, 2001; WEF, 2009). Another loss is incurred if the actual operating condition does not match the pump's *best efficiency point* (BEP). The various methods for controlling flow rate decrease system efficiency. Throttling valves to reduce the flow rate increases the pumping head, flow control valves burn head produced by the pump, recirculation expends power with no useful work, and VFDs produce a minor amount of heat. Of these methods, VFDs are the most flexible and efficient means to control flow despite the minor heat loss incurred. Table 17.1 summarizes typical pump system efficiency values; note that inefficiency in more than one component can add up quickly, resulting in a very inefficient pumping system.

TABLE 17.1

Pump System Efficiency

Pump System Component	Efficiency			
	Range (%)	Low (%)	Average (%)	High (%)
Pump	30–85	30	60	75[a]
Flow control[b]	20–98	20	60	98
Motor[c]	85–95	85	90	95
Efficiency of system	—	5	32	80

[a] For pumping wastewater; pump system efficiencies for clean water can be higher.
[b] Represents throttling, pump control valves, recirculation, and VFDs.
[c] Represents nameplate efficiency and varies by horsepower.

DID YOU KNOW?

The best efficiency point (BEP) is the flow rate (typically in gallons per minute or cubic meters per day) and head (in feet or meters) that give the maximum efficiency on a pump curve.

Inefficiencies in pumping often arise from a mismatch between the pump and the system it serves due to improper pump selection, changes in operating conditions, or the expectation that the pump will operate over a wide range of conditions. Signs of an inefficient pumping system include the following:

- Highly or frequently throttled control values
- Bypass line (recirculation) flow control
- Frequent on/off cycling
- Cavitation noise at the pump or elsewhere in the system
- A hot running motor
- A pump system with no means of measuring flow, pressure, or power consumption
- Inability to produce maximum design flow

The literature provides several examples of plants reducing pumping energy by as much as 5% through pump system improvements (SAIC, 2006). Energy savings result from lowering pumping capacity to better match system demands, replacing inefficient pumps, selecting more efficient motors, and installing variable speed controllers. Generally speaking, energy conservation measures for pumping are conventional and do not represent an area where recent technology innovation has played a part in improving energy conservation and efficiency. Pumping ECMs, however, are still extremely important to reducing and optimizing energy use at wastewater treatment plants.

17.2.1 Pumping System Design

Appropriated sizing of pumps is key to efficient operation of wastewater treatment plants. Pumps sized for peak flow conditions that occur infrequently or, worse, in the future toward the end of the pump's service life operate the majority of the time at a reduced flow that is below their BEP. Peak flow is typically several times greater than average daily flow and can be an order of magnitude different than minimum flow, especially for small systems or systems with significant inflow and infiltration (I&I). In some systems, these projected future flows are never reached during the design life of the pump. For existing treatment plants, utilities should evaluate the operation of existing pumps and identify opportunities for energy reduction. A good starting point is to determine the efficiency of existing pumping systems, focusing first on pumps that operate for the most hours and have potential problems, as identified earlier as pumping system inefficiencies. Plants should collect performance information on the flow rate, pressure, and delivered power to the pumps. Field measurements may be necessary if the plant does not regularly record this information. Pump and system curves can then be constructed to determine the actual operating points of the existing system. Operating points more than 10% different from the BEP signal room for improvement.

To improve efficiency, utilities should consider replacing or augmenting large-capacity pumps that operate intermittently with smaller capacity pumps that will operate for longer periods and closer to their BEP. When replacing a pump with a smaller unit, both the horsepower and efficiency change. A quick way to estimate the annual energy cost savings is to approximate cost before and after the improvement and determine the difference using Equation 17.1:

$$\text{Annual energy savings (\$)} = [hp_1 \times L_1 \times 0.746 \times hr \times E_1 \times C] \\ - [hp_2 \times L_2 \times 0.746 \times hr \times E_2 \times C]$$

(17.1)

where

hp_1 = Horsepower output for the larger capacity pump

hp_2 = Horsepower output for the smaller capacity pump

L_1 = Load factor of larger capacity pump (percentage of full load/100, determined from pump curve)

L_2 = Load factor of smaller capacity pump (percentage of full load/100, determined from pump curve)

hr = Annual operating hours

E_1 = Efficiency of the larger capacity pump

E_2 = Efficiency of the smaller capacity pump

C = Energy (electric power) rate ($/kWh)

See Box 17.1 for a summary of how the Town of Trumbull, Connecticut, was able to save more than $1500 per year by adding a small pump to one of its existing sewage pumping stations. When applied correctly, replacement of standard drives with VFDs can also yield significant improvements.

For greenfield plants or new pump stations, utilities should consider and plan for staging upgrades of treatment capacity as part of the design process; for example, multiple pumps can be specified to meet a future design flow instead of one large pump so that individual pumps can be installed as needed, say at year zero, year ten, and year twenty. The State of Wisconsin's Focus on Energy™ best practices guidebook (SAIC, 2006) estimates that staging of treatment capacity can result in energy savings between 10 and 30% of total energy consumed by a unit process.

17.2.2 Pump Motors

The induction motor is the most commonly used type of alternating-current (AC) motor because of its simple, rugged construction and good operating characteristics. It consists of two parts: the *stator* (stationary part) and the *rotor* (rotating part). The most important type of polyphase induction motor is the three-phase motor.

Important Note: A three-phase (3-θ) system is a combination of three single-phase (1-θ) systems. In a 3-θ balanced system, the power comes from an AC generator that produces three separate but equal voltages, each of which is out of phase with the other voltages by 120°. Although 1-θ circuits are widely used in electrical systems, most generation and distribution of AC current is 3-θ.

**BOX 17.1. TOWN OF TRUMBULL, CT, IMPROVED EFFICIENCY
AT RESERVOIR AVENUE PUMP STATION**

Background: Wastewater from the Town of Trumbull, in southwestern Connecticut, is collected and conveyed to a WWTP in Bridgeport via ten sewage pump stations. One of these is the Reservoir Avenue Pump Station, which consists of two 40-hp direct-drive pumps designed to handle an average daily flow of 236 gallons per minute (gpm). Each pump was operated at a reduced speed of 1320 rpm at 50.3 feet of total dynamic head (TDH) with a duty point of approximately 850 gpm. A bubbler-type level control system was used to turn the pumps off and on. One pump could handle the entire peak inflow (usually less than 800 gpm), with the second pump operating only during peak flow conditions.

Energy Efficiency Upgrades: To reduce energy use, the town installed a new 10-hp pump and modified the system control scheme. The new pump handles the same volume as the original pump but operates for a longer time between standby periods. In addition, the speed control was eliminated and the original pumps, when used, are run at full speed of 1750 rpm. This allowed the impellers of the original pumps to be trimmed from 11.25 inches in diameter to 10 inches. The original pumps are used for infrequent peak flows that cannot be handled by the new 10-hp pump. Under normal operating conditions, the operating point for the new pump is 450 gpm at 407 TDH compared to 850 gpm at 50.3 feet of head for the whole system. Improvements made to the lighting and control systems resulted in additional energy savings.

Energy Savings: Annual energy savings from modifying the pumping system have been 17,643 kWh. Total energy savings have been 31,875 kWh/yr, or approximately $2600/yr based on a rate of 8¢/kWh. Total implementation costs were $12,000, resulting in a simple payback of 4.6 years.

—EERE (2005c)

The driving torque of both DC and AC motors is derived from the reaction of current-carrying conductors in a magnetic field. In the DC motor, the magnetic field is stationary and the armature, with its current-carrying conductors, rotates. The current is supplied to the armature through a commutator and brushes. In induction motors, the rotor currents are supplied by electromagnet induction. The stator windings, connected to the AC supply, contain two or more out-of-time-phase currents, which produce corresponding magnetomotive force (mmf). This mmf establishes a rotating magnetic field across the air gap. This magnetic field rotates continuously at constant speed regardless of the load on the motor. The stator winding corresponds to the armature winding of a DC motor or to the primary winding of a transformer. The rotor is not connected electrically to the power supply.

The induction motor derives its name from the fact that mutual induction (or transformer action) takes place between the stator and the rotor under operating conditions. The magnetic revolving field produced by the stator cuts across the rotor conductors, inducing a voltage in the conductors. This induced voltage causes rotor current to flow; hence, motor torque is developed by the interaction of the rotor current and the magnetic revolving field.

The cost of running these electric induction motors (and other motors) can be the largest fraction of a plant's total operating costs. The Water Environment Federation (WEF) estimates that electrical motors make up 90% of the electric energy consumption of a typical wastewater treatment plant (WEF, 2009). Inefficient motors, operation outside of optimal loading conditions, and mechanical or electrical problems with the motor itself can lead to wasted energy at the plant and are opportunities for savings.

The percent energy savings resulting from replacing older motors with premium motors is modest, typically between 4 and 8% (NEMA, 2006). Savings can be higher when energy audits reveal that existing motors achieve very low efficiencies, or when existing motors are oversized or underloaded. Many plants have coupled motor replacements with upgrades from fixed-speed to variable-speed drives for significantly higher energy savings.

The bottom line on pump motors in wastewater treatment plant applications? Keep in mind that the upgrading of motors is a conventional ECM that has been practiced at wastewater treatment plants for some time.

17.2.2.1 Motor Efficiency and Efficiency Standards

Motor efficiency is a measure of mechanical power output compared to electrical power input, expressed as a percentage (WEF, 2009):

$$\text{Motor efficiency} = P_m/P_e \tag{17.2}$$

where

  P_m = Mechanical power output of the motor (in watts)

  P_e = Electrical power input to the motor (in watts)

No motor is 100% efficient; all motors experience some power loss due to friction, electrical resistance losses, magnetic core losses, and stray load losses. Smaller motors generally experience higher losses compared to larger motors.

The U.S. Congress, in the Energy Policy Act (EPACT) of 1992, set minimum efficiency standards for various types of electric motors manufactured in or imported to the United States. Minimum nominal, full-load efficiencies typically range from 80 to 95% depending on size (i.e., horsepower) and other characteristics. Motors manufactured since 1997 have been required to comply with EPACT standards and to be labeled with a certified efficiency value. The National Electrical Manufacturers Association (NEMA) premium efficiency standard has existed since 2001 (NEMA, 2006) as a voluntary industry standard and has been widely adopted due to it power (and thus cost) savings over EPACT 1992 compliance standards. The 2007 Energy Independence and Security Act raised efficiency standards of motors to NEMA premium efficiency levels and set new standards for motors not covered by previous legislation.

Submersible motors are commonly used in wastewater treatment plants. They serve specialized applications in environments that are not suited for NEMA motors. There is currently no efficiency stand for submersible motors, and their efficiency is less than NEMA motors. Additionally, their power factor is usually lower. Their selection is usually driven by the application, though some applications have alternatives that use NEMA motors. Efficiency should be considered in the evaluation of alternatives in these applications as it affects the life-cycle costs used in the selection process.

DID YOU KNOW?

Plants should consider buying new energy-efficient motors

- For new installations
- When purchasing new equipment packages
- When making major modifications to the plant
- Instead of rewinding older, standard efficiency units
- To replace oversized or underloaded motors
- As part of a preventive maintenance or energy conservation program

—EERE (1996)

Operating efficiency in the field is usually less than the nominal, full-load efficiency identified by the motor manufacturer. One reason for this is the operating load. As a rule of thumb, most motors are designed to operate between 50 and 100% of their rated load, with maximum efficiency occurring at about 75% of maximum load. A motor rated for 20 hp, for example, should operate between 10 and 20 hp and would have its best efficiency at around 15 hp. Larger motors can operate with reasonable efficiency at loads down to the 25% range (USDOE, 1996). Motors operated outside of the optimal loading lose efficiency. Other factors that reduce efficiency in the field include power quality (i.e., proper voltage, amps, and frequency) and temperature. Motors that have been rewound typically are less efficient compared to the original motor.

Accurately determining the efficiency of motors in service at a plant is challenging because there is no reliable field instrument for measuring mechanical output power. Several methods are available, however, to approximate motor efficiency.

17.2.2.2 Motor Management Programs

Wastewater utilities should consider purchasing new energy-efficient premium motors instead of rewinding older units when replacing equipment and when making major improvements at the plant. Motor replacement is best done as part of a plantwide motor management program. A first step in program development is to create an inventory of all motors at the plant. The inventory should contain as much information as possible, including manufacturers' specifications, nameplate information, and field measurements such as voltage, amperage, power factor, and operating speed under typical operating conditions. Following the data-gathering phase, plant managers should conduct a motor replacement analysis to determine which motors to replace now and which are reasonably efficient and can be replaced in the future or at time of failure.

A key input to any motor replacement analysis is economics. A simple approach is to calculate the annual energy savings of the new motor compared to the old unit and determine the payback period in years. (In other words, when will the cumulative energy savings exceed the initial costs?) The following simple equation can be used to determine annual energy savings:

$$\text{Annual energy savings (\$)} = hp \times L \times 0.746 \times hr \times C \times (E_p - E_e) \qquad (17.3)$$

where

　hp = Horsepower output of motor

　L = Load factor (percentage of full load/100)

　0.746 = Conversion from horsepower to kilowatts

　hr = Annual operating hours

　C = Energy (electric power) rate ($/kWh)

　E_e = Existing motor efficiency as a percentage

　E_p = Premium motor efficiency as a percentage

Simple payback in years can then be calculated as the new motor cost (capital plus installation) divided by the annual energy savings. When comparing buying a premium motor instead of rewinding an existing one, the cost of rewinding the existing motor should be subtracted from the motor cost. Any cash rebated from the local electric utility or state energy agency should also be subtracted from the cost of the new motor. When replacing pumps, motors, or control systems, upgrading the electrical service, wiring, transformers, and other components of the electrical system should be considered when calculating energy savings and life-cycle costs. Utilities should also consider the importance of reliability and environmental factors when making motor replacement decisions. More robust economic analysis such as net present value life-cycle cost analysis should be considered, especially for large expenditures.

The ENERGY STAR® Cash Flow Opportunity (CFO) calculator is an easy-to-use spreadsheet tool that can help plant managers calculate simple payback as well as cost of delay, which is the lost opportunity cost if the project is delayed 12 months or more. The last sheet of the workbook provides a summary that can be given to senior managers and decision makers to help convince them of the financial soundness of energy efficiency upgrades. The CFO calculator and other financial tools are available for free download at http:www.energystar.gov/index.cfm?c = assess_value.financial_tools.

The task of motor inventory management and replacement analysis is made significantly easier by publically available software tools. Developed by the U.S. Department of Energy Industrial Technologies Program, MotorMaster+ is a motor selection and management tool available free online at http://www1.eere.energy.gov/manufacturing/tech_deployment/software_motormaster.html. It includes inventory management features, maintenance logging, efficiency analysis, savings evaluation, and energy accounting. The site includes a catalog of 17,000 motors from 14 manufacturers, including NEMA Premium® efficiency motors, and motor purchasing information. In addition to the software, the sponsors of the Motor Decisions Matter℠ campaign developed a spreadsheet tool to assist plant managers with motor replacement or repair decision making. The tool is titled the "1*2*3 Approach to Motor Management" and is available for free download at http://www.motorsmatter.org/tools/123approach.html.

17.2.2.3 Innovative and Emerging Technologies

Siemens Energy and Automation, in cooperation with the Copper Development Association, has developed "ultra-efficient" copper rotor squirrel-cage-type induction AC motors. These motors exceed NEMA premium full-load efficiency standards by up to 1.4%; however, they are only currently available in outputs up to 20 hp. In addition to using high-conductivity copper rotors in place of aluminum, the new motors offer the following efficiency improvements (EERE, 2008):

- Optimized rotor and stator design
- Low-friction bearings
- Improved cooling system
- Polyurea-based grease
- Dynamically balanced rotors
- Precision-machined mating surfaces for reduced vibration
- The motor's insulation is designed to be compatible with VFDs

The U.S. Department of Energy (USDOE), in cooperation with Baldor Electric Company and other private partners, is developing a new grade of ultra-efficient and power-dense electric motors, with the goal of a 15% reduction in motor energy loss over NEMA premium motors. For example, if a NEMA premium motor with particular characteristics and output horsepower was 92% efficient and thus had 8% loss, this new grade of motor would reduce loss by 0.15 × 8% = 1.2%, for a new overall efficiency of 93.2%. The new grade of motor will also be 30% smaller in volume and 30% lower in weight, leading to decreased motor costs due to lower material costs (EERE, 2011).

17.2.3 Power Factor

As discussed in Chapter 5, power factor is important because customers whose loads have low power factor require greater generation capacity than what is actually metered. This imposes a cost on the electric utility that is not otherwise recorded by the energy and demand charges. Two types of power make up the total or *apparent power* supplied by the electric utility. Their relationship is shown in Figure 17.1. The first is *active* (also called *true*) *power*. Measured in kilowatts, it is the power used by the equipment to produce work. The second is *reactive power*. This is the power used to create the magnetic field necessary for induction devices to operate. It is measured in units of kilovolt-ampere reactive (kVAR). The vector sum of the active and reactive power is called *complex power*. Power factor is the ratio of the active power to the apparent power. The various types of power are summed up below:

- *Active power (true power) (P)*—Power that performs work, measured in watts (W)
- *Reactive power*—Power that does not perform work (sometimes called *wattless power*), measured in volt–ampere reactive (VAR)
- *Complex power*—The vector sum of the true and reactive power, measured in volt–ampere (VA)
- *Apparent power*—The magnitude of the complex power, measured in volt–ampere (VA)

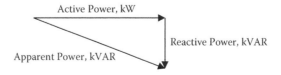

FIGURE 17.1
Vector relationship of AC power (power triangle).

The power factor of fully loaded induction motors ranges from 80 to 90%, depending on the type of motor and the speed of the motor. Power factor deteriorates as the load on the motor decreases. Other electrical devices such as space heaters and old fluorescent or high-intensity discharge lamps also have poor power factor. Treatment plants have several motors, numerous lamps, and often electric heaters, which, combined, lower the facility's overall power factor.

Power factor may be leading or lagging. Voltage and current waveforms are in phase in a resistive AC circuit; however, reactive loads, such as induction motors, store energy in their magnetic fields. When this energy gets released back to the circuit it pushes the current and voltage waveforms out of phase. The current waveform then lags behind the voltage waveform (see Figures 17.2 to 17.4).

Improving power factor is beneficial as it improves voltage, decreases system losses, frees capacity to the system, and decreases power costs where fees for poor power factor are billed. Power factor can be improved by reducing the reactive power component of the circuit. Adding capacitors to an induction motor is perhaps the most cost-effective means to correct power factor as they provide reactive power. Synchronous motors are an alternative to capacitors for power factor correction.

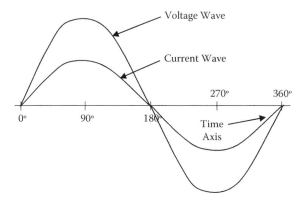

FIGURE 17.2
Voltage and current waves in phase.

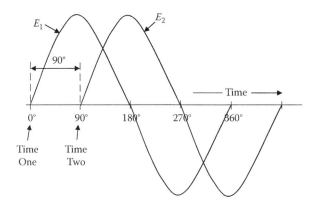

FIGURE 17.3
Voltage waves 90° out of phase.

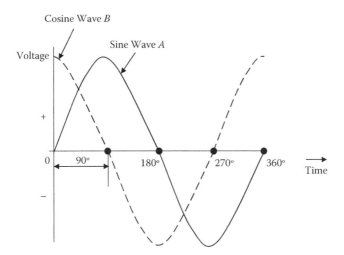

FIGURE 17.4
Wave B leads wave A by a phase angle of 90°.

Like induction motors, synchronous motors have stator windings that produce a rotating magnetic field; however, unlike the induction motor, the synchronous motor requires a separate source of DC from the field. It also requires special starting components. These include a salient-pole field with starting grid winding. The rotor of the conventional type of synchronous motor is essentially the same as that of the salient-pole AC generator. The stator windings of induction and synchronous motors are essentially the same.

In operation, the synchronous motor rotor locks into step with the rotating magnetic field and rotates at the same speed. If the rotor is pulled out of step with the rotating stator field, no torque is developed and the motor stops. Because a synchronous motor develops torque only when running at synchronous speed, it is not self-starting and hence requires some device to bring the rotor to synchronous speed. For example, the rotation of a synchronous motor may be started with a DC motor on a common shaft. After the motor is brought to synchronous speed, alternating current is applied to the stator windings. The DC starting motor now acts as a DC generator, supplying DC field excitation for the rotor. The load then can be coupled to the motor.

Synchronous motors can be run at lagging, unity, or leading power factor by controlling their field excitation. When the field excitation voltage is decreased, the motor runs at lagging power factor. This condition is referred to as *under-excitation*. When the field excitation voltage is made equal to the rated voltage, the motor runs at unity power factor. The motor runs at leading power factor when the field excitation voltage is increased above the rated voltage. This condition is *over-excitation*. When over-excited, synchronous motors can provide system power factor correction. Synchronous motors above 300 hp and below 1200 rpm are often less expensive than a comparable induction motor (Thumann and Dunning, 2008).

The feasibility of adding capacitors depends on whether the electric utility charges for low power factor. Corrective measures are infrequently installed, as many electric utilities do not charge small customers for poor power factor but rather price it into the electrical rates as a cost of business. A cost evaluation is necessary to determine the type of correction equipment to use. The evaluation should include motor type, motor starter, exciter (for synchronous motors), capacitors, switching devices (if needed), efficiency, and power factor fees (IEEE, 1990). Manufacturers should be consulted before installing capacitors to reduced voltage solid-state starters and VFDs, as there can be problems if they are not properly located and applied.

17.2.4 Variable-Frequency Drives

In Section 5.3, variable-frequency drives (VFDs) were discussed in detail. It was pointed out that VFDs are used to vary the speed of a pump to match the flow conditions. They control the speed of a motor by varying the frequency of the power delivered to the motor. The result is a close match of the electrical power input to the pump with the hydraulic power required to pump the water. This section expands on the earlier VFD discussion and addresses energy savings, applications, and strategies for wastewater pumping stations.

17.2.4.1 Energy Savings

Variable-frequency drives have been used by many wastewater utilities to conserve energy and reduce costs. A literature review found numerous success stories, with energy savings ranging from 70,000 kWh/yr for smaller WWTPs (average daily flows of 7 to 10 MGD) to 2,8000,000 kWh/yr for larger WWTPs (average daily flows of 80 MGD) (EERE, 2005b; Efficiency Partnership, 2009; EPRI, 1998). VFDs are now more available, and affordable; paybacks for VFDs range from 6 months to 5 years, depending on the existing level of control and annual hours of operations (SAIC, 2006). To approximate the potential energy savings, utilities should develop a curve of actual flow in hourly increments during a day. Using the curve, energy consumed by a constant-speed motor and throttling valve can be estimated and compared to energy consumed by a VFD system.

17.2.4.2 Applications

Variable-frequency drives can be installed at remote collection system pumping stations, at lift stations, on blowers, and on oxidation ditch aeration rotor drives. A common application of VFDs is for pumps that experience a large variation in diurnal flow, such as at wastewater pumping stations; however, if VFDs are not selected and applied correctly, they can waste energy. Operating below 75% for full load, VFDs can have very low efficiencies. When selecting a VFD, information should be obtained from the VFD manufacturer showing the efficiency at different turndown rates. VFDs are not applicable in all situations. VFDs may not be effective when a large static head must be overcome or where there is little variation in the flow rate (WEF, 2009). Additionally, some motors are not suited for use with VFDs. When the drive reduces the frequency to the motor, the voltage decreases, but the amperage increases which can generate heat. More commonly, voltage spikes that develop from the non-sinusoidal wave form produced by VFDs can damage motor insulation if not properly filtered. Conductors within the motor should be properly insulated, and the motors should be capable of dissipating the heat.

17.2.4.3 Strategies for Wastewater Pumping Stations

Variable-frequency drives can be costly to install in an existing pump station and require space in the electrical room. The range of flow, number of pumps, and hours of operation also have to be considered when evaluating the implementation of VFD control. Although equipping all pumps with VFDs provides maximum operation flexibility, this can be costly and, in retrofit projects, not always feasible. Often the rewards of having VFDs can be achieved at less cost with half of the pumps, or as few as one pump, being equipped. One VFD can be feasible in small stations where two pumps are run in duty/standby mode because the duty pump runs the majority of the time, reaping the savings with the VFD. In situations where both pumps are run in the lead/lag mode to cover the range of flow encountered, it is usually beneficial to have both pumps equipped with VFDs. This allows the pumps to alternate the lead position, which balances their hours, and it simplifies the controls as both pumps can be operated in the same manner.

In the case of larger stations with three or more pumps of the same size operated in lead/lag mode, the number of VFDs required depends on the range of flow and the space available. If one pump runs the majority of the time with infrequent assistance from the others, then one VFD would likely suffice; however, if the second pump operates frequently, then at least two VFDs are recommended. In the two-VFD scenario, when an infrequent peak flow is necessary, the third constant-speed pump can provide the base load while both VFD-driven pumps adjust to meet the demand. Depending on the size of the pumps, it could be more beneficial to install a smaller pump instead and run it with a VFD. This maximizes the efficiency of the system because when the large pumps are run they are near their BEP without the heat losses generated by VFDs.

Large stations with multiple pumps of different sizes need to be evaluated on a case-by-case basis. Typically, VFDs are placed on the smaller pumps so they can be used to fill in the peaks before another large pump is turned on. The controls are simple and sequencing is easy to maintain when a pump is down for service. Additionally, the cost is lower as small VFDs are less expensive than large ones.

It is important to run each pump periodically. Bearings in pumps that sit too long can be damaged from brinnelling. Brinnelling is a material surface failure caused by contact stress that exceeds the material limit; it occurs when just one application of a load is great enough to exceed the material limit, and the result is a permanent dent or "brinnell" mark. Also, stuffing boxes can dry out and leak. It is beneficial from an operations and maintenance (O&M) standpoint to exercise equipment at intervals recommended by the equipment manufacturer to ensure their reliability when called upon. Energy-wise, it is best to do this during off-peak electrical hours such as mornings or on weekends.

17.3 Design and Control of Aeration Systems

The aeration process can account for the largest energy demand of any operation at the facility. Although the demand is site specific and can vary widely from plant to plant, the fraction of energy used for aeration ranges from 25% to as much as 60% of total plant energy used (WEF, 2009). Because of the high energy use associated with aeration, energy savings can be gained by designing and operating aeration systems to match, as closely as possible, the actual oxygen demands of the process. Through improved understanding of the oxygen demands of a particular wastewater and how those demands fluctuate with time of day and season, wastewater treatment plants can build flexibility into their aeration systems so that operation can address real-time demands efficiently.

17.3.1 ECMs for Aeration Systems

Wastewater is aerated either by bubbling air or high-purity oxygen through it or by mixing it so that oxygen is transferred through contact with the atmosphere. The two most common types of aeration systems are *diffused aeration* and *mechanical surface aeration*. Hybrid systems that combine diffused air and mechanical mixing include jet systems, U-tube aerators, and submerged turbine aerators. The energy efficiency of an aeration system depends on several key factors:

- *Diffuser flux rate*—The rate of airflow per unit surface area of the diffuser (in standard cubic feet per minute per square foot of diffuser area). A minimum rate is typically required to uniformly distribute air to diffusers.

- *Oxygen transfer rate (OTR)*—The mass of oxygen dissolved in the mixed liquor per unit of time (in lb/hr). The OTR for clean water as determined by the manufacturer following standard test protocols is the *standard OTR, or SOTR*. The oxygen transfer rate under field conditions is designated by the subscript f (OTR_f).

- *Oxygen transfer efficiency (OTE)*—The mass of oxygen transferred to the liquid from the mass of oxygen supplied (expressed as a percentage). Similar to OTR, the transfer efficiency as determined by the manufacturer for clean water at a given gas flow rate and power input is called the *standard oxygen transfer efficiency, or SOTE*. The transfer efficiency for field conditions is often designated by the subscript f (OTE_f).

- *Alpha*—This is the ratio of oxygen transfer efficiency in wastewater vs. clean water.

- *Mixed liquor dissolved oxygen concentration*—Although not strictly a design factor, one of the most significant and controllable factors affecting aeration energy efficiency is the mixed liquor dissolved oxygen concentration. The closer the dissolved oxygen concentration is to saturation, the greater the resistance for dissolved oxygen dissolution and the lower the OTE.

Designers try to maximize the OTE_f under most operating conditions so that the plant will operate efficiently. OTE_f depends on a number of external factors, including water temperature and site elevation. It decreases with increasing concentration of solids and surfactants. Within the basin itself, it increases from the inlet to the outlet as organic material is biodegraded. It increases with decreasing flux rate and is generally higher for deeper basins. Although commercially available aeration equipment has a wide variety of SOTE values, fine-pore diffusers have the highest efficiency compared to any other diffused air or mechanical aeration system.

Many different basin configurations and a variety of aeration equipment can be used to improve aeration efficiency. No single approach is appropriate for every system. Life-cycle cost analysis should always be conducted to ensure that ECMs are appropriately factored into the decision-making process.

17.3.1.1 ECMs for Diffused Aeration Systems

Diffused aeration systems introduce air below the wastewater surface. Major components of diffused aeration systems are the air intake system, blowers, the air piping system, diffusers, and controls. Air intake systems are typically equipped with filters to protect blowers and diffusers from airborne particulates. Blowers are essentially lower pressure, high-volume air compressors. Common types are *positive displacement* and *centrifuge*. Air piping systems deliver air from blowers to the diffusers. Head loss within the air piping system is typically a small portion (<10%) of total system pressure (WEF/ASCE/EWRI, 2010). Aeration control systems are key to keeping the aeration system operating efficiently over the entire operating range of the system. Conventional ECMs for aeration systems include: (1) proper sizing of blowers, (2) dedicated blowers for channel aeration, and (3) configuration of diffusers within a basin. A fourth method emerging is intermittent aeration. Each is discussed separately below.

17.3.1.1.1 *Proper Sizing of Blowers*

Many blower configurations can operate efficiently, especially in larger plants. In general, blower systems should be designed for a minimum 5:1 turndown ratio, meaning that a system should be capable of operating at one-fifth of its full capacity (Cantwell et al., 2009). Some common arrangements that provide for efficient blower operation with backup are to design for four blowers at 33% each of design flow, or two blowers at 25% each of design flow plus two blowers at 50% each of design flow.

Many plants have more capacity in their aeration system than needed because the population growth projected during the design phase (a factor in designing treatment plant capacity for a 20-year future projected loading) has not yet occurred or will never occur. In other cases, changes in local industries or aggressive pretreatment programs may have resulted in decreased organic loading and reduced aeration requirements. In these situations and others, it may not be possible for the existing aeration system to operate efficiently to meet the existing load. Using an aeration system to supply parasitic loads, such as channel air and air lift pumps, further increases the design capacity of the system. These applications, which typically require lower pressure, can often be more efficiently served by a smaller blower. Determining the actual process air requirements without parasitic loads allows a more efficient system to be designed.

The energy savings associated with retrofits to increase blower turndown depends on many factors, including where the plant is within its design life and how closely project growth matches actual growth. If the original design was oversized, energy savings can be significant. Several methods are available to reduce energy use in these situations, such as replacing larger blowers with one or more smaller units or installing variable-frequency drives. Inlet throttling may be applicable, depending on the blower type, to modulate the air flow rate of existing blowers. Box 17.2 shows how one utility was able to reduce energy use by approximately 1,000,000 kWh/hr by installing smaller blowers.

17.3.1.1.2 *Dedicated Blowers for Channel Aeration or Air Lift Pumps*

The air for channel aeration or air lift pump processes is often tapped from the main aeration system air header; however, particularly for channel aeration, the pressure required is significantly lower than the main aeration system pressure. This excess pressure is usually reduced by throttling the air through a flow control valve. By providing a small blower rated at the specific pressure required of the service, the energy requirement can be reduced. This approach is usually viable only for large plants, as the payback period to implement it in smaller plants is too long to make the change feasible.

17.3.1.1.3 *Configuration of Diffusers*

The configuration of diffusers within an aeration basin should allow for maximum operational flexibility to respond to varied conditions and treatment goals. It is also important that their layout promotes plug flow within the basin and reduces short-circuiting. A common approach is to use tapered aeration to reduce the rate of oxygen supply along

---

BOX 17.2. CITY OF WAUKESHA REPLACED EXISTING BLOWERS WITH SMALLER UNITS TO IMPROVE EFFICIENCY

Background: The City of Waukesha is a medium-sized community of approximately 70,000 residents located 15 miles west of Milwaukee, Wisconsin. The City's conventional activated sludge wastewater treatment facility treats between 10 and 12 million gallons per day (MGD) and has a design flow of 18.5 MGD. The plant chemically removes phosphorus by adding ferric chloride in a tertiary treatment process prior to filtration, ultraviolet (UV) disinfection, and discharge to the Fox River. The plant's six aeration basins were equipped with ceramic fine-bubble diffusers. Five 700-hp, inlet-throttled centrifugal blowers provided air to the aeration system. Since the original plant design, the City began aggressively enforcing their industrial pretreatment program. In addition, some industries closed or moved out. The combined effect was a significant reduction in organic loading to the plant. With just one blower running, dissolved oxygen (DO) concentrations were high, between 4.5 and 8.0 milligrams per liter (mg/L). The facility could not achieve sufficient turndown of the blower prior to implementing energy efficiency improvements.

Energy Efficiency Upgrades: In July 2003, the plant replaced two 700-hp blowers with two 350-hp blowers so they could operate at a lower DO concentration. They also upgraded their DO probes with new membrane units and replaced the existing single-loop PID DO control system with an integrated direct-flow control system with most open valve (MOV) control logic. During the upgrade, they took three of the six basins out of service.

Energy Savings: Total energy savings from the blower replacement were substantial at approximately 1,000,000 kWh per year (approximately 9% of total plant energy use), with an associated annual energy cost savings of more than $65,000. Total implementation costs were approximately $200,000, resulting in a simple payback of about 3 years.

—Cantwell et al. (2009)

---

the length of a basin (WEF/ASCE/EWRI, 2010). It can be accomplished by placing more diffusers at the inlet to the basin where the organic loading is highest and decreasing the number of the diffusers along the basin's length. Tapered aeration better matches the oxygen demand across the basin by providing more air to the head of the basin where it is needed and less air near the end of the basin where the food-to-microorganism ratio (F/M) is lower, thereby saving energy.

If an existing aeration system is underperforming, utilities should examine the configuration of diffusers to identify possible causes and potential improvements. Changes in the number of diffusers and diffuser configuration could lead to increased energy efficiency. For example, the Waco Metropolitan Area Regional Sewer System (WMARSS) treatment facility in Waco, Texas, was not meeting its nitrification goals with the plant's existing fine-bubble aeration system. An analysis of the facility operations revealed that the aeration system was being operated in excess of the maximum air flow rate of the diffusers, producing coarse bubbles instead of fine bubbles, which reduced the oxygen transfer efficiency. The analysis also concluded that additional diffusers were required to effect nitrification. The utility installed 700 additional diffusers in each of the plant's five aeration basins, bringing the total number of diffusers in each basin to 3500. This modification, along with implementation of automated DO control, has reduced energy consumption by an average of 4,643,000 kWh per year (an average 33% reduction) and had a payback period of less than 3 years. See Appendix I for a summary of the treatment plant energy-use improvements and cost information for the Waco facility.

17.3.1.1.4 Intermittent Aeration

Intermittent aeration saves energy by reducing the number of hours that an aeration system operates or the aeration system capacity. It is not appropriate for all facilities, especially those at or near capacity, and must be evaluated on a case-by-case basis so as not to adversely impact the treatment process. The methodology involves momentarily stopping air flow to an aeration zone or cycling air flow from zone to zone. The cycle length can be controlled with DO concentration or can be strictly time based. When controlling with the DO concentration, air flow is turned off at a set high level and turned back on based on a lower limit. The cycle length on time-based systems is strictly controlled by a set maximum time. Many basins are limited by mixing, which must be considered when setting the maximum length of time that the air can be turned off. Additionally, settling of solids within the basin should be factored in the cycle length.

17.3.1.2 ECMs for Mechanical Surface Aerators

Mechanical surface aerators vigorously agitate the wastewater, transferring oxygen from the air by increasing the water–atmosphere interface. Common types of equipment include low-speed mechanical aerators, direct-drive surface aerators, and brush-type surface aerators. Slow-speed mechanical aerators are used in both pond systems and the activated sludge process. In ponds, they are mounted on floats and held in position using guy wires. Mechanical aerators are mounted above the wastewater on a platform in the middle of aeration basins. A shaft extends down through the platform into the tank to mix the wastewater. Brush aerators are used in oxidation ditches where, in addition to providing mechanical aeration, they impart the horizontal velocity necessary to keep the contents of the ditches moving and particles in suspension.

In general, ECMs for mechanical aerators are conventional retrofits. One ECM identified in the literature is the ability to adjust the submergence of fixed mechanical mixers through the use of adjustable weirs. Oxygen transfer can be improved and energy use is reduced by installing motor-operated weirs that change the submergence of the impeller based on the dissolved oxygen concentration (WEF, 2009). Thus, submergence of the impeller delivers more or less oxygen in response to real-time conditions, resulting in energy savings. In general, radial-flow, low-speed mechanical aeration systems can provide higher aeration efficiency than high-speed machines (WEF/ASCE/EWRI, 2010).

Cycling aerators off during nighttime hours can be effective in reducing aeration energy use in pond systems with multiple surface aerators. As the influent load to the plant decreases in the evening, the DO concentration rises. This is a potential opportunity to decrease surface aeration. Operationally, it is better to cycle the aerators so each aerator is off for only a short time before another is put in service. As ponds come in different shapes and sizes, the number of aerators and determination of which aerators to turn off must be carefully evaluated on a case-by-case basis to prevent settling and the generation of odor.

A new development in mechanical aerators is the use of multiple impellers. Single impeller mechanical aerators are limited in their turndown due to the need to keep the contents of the basin from settling. A dual-impeller aerator by Eimco Water Technologies includes a lower impeller near the bottom of the basin floor to augment the surface impeller. This provides additional mixing energy near the floor of the basin, permitting greater power turndown when a VFD is used and associated energy savings. Data from full-scale installations were not identified through a literature review; thus, potential energy savings have not been quantified and this technology remains classified as an emerging ECM.

17.3.2 Control of the Aeration Process

Control of the aeration process is critical to efficient operation of wastewater treatment plants as both over- and under-aeration have detrimental effects. The energy wasted on over-aeration mounts quickly, as the energy expended increases exponentially with increasing DO concentrations. The DO concentration required to maintain stable biological activity is site specific but usually ranges from 1.0 to 2.0 mg/L for activated sludge systems and can be as low as 1.0 mg/L for nitrification. As noted previously, operating at DO concentrations closest to saturation increases the resistance of dissolved oxygen to dissolution. This both lowers the oxygen transfer efficiency (OTE) and increases the energy expended to drive oxygen into solution. In addition to wasting energy, the following operation problems have been reported in association with excess dissolved oxygen:

- Poor biosolids settling
- Increased foam caused by filamentous organisms (can also occur at low DO)
- Negative impacts on the anoxic zone of a biological nitrogen removal system due to high DO levels in the recycle flow

Under-aeration can lead to underperformance of the activated sludge process, bulking issues, and, in some cases, issues with struvite (a phosphorus precipitate) formation in sludge processing resulting from unwanted biological phosphorus removal. The key point is to have good control over dissolved oxygen levels so the aeration system supplies only what is needed.

17.3.2.1 Automated DO Control

Automated control of the aeration process is an important ECM that can save a plant considerable energy by quickly adjusting to variable conditions within the basin. The oxygen required to maintain biological processes (i.e., oxygen demand) within the aeration basin is proportional to organic and ammonia loading in the influent wastewater. Oxygen demand for aeration, therefore, follows the same diurnal pattern, dipping in the middle of the night and peaking in the morning and evening. The ratio of peak to minimum oxygen demand can typically be 2:1 (Cantwell et al., 2009), although it can be much higher for small systems and resort communities. Intermittent discharge of ammonia-rich supernatant from

---

DID YOU KNOW?

Struvite is a phosphate mineral chemical compound (magnesium ammonium phosphate; chemical formula $NH_4MgPO_4 \cdot 6H_2O$) that forms as hard crystals inside the pipes in wastewater treatment plants, where it creates expensive maintenance problems.

biosolids dewatering operations can also dramatically increase the oxygen demand in the basin. Conversely, dilution from stormwater flow can reduce oxygen needs. In addition to fluctuating oxygen demand of the wastewater itself, the oxygen transfer efficiency in the basin also varies in response to changing air and water temperature and other wastewater characteristics such as concentrations of solids and surfactants.

In the past, wastewater treatment operators took field measurements to determine the DO concentration in the aeration basins. Based on the results, operation modifications were made (e.g., to blowers or aeration system valves) to increase or decrease the oxygen being delivered to the basins based on target setpoints. This was typically done only a few times (or once) per day and would not closely reflect diurnal variations in DO demand. In addition, a high safety factor was often applied to ensure that the DO level did not decrease below the target concentration should the influent wastewater characteristics change quickly.

To more closely match the air delivered to the biological process oxygen demand, utilities now commonly install automated control systems. Some new blowers come with automated control for reliable operations and enhanced energy savings (blower technologies are discussed in Section 17.4). Because the energy required increases exponentially as the DO concentration increases, energy savings from automated DO control can be significant. It has been estimated, for example, that tight control of DO in the aeration process can save a wastewater plant between 10 and 30% of total energy costs (WEF/ASCE/EWRI, 2006). Energy savings will be site specific and are highly dependent on the control system in place prior to the upgrade to automated process control. For medium to large WWTPs, the payback period for installing automated DO control is generally within a few years (WEF, 2009).

17.3.2.1.1 *How Automated DO Control Works*

Automated DO control system use real-time dissolved oxygen concentration readings from DO probes located within the aeration basins as inputs to a process controller. The process controller provides control output to the aeration system that responds by adjusting the brush rotor or blower speed, the position of variable vane diffusers on the blower, or the position of the drop-leg control valves at the basin to deliver the proper amount of air to maintain the target DO concentration. A simple control system might use one DO probe and one target DO concentration for all aeration basins. A more complex control strategy involves individual DO probes and air header control valves for each basin or stage within each basin. Individual target DO concentrations for each basin or stage can further increase energy savings. Major components of an automated DO control system include the following:

- *DO probes*—Typical configurations are membrane (most common), galvanic, or new optical technology (see below for a detailed discussion). Probes should be installed in each aeration basin near the center or close to the inlet of a plug-flow basin.

- *Blower air flow control*—The total air flow supplied to the system is controlled by modulating the air flow rate delivered by the blowers. The control mechanism depends on the type of blower. Positive displacement blowers can use VFDs to modulate air flow. Air flow for multistage centrifugal blowers is often controlled by inlet throttling; however, VFDs also can be used to improve efficiency and turndown. New single-stage centrifugal blowers used variable speed, inlet guide vanes, and variable discharge diffusers to modulate flow for enhanced energy efficiency. See Section 17.4 for more information on blower technology and air control.

- *Basin air flow control*—The total air flow supplied by the blowers is divided between multiple aeration tanks and multiple grids in each tank. The air flow in each zone should be proportional to process demand in each zone. In small facilities, basin air flow control is often done manually. In larger facilities, automatically controlled air flow valves can be used to continuously modulate air flow as DO concentrations change. In the largest facilities, automatic control may also be provided for individual zones. Most open valve (MOV) control can be used to automatically adjust header pressure so as to maintain a most open valve at an essentially full open position and minimize system pressure and energy.
- *Process control system*—The aeration system process controllers receive information from the DO probes, process results (i.e., compares the basin readings to setpoints), and send signals to air control mechanisms to make a change if needed. Most systems are composed of programmable logic controllers (PLCs), usually networked together by a supervisory control and data acquisition (SCADA) system. In larger facilities, a distributed control system (DCS) is sometimes used, combining local controllers and computer-based operator interface.

Automated DO control systems typically use some form of a feedback control loop, whereby blower and aeration basin air flow rates are manipulated in response to changes in the DO level in the aeration basin. Control strategies can be very simple, such as an on/off or setpoint control, or complex based on proprietary algorithms. A common strategy for automated DO control is a cascaded control system. In the first loop of cascade control, the process controller sends a signal to the basin air flow control loop based on the DO probe readings in a basin; for example, if the DO reading is below the target, the controller will require more air in the basin. In some cases, the basin flow control valve is manipulated directly by the DO control loop. In most cases, a flow meter and separate air flow controller are provided. In this type of system, the output of the DO control loop is the setpoint for the air flow controller. The second loop is established between a pressure transducer on the main header and the blower system. Pressure in the line will naturally increase or decrease based on modulations of the basin air flow control valve. If the basin valve is opened, the header pressure will decrease and the pressure control loop will send a signal to the blower controller to increase blower air flow. That is, the output of the pressure control loop is the setpoint of the blower air flow controller.

17.3.2.1.2 DO Measurement Equipment

Dissolved oxygen can be measured by membrane electrodes, galvanic electrodes, and optical DO technology (fluorescence or luminescence). Membrane electrodes, historically the most common DO measurement device, are composed of two metal electrodes separated from a test solution by a membrane. As oxygen permeates the membrane, the cathode reduces it and creates a potential that can be correlated to the amount of dissolved oxygen in the system. They are fairly reliable but must be calibrated frequently, typically monthly or weekly depending on the manufacturer and site conditions. The membranes must also be replaced fairly frequently (often quarterly but can be more frequently) (WEF/ASCE/EWRI, 2006). This can be a time-consuming and tedious activity for operators. Galvanic electrodes, such as the proprietary Zullig probe, apply a galvanic current to measure the oxygen. This type of probe requires significantly less maintenance than membrane-style probes and can obtain a slightly better energy savings as it maintains its accuracy longer.

17.3.2.1.3 Calibration and Maintenance on Membrane-Type Dissolved Oxygen Probes

Most dissolved oxygen probes can be field calibrated to match a known DO concentration reading. A laboratory unit can be used to measure the mixed liquor DO concentration, and this value can be entered into the transmitter using a menu setting. Membrane-type DO probes also have an air calibration available. In this method, the clean probe is exposed to air and the transmitter automatically adjusts the display and output to match the known concentration of oxygen in ambient air. In mixed liquor, all types of DO probes can accumulate deposits of biological growth or grease. These deposits cause inaccuracy in the DO concentration measurement. Frequency of cleaning varies from once per week to once per month, depending on site conditions. Cleaning of most DO probes is accomplished by removing the probe from the mixed liquor and wiping it with a damp cloth. In installations with a large number of DO probes, utilities should consider the use of self-cleaning probes. These employ air blasts, water spray, or mechanical wipers to periodically and automatically remove deposits from the face of the probe (USEPA, 2010).

The newest technology on the market, the optical DO probe, measures changes in light emitted by a luminescent or fluorescent chemical and relates the rates of change in the emission to the DO concentration in solution. They work on the principle that DO quenches both the intensity and duration of the luminescence or fluorescence associated with certain chemical dyes. Thus, the duration of the dye luminescence or fluorescence is inversely proportional to the dissolved oxygen concentration. Several manufacturers offer optical DO probes including Hach, Orion, YSI, Insite IG, Endress & Hauser, and Analytical Technologies, Inc.

The optical DO probe offers several advantages over the traditional membrane probe that make it a good candidate for automated DO control systems. The optical DO probe does not consume electrolyte and requires less frequent calibration. There are no membranes to replace, so maintenance requirement are low, with only the sensor cap requiring replacement approximately once per year (WEF/ASCE/EWRI, 2006). Accuracy and reliability are also generally greater for the optical DO probe compared to the membrane probe.

Using optical DO probes instead of traditional membrane probes in automated DO control systems is not considered an ECM itself; however, a more reliable and easy to use instrument could pave the way for increased automated DO control installations. Brogdon et al. (2008) reported on energy savings realized by a Tennessee Valley Authority (TVA) demonstration project to advance the use of optical DO probes and variable-speed drives for automated DO control among small- to medium-sized utilities. Energy savings associated with the projects ranged from 14 to 40%.

Upgrading a system with optical DO probes is often combined with other aeration system upgrades to ensure reliable operation. The Bartlett Wastewater Treatment Plant #1 in Tennessee implemented VFD control on one of the two operating rotors in each of the plant's oxidation ditches using DO readings from optical probes. Prior to implementing this modification, one rotor in each ditch was operated at constant full speed and the second rotor in each ditch was manually activated during peak flow periods. Following implementation of the ECM, the need to run the second rotor during peak flow conditions was eliminated. The ECM reduced total plant electrical energy use by approximately 13% and saved the utility more than $9000 per year. A case study of the Bartlett ECM project is presented in Appendix D.

17.3.2.1.4 Advances in DO Control Strategies

Although automated DO control is a significant improvement over manual control, it has some limitations. It takes time for the DO concentration in the basin to change in response to a change in organic or ammonia loading (at least a few minutes, but it can be much longer). It also takes time for the process to reach equilibrium after the airflow is increased or decreased. This can cause the valve position to repeatedly open and close before DO in the basin has stabilized. For example, an increase in airflow could overshoot the DO target, causing a second manipulation of valve position and airflow rate to reach the target. Repeated adjustments to find the DO setpoint is commonly referred to as *hunting*. Operators tune the control system to reduce hunting; however, conservative tuning can make the system unresponsive to changes within the basin. Tuning is made all the more challenging by the nonlinear relationship between DO concentration and air flow to the basin. New advances in DO control algorithms attempt to address these issues. Two proprietary ECMs that are emerging for automated DO control are discussed below: integrated air flow control and automated SRT/DO control. Also provided is a description of the most open valve (MOV) control methodology. Alternatives to DO-based control are discussed in Section 17.3.3.

17.3.2.1.4.1 Most Open Valve Control   The goal of MOV control is to avoid excessive throttling on the discharge side of the blowers because it is not energy efficient to build pressure and then waste it across throttling valves. The amount of throttling should be limited to what is required to properly split the air flow. This is accomplished by ensuring that the control butterfly valve serving the zone with the highest oxygen demand is essentially fully open. MOV is now commonly integrated into new aeration control systems (see Appendix F). It may not be cost effective as an add-on to an existing control system, but utilities should consider specifying for it when upgrading blowers or aeration controls.

17.3.2.1.4.2 Integrated Air Flow Control   Integrated air flow control is a proprietary aeration control system that was developed by ESCOR (Energy Strategies Corporation, now part of Dresser, Inc.) that eliminates the pressure control loop common in many automatic DO control systems. Particularly in smaller systems, the pressure control loop can cause instability in the operation of the blowers and control valves (cyclic oscillation, or hunting) as the control system attempts to adjust air flow and pressure in response to changes in the process and ambient air conditions. Full-scale implementation of integrated air flow control has resulted in better stability and simplified tuning of the aeration process, leading to more efficient blower operation. The Narragansett Bay Commission's Bucklin Point facility implemented an ESCOR aeration control system following an upgrade of their 46-MGD facility in 2005. As part of the upgrade, the aeration system was reconfigured to improve the plant's Modified Ludzack–Ettinger (MLE) process to help meet biological nutrient (nitrogen) removal goals. The upgraded plant, operating with two of the plant's three 600-hp blowers (one in standby/spare status), had difficulty maintaining consistent nitrogen removal primarily due to inadequate air supply control. Implementation of the ESCOR integrated air flow control system, unlike the system's original pressure-based control system, employs direct flow control of the blowers. As the DO in the aeration basin varies from setpoint, the required incremental changes in air flow are used to modify both aeration drop leg air flow and blower air flow. The control system's MOV logic directly manipulates basin air flow control valve positions to ensure that at least one valve is always at maximum open position, thereby minimizing system pressure without using a pressure control setpoint. The reduced complexity compared to the pressure control system results in more robust and accurate control, and elimination of the pressure control loop minimizes tuning. Implementation of the integrated air flow control provided the required DO control to meet the plant's total nitrogen discharge requirements and reduced electricity consumption at the facility an average of approximately 1,247,000 kWh per year (an average reduction of 12%) in the first 3 years of operation following commissioning at the end of 2006 (a savings of nearly $236,000). This energy saving was the result of eliminating the need to constantly run the second of the two plant blowers.

17.3.2.1.4.3 Automatic SRT/DO Control   Dissolved oxygen and sludge age (solids retention time, SRT) are two of the most important operating parameters in activated sludge treatment. Although reducing DO in the aeration process affects energy savings (i.e., less DO lowers the energy consumption of the blowers), it often requires increasing SRT to compensate for the deterioration in process performance. Increasing the sludge age in an activated sludge process, however, can lead to an increase in the sludge settling volume index (SVI), which can increase the plant's effluent total suspended solids (TSS). Ekster Associates developed a proprietary algorithm (see Box 17.3), OPTIMaster™, which is based on activated sludge modeling, plant historical data, and statistical process control. It provides setpoint optimization for sludge age and DO and automates control of these parameters (through automatic sludge wasting and blower output adjustment) to optimize aeration. The algorithm selects sludge age and a range of mixed liquor suspend solids (MLSS) and DO concentrations to maintain the proper SVI at minimum aeration. Oxnard, California, implemented the OPTIMaster™ system in 2006 and reported a reduction of approximately 20% in the total plant's electrical energy use. A case study of the Oxnard plant's implementation of the OPTIMaster™ system is provided in Appendix E.

17.3.2.2 Emerging Technologies Using Control Parameters Other Than DO

Instead of monitoring and control based on DO concentrations in the aeration basin, another innovation is to take alternative measurements of biological activity and use this information for process control. This section describes three emerging ECMs for automated control of the aeration process using a measurement parameter other than dissolved oxygen: respirometry, critical oxygen point control, and off-gas monitoring.

17.3.2.2.1 Respirometry

Respirometry involves measuring the oxygen uptake rate (OUR) by a biological treatment culture. In bench-scale respirometry experiments, a sample of mixed liquor representing the biomass in the aeration basin, possibly amended with an organic substrate or ammonia, is placed in a sealed vessel. The rate of oxygen consumption within the vessel is monitored over time. A review of the literature suggests that online control of aeration using respirometry is possible but has not been successfully implemented on a full-scale basis. Online respirometers also require a representative sample of biomass from the aeration basin. This source is typically a fresh sample from the mixed liquor, the return activated sludge line, or an offline pilot reactor (Love et al., 2000). The sample is contained in a well-mixed batch reactor or flow-through system. Oxygen consumption is measured over time (either for the liquid phase or a sealed gas phase). OUR is based on a mass balance on either the liquid phase or both the gas phase and the liquid and gas phases within the respirometers. It is important that sufficient oxygen be present in the liquid or gas phase to prevent oxygen limited conditions (Love et al., 2000).

BOX 17.3. ALGORITHMS: WHAT ARE THEY?

An *algorithm* is a specific mathematical calculation procedure. More specifically, "an algorithm is any well-defined computational procedure that takes some value, or set of values, as input and produces some value, or set of values, as output" (Cormen et al., 2002). In other words, an algorithm is a recipe for an automated solution to a problem. A computer model may contain several algorithms. The word *algorithm* is derived from the name al-Khowarizmi, who was a ninth-century Persian mathematician.

Algorithms should not be confused with computations. Whereas an algorithm is a systematic method for solving problems and computer science is the study of algorithms (although the algorithm was developed and used long before any device resembling a modern computer was available), the act of executing an algorithm—that is, manipulating data in a systematic manner—is called *computation*.

The following algorithm for finding the greatest common divisor of two given whole numbers, attributed to Euclid (c. 300 BC) may be stated as follows:

- Set *a* and *b* to the values *A* and *B*, respectively.
- Repeat the following sequence of operations until *b* has value 0:
  - Let *r* take the value of *a mod b*.
  - Let *a* take the value of *b*.
  - Let *b* take the value of *r*.
- The greatest common divisor of *A* and *B* is the final value of *a*.

The operation *a mod b* gives the remainder obtained by dividing *a* by *b*.

Note that the problem, that of finding the greatest common divisor of two numbers, is specified by stating what is to be computed; the problem statement itself does not require that any particular algorithm be used to compute the value. Such method-independent specifications can be used to define the meaning of algorithms; in other words, the meaning of an algorithm is the value that it computes. Moreover, there are several methods that can be used to compute the required value; Euclid's method is just one. The chosen method assumes a set of standard operations (such as basic operations on the whole number and a means to repeat an operation) and combines these operations to form an operation that computes the required value. Also, it is not at all obvious to the vast majority of people that the proposed algorithm does not actually compute the required value. That is one reason why a study of algorithms is important—to develop methods that can be used to establish what a proposed algorithm achieves.

17.3.2.2.2 *Mass Balance and Measuring Plant Performance*

The simplest way to express the fundamental engineering principle of *mass balance* is to say, "Everything has to go somewhere." More precisely, the *law of conservation of mass* says that when chemical reactions take place matter is neither created nor destroyed. This important concept allows us to track materials (e.g., pollutants, chemicals, microorganisms) from one place to another. The concept of mass balance plays an important role in treatment plant operations (especially wastewater treatment) where we assume a balance exists between the material entering and leaving the treatment plant or a treatment process: "What comes in must equal what goes out." The concept is very helpful in evaluating biological systems, sampling and testing procedures, and many other unit processes within the treatment system.

Online respirometric control has a theoretical advantage over traditional automated dissolved oxygen control. Whereas DO is essentially an "after-the-fact" analysis once the oxygen needs of the biomass have been met, OUR as measured by a respirometer is a more direct measure of biomass needs and can be used to predict oxygen requirements for wastewater as it enters the basin. A study at the James C. Kirie Water Reclamation Plant in Chicago showed that respirometric control is technically feasible using a feed-forward control strategy (Tata et al., 2000). Online respirometers were installed in one aeration basin to determine OUR. Researchers used plant data and literature values to develop two semi-theoretical mathematical models to predict the aeration rate as a function of average basin OUR. Side-by-side experiments were conducted to compare the plant's existing automated DO control strategy to an experimental control algorithm based on OUR. Reduced airflow based on the OUR control strategy caused a reduction in effluent quality, leading Tata et al. (2000) to conclude that, even though online respirometric control is technically feasible, more work would be needed to configure an optimal control system.

Although respirometry has been used widely in the United States to determine the kinetics of aerobic biological processes, online respirometric control of aerobic treatment processes is not common. Trillo et al. (2004) evaluated the use of respirometry and noted the following limitations:

- Most respirometry analytical devices do not provide true, real-time measurements but rely instead on cyclic sampling and analysis.
- The devices require high maintenance because they utilize sampling pumps and require replenishment of chemical reagents.
- The technique requires conditioning of samples or changes in mixed liquor conditions that may lead to results that are not representative of actual process conditions.

The most appropriate applications may be sequencing batch reactors and oxidation ditches (WEF/ASCE/EWRI, 2006). More commonly, respirometric measurements have been used to create diurnal load profiles, which are then used as an input to more common DO control strategies. In 2002, the International Water Association (IWA) published a report entitled "Respirometry in Control of the Activated Sludge Process: Benchmarking Control Strategies" (Copp et al., 2002). This report contains an evaluation of current control devices and a protocol for evaluating aeration control strategies using respirometry. The reader is directed to this report for detailed analysis and recommendations for online respirometric measurement.

17.3.2.2.3 *Critical Oxygen Point Control Determination*

Critical oxygen point control is a control method based on respirometric measurements. The theory is as follows: Bacteria respire by diffusion of oxygen across their cell wall. Oxygen diffuses from a high concentration external to the bacterial cell wall to the low concentration internal to the bacterial cell. Diffusion will only take place when the oxygen concentration differential across the cell wall is sufficient to drive the oxygen through it. The minimum concentration at which this occurs is called the *critical oxygen point*. Below the critical oxygen point, the biodegradation rate will rapidly decrease. At the critical oxygen point, the biodegradation rate will be at a maximum for the available food source (i.e., organic compounds and ammonia in the wastewater being treated). Exceeding the critical oxygen point will not materially affect the biodegradation rate. For carbonaceous bacteria, this critical oxygen point is very distinct. Accurately knowing the critical oxygen point for the active biomass allows the optimal DO setpoint to be determined. Strathkelvin Instruments (North Lanarkshire, Scotland) has developed a proprietary software upgrade to their Strathtox line of respirometers that, in real time, determines the critical oxygen point of the wastewater under aeration and utilizes that data to change the DO setpoint to control optimum delivery of oxygen in the aeration basins.

17.3.2.2.4 *Off-Gas Analysis*

Off-gas testing is a standard test for determining in-process oxygen transfer efficiency (OTE) based on a gas-phase mass balance of oxygen entering the aeration basin and oxygen leaving the basin at the wastewater surface. It has been historically used for evaluating aeration system performance but has recently received attention as a parameter for aeration system control. A feed-forward, off-gas monitoring and control system was tested successfully at the Grafton WWTP in Wisconsin (Trillo et al., 2004). The Grafton WWTP treats 1.1 MGD on average, using two parallel aeration basins equipped with fine pore diffusers. Multistage centrifugal blowers provide the air flow to the system. The off-gas control system consists of a stainless steel hood for collecting a representative sample of the aeration system off-gas, a sample conditioning and transport system, gas sensors, and a programmable logic controller (PLC). In 2001, the plant began operating the new off-gas control system in one of its two aeration basins. Trillo et al. (2004) reported the following advantages of the off-gas control system compared to operation of the conventional feedback-based DO control system:

- It resulted in smaller aerations in basin DO (standard deviation of 0.12 mg/L compared to 0.36 mg/L for feedback-based DO control for typical 6-day performance).
- The effluent DO setpoint was reduced from 2.0 mg/L to 1.75 mg/L.
- Recovery time after power loss was reduced by 50%.

The authors did not present a side-by-side comparison of energy use for the two treatment trains but postulated that the feed-forward off-gas control could reduce energy use by more than 20% compared to conventional feedback-based DO control systems. See Trillo et al. (2004) for additional information.

17.3.3 Innovative and Emerging Control Strategies for Biological Nitrogen Removal

The bacteria responsible for biological nitrification (referred to as *nitrifiers*, or nitrifying bacteria) exhibit significant biological diversity. Many can operate at low DO concentrations, particularly following alternating anoxic and aerobic environments (Littleton et al., 2009). Thus, relying on automation of DO alone may not result in the most energy-efficient system. Although advanced control of nitrification using multiple measurement parameters such as ammonia and nitrate and nitrite has been growing overseas, there are still few full-scale applications in the United States. Still, several proprietary control systems are on the market and have been tested at full-scale WWTPs. The two described in this section are the SymBio® process and the Bioprocess Intelligent Optimization System (BIOS).

17.3.3.1 SymBio®

The SymBio® process by Eimco Water Technologies uses online monitoring of nicotin-amide adenine dinucleotide (NADH) to determine changes in biological demands. Based on the results, air flow to the basin is controlled to promote simultaneous nitrification and denitrification (SNdN) of wastewater. SNdN refers to a condition in an activated sludge or biofilm process in which the positive bulk liquid DO concentration is low enough (typically below 1.0 mg/L) that the DO diffusing into the floc is removed before it can penetrate the entire floc depth. Thus, nitrification is occurring on the exterior portions of the floc and denitrification is occurring in the anoxic, interior portion, allowing for total nitrogen removal. The monitoring device, the NADH sensor, uses a fluorescence sensor to detect changes in NADH and provide information on the status of biological wastewater treatment processes. Weerapperuma and de Silva (2004) reported that the NADH sensor requires minimal maintenance and can provide real-time information for process control. The manufacturer claims a 25 to 30% energy savings compared to nitrifying plants without this control technology; however no independent data from full-scale facilities has been published to verify these claims.

17.3.3.2 Bioprocess Intelligent Optimization System (BIOS)

BIOS is a proprietary control algorithm, online process simulation program originally developed by Biochem Technology, Inc., to optimize the operation of a Modified Ludzack–Ettinger (MLE) biological nitrogen removal process. Because the denitrification and nitrification sections of the MLE process are an integral part of many other biological nutrient removal processes (e.g., four- and five-stage Bardenpho and A^2O processes), the BIOS control system can be applied to other processes having the MLE component.

BIOS is a feed-forward optimization that conducts simulation calculations based upon online measurement of temperature, ammonia, nitrate, and influent wastewater flow rate, integrating these process measurements with laboratory analytical results for mixed liquor suspended solids (MLSS) as inputs to the algorithm. The BIOS simulation provides a continuous output of DO setpoints for the biological treatment process according to the load entering the bioreactor. Additionally, the internal recycle flow (IRQ) rate from the aerobic zone to the upstream anoxic zone in the MLE process or multi-zone biological nutrient removal (BNR) process is controlled to achieve optimal total nitrogen removal. Using BIOS to control the biological nitrogen removal process produces low effluent total nitrogen concentration while minimizing aeration energy consumption.

Specifically, an ammonia analyzer located in the anoxic zone provides the control system with the ammonia concentration in the aerobic zone influent, and a nitrate analyzer located at the end of the aerobic zones provides the control system with the nitrate concentration in the internal recycle flow (IRQ) stream. BIOS conducts iterative biological and hydraulic simulations that predict the nitrification reaction rates in the aerobic zones and the denitrification reaction rates in anoxic zones under different DO and IRQ conditions. The simulation iterative calculations take into account that the IRQ will dilute the ammonia concentration in the anoxic zone and decrease one-pass hydraulic retention times in both the anoxic zone and aerobic zones. As a result, the simulation provides optimal DO setpoints (for controlling and optimizing the aeration rate) and IRQ (for controlling the process recirculation pump rates) in real-time based on the changing characteristics of the wastewater. See Figure 17.5 for a diagram of the BIOS process.

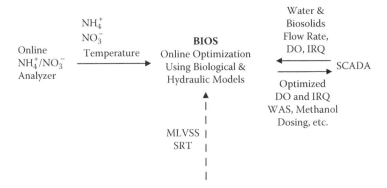

FIGURE 17.5
Representation of the BIOS process. (Adapted from Lui et al., in *Proceedings of 2005 WEFTEC Annual Conference*, Water Environment Federation, Alexandria, VA, 2005.)

17.4 Blowers

As noted earlier, the aeration process can account for 25% to as much as 60% of total plant energy use (WEF, 2009). This section builds on energy conservation measures (ECMs) for aeration system design and operating presented in the previous section by providing technical information and cost/energy data for ECMs related to innovation and emerging blowers and diffuser equipment. Unlike other ECMs described in this chapter, blower and diffuser designs are often unique to manufacturers; hence, this section contains information on proprietary systems as examples. The wastewater industry is constantly evolving, and new equipment not identified in this book may be available now or will be soon. When evaluating new equipment, design engineers and plant owners should work closely with their state regulatory agency to assess operating principles and potential energy savings

Blowers are an integral piece of the aeration system. There are many configurations, but all consist of lobes, impellers, or screws mounted on one or more rotating shafts powered by a motor. As the shaft turns, the blower pulls in outside air and forces it through distribution pipes into aeration basins at pressures typically between 5 and 14 pounds per square inch gauge (psig). The energy consumption of blowers is a function of air flow rate, discharge pressure, and equipment efficiency (WEF, 2009). Blower efficiency varies with flow rate, speed, pressure, inlet conditions, and actual design.

Blowers can be categorized as either *positive displacement blowers*, which provide a constant volume of air at a wide range of discharge pressures, or *centrifugal blowers*, which provide a wide range of flow rates over a narrow range of discharge pressure. Centrifugal blowers are either multi-stage and have a sequence of impellers mounted along a single shaft directly connected to a motor with a flexible coupling, or single-stage, typically with one impeller with speed-increasing gears or a variable-frequency drive. Single-stage centrifugal blowers can be conventional, integrally geared blowers or gearless (also known as high-speed "turbo") blowers. Positive displacement or centrifugal blowers (multi-stage or new high-speed turbo blowers) are well suited for small plants. Large plants more often use multi- or single-stage centrifugal blowers, as high-speed turbo blowers are not yet available in capacities suitable for large plants. Table 17.2 lists the types of blowers, describes their operation, and provides information on their advantages and disadvantages.

TABLE 17.2

Overview of Blower Types for Aeration of Wastewater

Category	Description and Operation	Types	Advantages	Disadvantages
Positive displacement	Provides fixed volume of air for every shaft revolution. Operates over a wide range of discharge pressures.	Most common is two counter-rotating shafts (rotary) with two- or three-lobed impellers on each shaft.	Low capital cost; economical at small scale. Can achieve higher output pressure at same air flow rates. Simple control scheme for constant flow applications.	Difficult to operate at variable flow rates without VFD. Can be noisy (enclosures are commonly used for noise control). Requires more maintenance than other types. Typically least energy efficient.
Centrifugal multi-stage	Uses a series of impellers with vanes mounted on rotating shaft (typically 3600 rpm). Each successive impeller increases discharge pressure. Individual units operate at narrow range of discharge pressures at wide range of flow rates.	Number of stages dictates discharge pressure.	Can be more energy efficient than positive displacement. Lower capital costs compared to single-stage centrifugal blowers. Can be quieter than single-stage units.	Can be less energy efficient than single-stage centrifugal. Efficiency decreases with turndown.
Centrifugal single-stage integrally geared	Similar to multi-stage but uses a single impeller operating at high speed (typically 10,000–14,000 rpm) to provide discharge pressure. Uses gearing between motor and blower shaft.	Differences are in speed and type of control (e.g., one or two sets of variable vanes).	Can be more energy efficient than multi-stage or positive displacement. Can maintain good efficiency at turndown. Typically comes with integral control system for surge protection.	More moving parts than multi-stage units. Surge can be more damaging. Can be noisy (enclosures are commonly used for noise control). Higher capital costs compared to multi-stage or positive displacement.
Centrifugal single-stage gearless (high-speed turbo)	Centrifugal single-stage blower uses special low-friction bearings to support shaft (typically ~40,000 rpm). Uses a single or dual impeller.	Magnetic or air bearing.	Small footprint. Efficient technology for lower air flow capacity ranges. Can maintain good efficiency at turndown. May come with integrated control systems to modulate flow and for surge protection. Can be easy to install (place, plumb, and plug in).	Typically higher capital costs compared to multi-stage or positive displacement blowers (although likely less expensive than integrally geared). Limited experience (new technology). More units required for larger plants (will change as manufacturers expand air flow range).

Source: Adapted from WEF, *Energy Conservation in Water and Wastewater Facilities*, MOP No. 32, McGraw Hill, New York, 2009; WEF/ASCE/EWRI, *Design of Municipal Wastewater Treatment Plants*, 5th ed., MOP No. 8, ASCE Manuals and Reports on Engineering Practice No. 76, McGraw Hill, New York, 2010.

Abbreviations: psi, pounds per square inch; rpm, revolutions per minute; scfm, standard cubic feet per minute; VFD, variable-frequency drive.

TABLE 17.3

Typical Blower Efficiencies

Blower Type	Nominal Blower Efficiency (%)	Nominal Turndown (% of rated flow)
Positive displacement (variable speed)	45–65	50
Multi-stage centrifugal (inlet throttled)	50–70	60
Multi-stage centrifugal (variable speed)	60–70	50
Single-stage centrifugal, integrally geared (with inlet guide vanes and variable diffuser vanes)	70–80	65
Single-stage centrifugal, gearless (high-speed turbo)	70–80	50

Source: Adapted from Gass, J.V., Black & Veatch, *Scoping the Energy Savings Opportunities in Municipal Wastewater Treatment*, presented at CEE Partner's Meeting, September 29, 2009.

Table 17.3 presents typical ranges of isentropic (nominal) energy efficiency and turndown (low-flow) for different blower types. Note that there is significant variation from small to large blowers of any type; the values presented are general rules of thumb and may vary with the application.

Controlling positive displacement blowers is typically done by varying blower speed with a variable-frequency drive or by the use of multiple blowers operating in parallel. Throttling air flow through the machine is not possible for this type of blower. Multi-stage centrifugal blowers can be controlled through a variety of techniques, the most efficient being VFDs followed by suction air flow throttling using inlet butterfly valves. VFD operation of multi-stage centrifugal blowers can be 15 to 20% more efficient than throttling (WER, 2009). This section identifies several innovative and emerging ECMs related to blower and diffuser equipment:

- Turbo blowers are a significant area of innovation in blower design and offer energy savings for the wastewater industry. They emerged in the North American market around 2007 and have been or are being tested and installed at many plants. Section 17.4.1 provides detailed information on turbo blower technology as an innovative ECM.

- Single-stage centrifugal integrally geared blowers are controlled using inlet guide vanes and variable diffuser vanes. This control technique has the advantages of managing air flow and pressure independently. See Section 17.4.2 for discussion of new single-stage centrifugal blower technology.

- Fine-bubble diffusers were once considered the standard for energy efficiency, but new materials and configurations capable of producing "ultra-fine" bubbles (1 mm or less) are now available. See Section 17.4.3 for a discussion of emerging diffuser ECMs.

- Technological advances are also progressing in the area of diffuser cleaning. See Section 17.4.4 for recommendations for preventing diffuser fouling.

A very new technology is the rotary screw compressor. The technology was released to the U.S. market in the summer of 2010. The manufacturers claim significant energy savings of up to 50% compared to rotary lobe blower technology. Units are being manufactured by Atlas Copco, AERZEN, Inc., and Dresser Roots.

17.4.1 High-Speed Gearless (Turbo) Blowers

High-speed gearless (turbo) blowers use advanced bearing design to operate at higher speeds (upwards of 40,000 rpm) with less energy input compared to multi-stage and positive displacement blowers. Some turbo blowers come in package systems with integrated VFDs and automated control systems to optimize energy efficiency at turndown. Turbo blowers are available in two primary configurations based on the manufacturer: (1) air bearing or (2) magnetic bearing. In an air bearing turbo blower, an air film is formed between the impeller shaft and its bearings as the shaft rotates at high speed, achieving friction-free floating of the shaft. Air bearing technology is offered by several manufacturers, including K-Turbo, Neuros, Turblex, and HSI. In a magnetic bearing design, the impeller shaft is magnetically levitated to provide friction-free floating of the shaft. Turbo blowers featuring magnetic bearing design are offered by ABS Group, Atlas Copco, and Piller TSC. A magnetic bearing high-speed turbo blower is also being developed by Dresser Roots. The friction-free bearing design coupled with high-efficiency motors contributes to the comparative high energy efficiency of the turbo blower technology. Turbo blowers have many practical advantages (Gass, 2009; Jones and Burgess, 2009):

- They are typically 10 to 20% more energy efficient than conventional multi-stage centrifugal or positive displacement for their current size range based on manufacturers' data. They offer good turndown capacity (up to 50%) with little drop in efficiency. It is important to note that efficiencies of turbo blowers at turndown are not yet well documented because the technology is so new.

- Some include a dynamic control package with integrated variable speed drive, sensors, and controls that automatically adjust blower output based on real-time dissolved oxygen demand in the aeration basin.

- They have a small footprint and are lightweight.

- They are quiet, with low vibration. Sound enclosures are standard equipment.

- They have few moving parts and low maintenance requirements.

Disadvantages of the turbo blower are that it is a new technology with relatively few installations, capital costs tend to be higher compared to other blower types, and multiple units may be needed for larger installations. Moreover, testing methods are not consistent among different manufacturers, and some efficiency claims are not yet well documented.

17.4.2 Single-Stage Centrifugal Blowers with Inlet Guide Vanes and Variable Diffuser Vanes

Single-stage centrifugal blowers equipped with inlet guide vanes pre-rotate the intake air before it enters the high-speed blower impellers. This reduces flow more efficiently than throttling. Blowers that are also equipped with variable outlet vane diffusers have improved control of the output air volume. Utilizing inlet guide vane and discharge diffusers on a single-stage centrifugal blower makes it possible to operate the blower at its highest efficiency point, not only at the design condition but also within a greater range outside of the design condition. PLCs can be used to optimize inlet guide vane operation (i.e., positioning) based on ambient temperature, differential pressure, and machine capacity. Automated DO and variable header pressure control can increase efficiency.

17.4.3 New Diffuser Technology

The development of fine-bubble diffuser technology in the 1970s led to significant reductions in aeration energy consumption over mechanical and coarse-bubble aeration due to the increased oxygen transfer rates afforded by the high surface area of the fine bubbler. It has been estimated that using fine-bubble diffusion can reduce aeration energy from 25% to as high as 75% (SAIC, 2006). Estimated energy savings of 30 to 40% are common (Cantwell et al., 2009; USEPA, 1999a). There are many different types of fine-bubble diffusers available, including ceramic/porous plates, tubular membranes, ceramic disks, ceramic domes, and elastomeric membrane disks, each with distinct advantages and disadvantages. In general, most diffusers are one of two types: (1) rigid ceramic material configured in discs, or (2) perforated membrane material. Ceramic media diffusers have been in use for many years and are considered the standard against which new, innovative media are compared. Membrane diffusers consist of a flexible material with perforated pores through which air is released. Most often configured in tubes, discs, or panels, they comprise the majority of new and retrofit installations.

Fine-bubble aeration has been implemented at many WWTPs and is considered a common conventional ECM. The focus of this section is ECMs related to new diffuser equipment that can achieve enhanced energy reduction over fine-bubble technology. Recent advances in membrane materials have led to ultra-fine-bubble diffusers, which generate bubbles with an average diameter between 0.2 and 1.0 mm. The primary appeal of ultra-fine-bubble diffusion is improved oxygen transfer efficiency (OTE). Additionally, some composite materials used in the manufacture of ultra-fine-bubble diffusers are claimed to be more resistant to fouling, which serves to maintain the OTE and reduce the frequency of cleaning. Concerns about ultra-fine-bubble diffusion include slow rise rates and the potential for inadequate mixing. Two proprietary ultra-fine-bubble diffuser designs, panel diffusers by Parkson and Aerostrip® diffusers by the Aerostrip Corporation, are discussed below.

Panel diffusers are membrane-type diffusers built onto a rectangular panel. They are designed to cover large areas of the basin floor and lay close to the floor. Panel diffusers are constructed of polyurethane and generate a bubble with a diameter of about 1 mm. OTE is a function of floor diffuser coverage, which translates to improved efficiency for panel diffusers. The advantages of panel diffusers include the increased OTE and the even distribution of aeration. Disadvantages include higher capital costs, a higher head loss across the diffuser, increased air filtration requirements, and a tendency to tear when over-pressured.

Aerostrip® is a proprietary diffuser design manufactured in Austria. The device is a long strip diffuser with a large aspect ratio. According to the manufacturer, "it is a homogeneous thermoplastic membrane held in place by a stainless steel plate." The Aerostrip® diffuser provides many of the same advantages and disadvantages as panel diffusers; however, it appears to be less prone to tearing. Also, the smaller strips allow tapering of the diffuser placement to match oxygen demand across the basin. Aerostrip®s may be mounted at floor level or on supports above the floor. Manufacturer claims regarding the strip membrane diffuser include the following (USEPA, 2010):

- Energy efficiencies are between 10 and 20% greater than the traditional ceramic and elastomeric membrane diffuser configurations.
- Uniform bubbles are released across the membrane surface.
- Bubbles resist coalescing.

- Membranes are not prone to clogging.
- Diffusers are self-cleaning, although Aerostrip® panels have been reported to be susceptible to frequent fouling, requiring bumping and flexing of the membrane to dislodge.

17.4.4 Preventing Diffuser Fouling

Diffuser fouling can reduce OTE and thereby increase the energy required to operate the aeration system. In general, fine-bubble diffusers have been shown to be more susceptible to fouling than coarse-bubble diffusers. Most require periodic maintenance, and some must be replaced regularly. Ceramic diffusers require periodic pressure washing or acid cleaning depending on the severity of the fouling. Pressure washing is often sufficient to remove fouling, including chemical precipitates, and can restore the diffuser to near-new condition. Occasionally, acid cleaning is needed to remove precipitates. Intrusion of mixed liquor into the body of ceramic diffusers nearly always necessitates their replacement. Membrane diffusers attract slime and precipitates. Where slime can be scrubbed off, removal of precipitates requires an acid bath.

Some manufacturers of perforated membrane diffusers claim their products are more resistant to fouling than porous plastic or ceramic diffusers. The susceptibility to fouling is impacted by the membrane material used. A commonly used membrane material is ethylene propylenediene monomer (EPDM) rubber, which has been shown to be susceptible to biological fouling, while polyurethane or silicone materials appear to be more resistant (Wagner and Von Hoessle, 2004). New PTFE composite membranes made by Ott, SSI, and EDI were developed to minimize fouling; however, because these materials are relatively new, insufficient data are available to support this claim.

Sanitaire® by ITT Water and Wastewater is an in-place gas cleaning system that can be used to clean ceramic fine-bubble diffusers without interruption of process or tank dewatering. It can be added as a retrofit or included as part of new installations. The system is designed to inject anhydrous HCL gas into the process air stream. At the gas/liquid interface inside the diffusers, the anhydrous HCL combines with water to form hydrochloric acid. The acid mixture reacts with and dissolves soluble minerals and removes biological foulants by decreasing the pH. Although many Sanitaire® clean-in-place systems exist, their use is limited to existing ceramic diffusers.

A recent publication by Southern California Edison and the University of California, Los Angeles, documented the development of a new monitoring device to help predict cleaning when diffused air systems require cleaning (Larson, 2010). The device measures oxygen transfer efficiency and is characterized by the study as low-cost ($3000 to $5000) and easy to use. It is auto-calibrated and does not require trained experts. Prototype analyzers were installed and tested at a 10-MGD WWTP, and plans are in place to install additional devices at several other plants in California. Larson (2010) estimated an average energy efficiency improvement of 15% with the installation of an online analyzer.

17.4.5 Innovative and Emerging Energy Conservation Measures

Unlike energy conservation measures for aeration and pumping, ECMs for advanced treatment technologies such as ultraviolet (UV) disinfection and membrane bioreactors (MBRs), and for other process functions such as anoxic zone mixing, are emerging and generally not yet supported by operating data from full-scale installations. They are very important, however, because wastewater treatment plants (WWTPs) are increasingly employing these technologies. The following discussion of ECMs for advanced technologies presents full-scale plant test results where available. Where operation data are not available the manufacturer's information is provided.

17.4.6 UV Disinfection

Although ultraviolet disinfection was recognized as a method for achieving disinfection in the late 19th century, its application virtually disappeared with the evolution of chlorination technologies. In recent years, however, there has been a resurgence in its use in the wastewater field, largely as a consequence of concerns related to security, safe handling, and effluent toxicity associated with chlorine residual. Even more recently, UV has gained more attention because of the tough new regulations on chlorine use imposed by both OSHA and USEPA. Because of this relatively recent increased regulatory pressure, many facilities are actively engaged in substituting chlorine for other disinfection alternatives. Moreover, UV technology itself has made many improvements, which now makes UV attractive as a disinfection alternative. As of 2007, approximately 21% of municipal WWTPs were using UV for disinfection (Leong et al., 2008). That number is only expected to rise as manufacturers continue to improve UV equipment designs and decrease costs, and as more and more WWTPs gain experience with the technology.

Ultraviolet radiation at certain wavelengths (generally between 220 to 320 nanometers) can penetrate the cell walls of microorganisms and interfere with their genetic material. This limits the ability of microorganisms to reproduce and, thus, prevents them from infecting a host. UV radiation is generated by passing an electrical charge through mercury vapor inside a lamp. Low-pressure, low-intensity lamps, which are most common at WWTPs, produce most radiation at 253.7 nm. Medium-pressure, high-intensity lamps emit radiation over a much wider spectrum and have 15 to 20 times the UV intensity of low-pressure, low-intensity lamps. Although fewer lamps are required as compared to low-pressure systems, medium-pressure lamps require more energy. The effectiveness of the UV process is dependent on the following:

- UV light intensity
- Contact time
- Wastewater quality (turbidity)

The Achilles' heel of ultraviolet for disinfecting wastewater is turbidity. If the wastewater quality is poor, the UV light will be unable to penetrate the solids and the effectiveness of the process decreases dramatically. For this reason, many states limit the use of UV disinfection to facilities that can reasonably be expected to produce an effluent containing ≤30 mg/L or less of BOD_5 and total suspended solids.

In the operation of UV systems, UV lamps must be readily available when replacements are required. The best lamps are those with a stated operating life of at least 7500 hours and those that do not produce significant amounts of ozone or hydrogen peroxide. The lamps must also meet technical specifications for intensity, output, and arc length. If the UV light tubes are submerged in the wastestream, they must be protected inside quartz tubes, which not only protect the lights but also make cleaning and replacement easier.

Contact tanks must be used with UV disinfection. They must be designed with the banks of UV lights in a horizontal position, either parallel or perpendicular to the flow or with banks of lights placed in a vertical position perpendicular to the flow.

Note: The contact tank must provide, at a minimum, 10-second exposure time.

It was stated earlier that turbidity problems have been a major problem with UV use in wastewater treatment; however, if turbidity is its Achilles' heel, then the need for increased maintenance (as compared to other disinfection alternatives) is the toe of the same foot. UV maintenance requires that the tubes be cleaned on a regular basis or as needed. In addition, periodic acid washing is also required to remove chemical buildup. Routine monitoring of UV disinfection systems is required. Checking on bulb burnout, buildup of solids on quartz tubes, and UV light intensity is also necessary (Spellman, 2009).

Note: Ultraviolet light is extremely hazardous to the eyes. Never enter an area where UV lights are in operation without proper eye protection. Never look directly into the ultraviolet light.

Energy requirements for UV depend on the number, type, and configuration of lamps used to achieve the target UV dose for pathogen inactivation. One of the most important factors affecting UV dose delivery is the UV transmittance (UVT) of the water being disinfected. UVT is defined as the percentage of light passing through a wastewater sample over a specific distance (1 centimeter). It takes into account the scattering and adsorption of UV by suspended and dissolved material in the water. UVT is affected by the level of pretreatment—filtered wastewater has a much higher UVT than unfiltered water. Microorganisms that move quickly through the reactor far from the lamp will receive a lower dose than microorganisms that have longer exposure to the UV radiation and are closer to the lamp. Other factors affecting UV dose delivery are temperature, lamp age, and lamp fouling. Because UV disinfection is complex and based on many factors, dose estimation methods are complicated and typically involve computational fluid dynamic modeling or bioassays. Dose can be maintained at a minimum level or can be controlled based on water quality (i.e., lowered during periods of improved quality) which can save energy.

A study funded by Pacific Gas & Electric found that the energy consumed by UV disinfection can account for approximately 10 to 25% of total energy use at a municipal wastewater treatment plant (PG&E, 2001). These findings were based on a detailed evaluation of seven wastewater treatment plants ranging in flow rate from 0.4 to 43 MGD. Data included plants with low-pressure, low-intensity lamps and higher pressure, high-intensity lamps in a variety of configurations. Energy required for low-pressure lamps ranged from approximately 100 to 250 kWh per million gallons (MG). Energy required for medium-pressure systems ranged from 460 to 560 kWh/MG, with one plant requiring 1000 kWh/MG to achieve a very high level of coliform inactivation. PG&E (2001) reported that UV disinfection performance in relation to input energy is not linear. An increasing amount of energy is required to obtain marginal reductions in most probable number (MPN) per milliliter for total coliforms.

DID YOU KNOW?

The effectiveness of a UV disinfection system depends on the characteristics of the wastewater, the intensity of UV radiation, the amount of time the microorganisms are exposed to the radiation, and the reactor configuration. For any one treatment plant, disinfection success is directly related to the concentration of colloidal and particulate constituents in the wastewater (USEPA, 1999b).

The ECMs for UV disinfection are fairly new, and energy savings/cost data are not well documented in the literature. Still, growing experience with UV disinfection has revealed practical design, operation, and maintenance strategies that can reduce the energy use of UV disinfection. The following sections summarize these ECMs and provide detailed information on upgrades and associated energy savings for several WWTPs as reported in the literature.

17.4.6.1 Design

17.4.6.1.1 Pretreatment

Pretreatment to remove suspended solids from wastewater, such as tertiary sand filtration or membranes, can increase UVT and allow a plant to reach the same level of treatment at a lower UV dose, thereby saving energy. If a plant uses iron or aluminum compounds for chemical precipitation of phosphorus, it is important to minimize residual iron and aluminum concentrations to prevent acceleration of UV lamp fouling (Leong et al., 2008).

17.4.6.1.2 Lamp Selection

Medium-pressure lamps require two to four times more energy to operate than low-pressure, low-intensity lamps. In some cases, WWTPs can save on energy costs by specifying low-pressure, low-intensity lamps. Tradeoffs are (1) a larger footprint of the same disinfection level, which can be significant because as many as 20 low-pressure, low-intensity lamps are needed to produce the same disinfecting power as one medium-pressure lamp, and (2) higher operating costs for maintenance and change-out of additional lamps. Low-pressure, high-output lamps are similar to low-pressure, low-intensity lamps except that a mercury amalgam is used instead of mercury gas so they can operate at higher internal lamp pressures. Thus, the UV output of a low-pressure, high-output lamp is several times that of a low-pressure, low-output lamp (Leong et al., 2008). Low-pressure, high-output lamps offer significant advantages, including reducing lamp requirements (i.e., quantity) compared to traditional low-intensity lamps, and reducing energy requirements compared to medium-pressure lamps.

The energy demand for low-pressure, high-output systems is similar to that of low-pressure, low-intensity systems (Leong et al., 2008). Thus, low-pressure, high-output lamps may be a good option for reducing the number of lamps and footprint while keeping the energy requirements low. Salveson et al. (2009) presented results of a pilot test at the Stockton, California, WWTP comparing design conditions and operation of medium-pressure and low-pressure, high-output lamps. The power draw for the low-pressure, high-output lamps was between 20 and 30% of the power draw (based on operational values, kW) for the medium-pressure lamps, reducing annual O&M costs significantly. These results are similar to information reported from one manufacturer for a 30-MGD plant treating secondary effluent. A low-pressure, high-output system would use 60 kW at peak flow compared to 200 kW for a medium-pressure system.

17.4.6.1.3 System Turndown

Similar to the design of blowers for aeration systems, it is important that designers allow for sufficient UV system turndown to respond to changes in flow and wastewater quality. Flexibility and control in design are key factors in operating efficiently from the day the technology is commissioned until the end of its design life. The configuration of the lamps dictates the approach for lamp turndown. In systems with vertical lamp configurations, the water level can vary during operation (with respect to the submerged portion of the lamp), whereas in a horizontal lamp configuration the water levels should remain relatively constant (with respect to lamp submergence). Individual rows of lamps can be turned on and off in vertical configurations. In horizontal arrangements, channel control is more typically used to respond to varying flows (Leong et al., 2008). Regardless of configuration, the number of channels should be selected to maintain a velocity that has been tested and is known to provide the required dose delivery.

17.4.6.1.4 System Hydraulics to Promote Mixing

As noted previously, UV dose delivery inside a UV reactor depends on the hydraulics. Optimized longitudinal and axial mixing of the water is critical to maintain a minimum UV dose throughout the reactor. In general, this is achieved by operating at a sufficiently high approach velocity to ensure turbulent flow conditions (WEF/ASCE/EWRI, 2010). WWTPs should conduct full-scale, pilot testing before installation to ensure that mixing effects are addressed in the design. Flow equalization prior to the UV reactor can also stabilize hydraulic conditions and prevent high or low flows from causing reduced UV disinfection performance. It is important to note that mixing is a balancing act. Extreme agitation of the wastewater can create bubbles that shield pathogens from exposure to the UV radiation.

17.4.6.2 Operations and Maintenance

17.4.6.2.1 Automation

Automation can reduce the number of lamps and channels operating based on real-time flow and wastewater characteristic data, thereby reducing energy use and also extending UV lamp life. Controls can be designed to turn off lamps or divert flow to a few operating channels depending on the UV system design. Control is most commonly flow-paced control or dose-paced control. Flow-paced is the simplest, with the number of lamps and channels in service based strictly on influent flow rate. Dose-paced control is based on the calculated dose, which is derived from the following online monitoring data (Leong et al. 2008):

- Flow rate
- UV transmittance (UVT)
- Lamp power (including lamp age and on-line intensity output data)

Dose-paced control more closely matches the UV dose delivered to wastewater conditions. For example, during periods of high solids removal, UVT will increase and UV output can be decreased to achieve the same dose. During wet weather events or other periods of low effluent quality, lamp output can be increased in response to reduced UVT. At the University of California, Davis, WWTP, process controls were implemented to divert flow automatically to one of two channels during low flow conditions (Phillips and Fan, 2005). This change provided the flexibility to operate at 33, 50, 67, and 100% of maximum power. The original design limited operation to 67% and 100% of maximum power. The annual energy use at the WWTP is expected to decrease by 25% once the process changes are fully implemented.

17.4.6.2.2 Lamp Cleaning and Replacement

The effectiveness of UV disinfection systems depends on the intensity of the ultraviolet radiation to destroy the microorganism in the treated wastewater. Two factors that affect UV intensity during operation are lamp age and quartz sleeve fouling. After an initial burn-in period, the lamp output decreases gradually toward the end of lamp life. The end of lamp life is defined by the manufacturer and is the operating hours at which the lamp reaches a specified minimum output. The operating lives of UV lamps are provided below (WEF/ASCE/EWRI, 2010):

- Low-pressure, low-intensity lamps—7500 to 8000 hours
- Low-pressure, high-intensity lamps—12,000 hours
- Medium-pressure lamps—5000 hours

WWTPs can provide a relatively consistent level of lamp output by establishing a schedule for staging lamp replacements. Algal growth, mineral deposits, and other materials can foul the lamp sleeve and subsequently decrease UV intensity and disinfection efficiency. Cleaning and maintaining quartz sleeves are critical to ensuring the optimum performance of UV disinfection and can result in substantial energy savings. Most equipment suppliers provide automatic cleaning mechanisms, which consist of chemical cleaning, mechanical cleaning, or both. One study found that a combination of mechanical and chemical cleaning was superior to mechanical cleaning alone (Leong et al., 2008; Peng et al., 2005).

The Efficiency Partnership (2001) presented an example of energy savings due to increased attention to UV system cleaning and lamp replacement. At the Central Contra Costa Sanitary District (CCCSD), lamp cleaning and maintenance are particularly important because the facility is disinfecting secondary effluent with fairly low water quality. CCCSD found that increased maintenance of the UV lamps (i.e., cleaning and replacement of UV bulbs) at its wastewater treatment plant resulting in a reduction in the number of UV banks required for the disinfection system, from nine to six banks. This new maintenance strategy resulted in a power savings of 105 kW.

17.4.7 Membrane Bioreactors

Membrane bioreactors (MBRs) are becoming more common as WWTPs are required to meet increasingly stringent effluent limits and, in some cases, reuse requirements in smaller footprints. The unique feature of MBRs is that, instead of secondary clarification, they use membrane treatment, either as vacuum-driven systems immersed in a biological reactor or pressure-driven membrane systems located external to the bioreactor for solids separation. Membranes are typically configured hollow-tube fibers or flat panels and have pore sizes ranging from 0.1 to 0.4 microns. Although MBRs have many operational advantages, they use more energy than conventional processes in order to move water through the membrane and for membrane scouring and cleaning. The energy requirements of MBR systems may be twice that of conventional activated sludge systems (WEF, 2009). Because the technology is not widespread, ECMs for MBRs are emerging. The emerging ECM identified in this report is membrane air scour alternatives.

Membrane fouling has been identified as the most significant technical challenge facing this technology (Ginzburg et al. 2008). Fouling occurs when the membrane pores become obstructed with the mixed liquor suspended solids being filtered, causing a loss in permeability. The main causes of membrane fouling are initial pore blocking, where particles smaller than the membrane pore size plug the openings, followed by cake fouling, where particles accumulate on the membrane over time, forming stratified "cake" layers (Lim and Bai, 2003; Peeters et al., 2007). Although different membrane manufacturers use different techniques to control for fouling, the primary method to address cake fouling is aeration along with periodic chemical cleaning. Peeters et al. (2007) reported that membrane aeration to control fouling accounts for 35 to 40% of total power consumption of an MBR.

In recent years, several membrane manufacturers have modified operational strategies to reduce air scour fouling control requirements (Wallis-Lage and Levesque, 2009). For example, Kubota varies the volume of air used for aeration based on the flux (e.g., lower air scour rates are used for lower flux values). The manufacturer of the Huber system claims reduced energy consumption for air scour due to a centrally positioned air intake and lower pressure. Siemens uses a combination of air and water to scour the membrane (Wallis-Lage and Levesque, 2009). General Electric (GE) implemented cyclic air scour, whereby aeration is turned on and off in 10-second intervals. A newer innovation is GE's 10/3 eco-aeration system where the membrane is scoured for 10 seconds on, 30 seconds off during non-peak flow conditions. GE claims that the system can reduce energy consumption by up to 50% compared to standard 10/10 aeration protocol (Ginzburg et al., 2008).

The literature includes pilot- and full-scale test data for a membrane fouling controller and algorithm used to clean the GE Zenon ZeeWeed MBR. The system uses real-time analysis of the membrane's filtration operating conditions to determine the fouling mechanism present in the MBR system. The information obtained from the algorithm dictates the implementation of specific control actions to respond to the particular fouling mechanism (e.g., membrane aeration, backwash, chemical cleaning—the biggest impact on energy consumption being membrane aeration). When aeration is identified as the control action, the fouling controller/algorithm gives the MBR programmable logic controller (PLC) system the information to select between the traditional 10/10 (air scour on/off) protocol and the 10/30 eco-aeration energy saving protocol. The algorithm was piloted and later full-scale tested at a 3,000,000-MGD plant in Pooler, Georgia. Ginzburg et al. (2008) concluded that additional research is required to further develop the online fouling controller to include additional control parameters such as membrane aeration flow rate, backwash flow rate, and backwash duration.

17.4.8 Anoxic and Anaerobic Zone Mixing

Many WWTPs are implementing biological nutrient removal (BNR) for nitrogen and phosphorus to protect receiving waters and prevent eutrophication, particularly in coastal regions. Biological nitrogen removal is a two-step process consisting of nitrification to convert ammonia to nitrate (NO_3) followed by denitrification to convert nitrate to nitrogen gas. Nitrification of ammonia is an aerobic process and can occur in the aerated zone with sufficient solids retention time (SRT). Significant energy can be required for complete nitrification of ammonia. Denitrification is an anoxic process accomplished in the absence of dissolved oxygen so the microorganisms will use nitrate as their oxygen source. Denitrification can be accomplished in a denitrifying filter, but most often it occurs in a suspended-growth anoxic zone where the denitrifying microorganisms can use organic material present in the wastewater instead of or in addition to an external carbon source. A common configuration of the suspended growth nitrification–denitrification process is the Modified Ludzack–Ettinger (MLE) process, which has an initial anoxic zone followed by an aerobic zone. Nitrification occurs in the aerobic zone from which pumps recycle nitrate-rich mixed liquor to the anoxic zone for denitrification.

Biological phosphorus removal works by exposing the biomass first to anaerobic conditions. As long as a sufficient food source (i.e., volatile fatty acids) is present, microorganisms referred to as *phosphate accumulating organisms* (PAOs) will release stored phosphorus in the anaerobic zone, which conditions them to uptake large amounts of phosphorus when they enter the aerobic zone. Phosphorus is removed when biomass is wasted from the aerobic zone.

It is important to mix the wastewater in the anoxic zone to maintain suspension of solids and ensure that denitrifying microorganisms come into contact with nitrate. Similarly, it is important to mix the wastewater in the anaerobic zone to maintain suspension of the solids and PAOs. The mixers, however, cannot impart oxygen to the water (this would cause them to use oxygen as their electron exceptor instead of nitrate). Similarly for the anaerobic zone in biological phosphorus removal systems, mixers are needed to contact waste and microorganisms but they must not transfer oxygen to the water (oxygen would promote growth of microorganisms other than PAOs that would compete with them for the food source). Low-speed submersible mixers are commonly used for these processes. Two emerging ECMs have been identified to reduce the energy required to mix anoxic and anaerobic zones: hyperbolic mixers and pulsed large-bubble mixing.

17.4.8.1 Hyperbolic Mixing

A new hyperboloid mixer that has undergone full-scale testing at two large wastewater treatment plants in the United States has shown significant energy savings compared to traditional submersible mixers. The mixer is a vertical-shaft type of mixer with a hyperboloid-shaped stirrer located close to the bottom of the tank. The stirrer is equipped with transport ribs that cause acceleration of the wastewater in a radial direction to promote complete mixing (see Figure 17.6). The hyperboloid mixer has been used in Europe for more than 10 years with installations in Germany, Holland, and Belgium (Gidugu et al., 2010).

A recent study at the Bowery Bay Water Pollution Control Plant (WPCP) in New York City compared the performance of traditional submersible mixers (specifically, two-blade propeller mixers mounted to the side of the tank) to a hyperbolic mixer, the HYPERCLASSIC HC RKO 2500® (Fillos and Ramalingam, 2005), for anoxic zone mixing. Researchers evaluated the two mixers based on their ability to

- Sustain uniform distribution of suspended solids in the basin.
- Maintain a low DO concentration (<0.3 mg/L).
- Maintain a hydraulic profile supportive of denitrification (as determined using tracer tests).

Although both mixers at the Bowery Bay WPCP were able to achieve good distribution of solids with low DO, the HYPERCLASSIC mixer had a superior hydraulic profile. Moreover, the authors reported lower energy needs for the HYPERCLASSIC mixer due to its design: 2.2 brake horsepower (bhp) for the HYPERCLASSIC mixer compared to 6.0 bhp for the submersible mixer. The authors reported a total energy cost of $1131 for the HYPERCLASSIC mixer compared to $3075 for the submersible mixer per anoxic zone per year, for a savings of close to $2000 based on the energy rate of $0.039/kWh. Based on this information, energy use for the HYPERCLASSIC mixer would be 29,000 kWh/yr compared to 78,850 kWh/yr for the submersible mixer, for an energy savings of 49,850 kWh/yr per anoxic zone

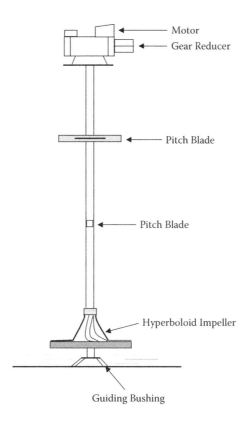

Motor

Gear Reducer

Pitch Blade

Pitch Blade

Hyperboloid Impeller

Guiding Bushing

FIGURE 17.6
Schematic of hyperboloid mixer. (Adapted from Gidugu, S. et al., *New England Water Environment Association Journal*, 44(1), 2010.)

(Fillos and Ramalingam, 2005). The capital cost of the HYPERCLASSIC mixer is approximately $10,000 more than the uniprop mixer, so simple payback would be approximately 5 years per anoxic zone.

Gidugu et al. (2010) reported results of side-by-side testing of the new hyperboloid mixer and a conventional hydrofoil mixer at the Blue Plains WWTP in Washington, DC. The hydrofoil mixer, which is widely used in the United States, has a vertical shaft and a hydrofoil impeller with four angled stainless steel blades (Figure 17.7) (Gidugu et al., 2010). Two 20-hp hydrofoil mixers were installed in one of the anoxic zones at the Blue Plains WWTP in October 2004 for evaluation. Six 10-hp hyperboloids mixers were installed in three anoxic zones (two per zone) for testing in October 2008. Researchers collected data to create DO and TSS profiles in all four anoxic zones in June 2008 to evaluate mixing.

Results showed results similar to those for the Bowery Bay WPCP, with the hyperboloid mixer achieving good distribution of solids with low DO. TSS concentrations within the hyperboloid mixer were spread out over a smaller range of values than within the traditional hydrofoil mixer, indicating more uniform mixing. Gidugu et al. (2010) compared energy use and reported 9.7 bhp per unit for the 10-hp hyperboloid mixer compared to 17.3 bhp for the hydrofoil mixer. Based on an electricity cost of $0.10/kWhr, they estimated energy savings potential of over $5000 per year per mixer. At a cost difference of only $2000 more for the hyperboloid mixer compared to the hydrofoil mixer, simple payback would be less than one year.

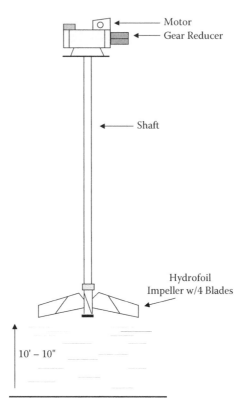

FIGURE 17.7
Schematic of conventional hydrofoil mixer. (Adapted from Gidugu, S. et al., *New England Water Environment Association Journal*, 44(1), 2010.)

17.4.8.2 Pulsed Large-Bubble Mixing

An innovative mixing technology by Enviromix called Bio^{M}x reduces energy required for anoxic or anaerobic zone mixing by firing short bursts of compressed air into the zone instead of mechanically mixing it. Uniquely designed nozzles produce a mass of large air bubbles, ranging from marble to softball size, which mix the water as they rise to the surface. The large air bubbles, much larger than those made by coarse-bubble diffusers, are designed to minimize oxygen transfer and maintain anoxic or anaerobic conditions. The system includes a PLC to manage the timing of the air control valve firing, which gives the operator flexibility to respond to different conditions within the tank. The manufacturer reports that the system has nonclogging, self-cleaning in-tank components that require no maintenance. An independent study at the F. Wayne Hill Water Resources Center in Gwinnett County, Georgia, compared the performance and energy use of Bio^{M}x to submersible propeller mixers. The plant, treating 30 MGD on average with a design flow of 60 MGD, operates up to ten parallel treatment trains each with anaerobic, anoxic, and aerobic zones for biological nitrogen and phosphorus removal. In the spring of 2009, the Bio^{M}x system was installed in two anaerobic cells of one treatment train. The system consisted of an Ingersoll Rand 5- to 15-hp variable-speed rotary screw compressor, piping, controls, and floor-mounted nozzles. Findings from the technology evaluation performed in January 2010 are summarized below:

- Dye tracer tests showed similar mixing for the Bio^{M}x and submersible mixer systems.

- Total suspended solids (TSS) profiles showed that the Bio^{M}x unit is capable of mixing to homogeneity similarly to the submersible mixing units, although variability in the Bio^{M}x cells was slightly higher.

- Continuous oxidation reduction potential (ORP) measurements over periods of 12 to 28 hours showed 95th-percentile ORP values of less than –150 millivolts (mV), which is indicative of anaerobic environments. Given the success in anaerobic environments (< –100 mV), the technology is also applicable for use in anoxic environments.

- Power analyzer readings taken simultaneously showed that the energy (in kilwatts) required to mix one anaerobic cell using the Bio^{M}x system was 45% less than the energy required by a submersible mixer. When operated in three cells using the same compressor, 60% less energy was required (0.097 hp/1000 cf).

Test results obtained by the manufacturer from April 2009 through February 2010 are available online (http://www.enviro-mix.com/documents/FWayneHillEnergySuccessStory 2009-091001.pdf).

References and Recommended Reading

Brogdon, J., McEntyre, C., Whitehead, L., and Mitchell, J. (2008). Enhancing the energy efficiency of wastewater aeration, in *Proceedings of the Water Environment Federation (WEFTEC 2008)*, Water Environment Federation, Alexandria, VA, pp. 444–451.

Burton, F.L. (1996). *Water and Wastewater Industries: Characteristics and Energy Management Opportunities*, Report CR106941, prepared for Electric Power Research Institute, Palo Alto, CA, by Burton Environmental Engineering, Los Altos, CA.

Cantwell, J., Newton, J., Jenkins, T., Cavagnaro, P., and Kalwara, C. (2009). *Running an Energy Efficient Wastewater Utility: Modifications That Can Improve Your Bottom Line* [webcast], Water Environment Federation, Alexandria, VA.

Copp, J.B., Spanjers, H., and Vanrolleghem, P.A. (2002). *Respirometry in Control of the Activated Sludge Process: Benchmarking Control Strategies*, Scientific and Technical Report No. 11, IWA Publishing, London.

Cormen, T.H., Leiserson, C.E., Rivest, R.L., and Stein, C. (2002). *Introduction to Algorithms*, 2nd ed., Prentice-Hall, New Delhi.

EERE. (1996). *Fact Sheet: Buying an Energy-Efficient Electric Motor*, DOE/GO-10096-314, Energy Efficiency and Renewable Energy, U.S. Department of Energy, Washington, DC (www1.eere. energy.gov/industry/bestpractices/pdfs/mc-0382.pdf).

EERE. (2005a). *Energy Tips—Motor Systems Tip Sheet #2: Estimating Motor Efficiency in the Field*, DOE/GO-102005-2021, Energy Efficiency and Renewable Energy, U.S. Department of Energy, Washington, DC (http://www1.eere.energy.gov/manufacturing/tech_deployment/pdfs/ estimate_motor_efficiency_motor_systemts2.pdf).

EERE. (2005b). *Performance Spotlight: Onondaga County Department of Water Environment Protection: Process Optimization Saves Energy at Metropolitan Syracuse Wastewater Treatment Plant*, DOE/ GO-102005-2136, Energy Efficiency and Renewable Energy, U.S. Department of Energy, Washington, DC (http://www1.eere.energy.gov/manufacturing/tech_deployment/pdfs/ onondaga_county.pdf).

EERE. (2005c). *Case Study—The Challenge: Improving Sewage Pump System Performance*, Energy Efficiency and Renewable Energy, U.S. Department of Energy, Washington, DC (http://www1. eere.energy.gov/manufacturing/tech_deployment/case_study_sewage_pump.html).

EERE. (2008). New motor technologies boost system efficiency, *Energy Matters*, Summer (http:// www1.eere.energy.gov/manufacturing/tech_deployment/summer2008.html#a284).

EERE. (2011). *Ultra-Efficient and Power-Dense Electric Motors*, Energy Efficiency and Renewable Energy, U.S. Department of Energy, Washington, DC (http://www1.eere.energy.gov/industry/intensiveprocesses/pdfs/electric_motors.pdf).

EERE. (2012). *Case Study—The Challenge: Improving Sewage Pump System Performance, Town of Trumbull*, Energy Efficiency and Renewable Energy, U.S. Department of Energy, Washington, DC (http:// www1.eere.energy.gov/manufacturing/tech_deployment/case_study_sewage_pump.html).

Efficiency Partnership. (2001). *Water/Wastewater Case Study: Central Contra Costa Sanitary District*, Flex Your Power, http://www.fypower.org/pdf/CS_Water_CCCSD.pdf.

Efficiency Partnership. (2009). *Water/Wastewater Case Study: South Tahoe Public Utility District*, Flex Your Power, http://www.fypower.org/pdf/CS_South_Tahoe.pdf.

EPRI. (1998). *Quality Energy Efficiency Retrofits for Wastewater Systems*, CR-109081, Electrical Power Research Institute, Charlotte, NC.

Fillos, J. and Ramalingam, K. (2005). *Evaluation of Anoxic Zone Mixers at the Bowery Bay WPCP*, New York City Department of Environmental Protection, New York.

Gass, J.V., Black & Veatch. (2009). *Scoping the Energy Savings Opportunities in Municipal Wastewater Treatment*, presented at CEE Partner's Meeting, September 29, 2009 (www.cee1.org/cee/ mtg/09-09mtg/files/WWWGass.pdf).

Gidugu, S., Oton, S., and Ramalingam, K. (2010). Thorough mixing versus energy consumption, *New England Water Environment Association Journal*, 44(1).

Ginzburg, B., Peeters, J., and Pawloski, J. (2008). On-line fouling control for energy reduction in membrane bioreactors, in *Proceedings of the Water Environment Federation, Membrane Technology 2008*, Water Environment Federation, Alexandria, VA, pp. 514–524.

Gusfield, D. (1997). *Algorithms on Strings, Trees, and Sequences: Computer Science and Computational Biology*, Cambridge University Press, Cambridge, UK.

IEEE. (1990). *Recommended Practices for Electric Power Distribution for Industrial Plants* (*IEEE Red Book*), ANSI/IEEE Std 141-1986, Institute of Electrical and Electronic Engineers, New York.

Jones, T. and Burgess, J. (2009). Municipal Water-Wastewater Breakout Session: High Speed "Turbo" Blowers, paper presented at Consortium for Energy Efficiency Program Meeting, June 3, 2009 (www.cee1.org/cee/mtg/06-09mtg/files/WWW1JonesBurgess.pdf).

Lafore, R. (2003). *Data Structures and Algorithms in Java*, 2nd ed., Sams Publishing, Indianapolis, IN.

Larson, L. (2010). *A Digital Control System for Optimal Oxygen Transfer Efficiency*, Report CEC-500-2009-076, California Energy Commission, Sacramento, CA (www.energy.ca.gov/2009publications/ CEC-500-2009-076/).

Leong, L.Y.C., Kup, J., and Tang, C. (2008). *Disinfection of Wastewater Effluent—Comparison of Alternative Technologies*, Water Environment Research Foundation, Alexandria, VA.

Lim, A.L. and Bai, R. (2003). Membrane fouling and cleaning in microfiltration of activated sludge wastewater, *Journal of Membrane Science*, 216, 279–290.

Littleton, H.X., Daigger, G.T., Amad, S., and Strom, P.F. (2009). Develop control strategy to maximize nitrogen removal and minimize operation cost in wastewater treatment by online analyzer, in *Nutrient Removal 2009 Conference Proceedings*, Water Environment Federation, Alexandria, VA.

Liu, W., Lee, G., Schloth, P., and Serra, M. (2005). Side by side comparison demonstrated a 36% increase of nitrogen removal and 19% of aeration requirements using a feed forward online optimization system, in *Proceedings of 2005 WEFTEC Annual Conference*, Water Environment Federation, Alexandria, VA.

Love, N.G. et al. (2000). *A Review and Needs Survey of Upset Early Warning Devices*, Report 99-WWF-2, Water Environment Research Foundation, Alexandria, VA.

Mitchell, T.M. (1997). *Machine Learning*, McGraw-Hill, New York.

NEMA. (2006). *NEMA Premium™: Product Scope and Nominal Efficiency Levels*, National Electrical Manufacturers Association, Rosslyn, VA.

Peeters, J., Pawloski, J., and Noble, J. (2007). The evolution of immersed hollow fiber membrane aeration for MBR, in *Proceedings of IWA 4th International Membrane Technologies Conference*, Harrogate, UK, May 15–17.

Peng, J., Qiu, Y., and Gehr, R. (2005). Characterization of permanent fouling on the surfaces of UV lamps used for wastewater disinfection, *Water Environment Research*, 77(4), 309–322.

PG&E. (2001). *Energy Benchmarking Secondary Wastewater Treatment and Ultraviolet Disinfection Processes at Various Municipal Wastewater Treatment Facilities*, Pacific Gas & Electric Company, San Francisco, CA.

Phillips, D.L. and Fan, M.M. (2005). *Automated Channel Routing to Reduce Energy Use in Wastewater UV Disinfection Systems*, University of California, Davis.

Poynton, C. (2003). *Digital Video and HDTV Algorithms and Interfaces*, Morgan Kaufmann, Burlington, MA.

SAIC. (2006). *Water & Wastewater Industry Energy Best Practice Guidebook*, prepared for Focus on Energy™ by Science Applications International Corporation, McLean, VA.

Salveson, A., Wade, T., Bircher, K., and Sotirakos, B. (2009). High Energy Efficiency and Small Footprint with High-Wattage Low Pressure UV Disinfection for Water Reuse, paper presented at the Ozone and Ultraviolet Technology Conference and Exposition on North American Water, Wastewater, and Reuse Applications, Cambridge, MA, May 4–6.

Spellman, F.R. (2009). *Handbook of Water and Wastewater Treatment Plant Operations*, 2nd ed., CRC Press, Boca Raton, FL.

Spellman, F.R. and Drinan, J. (2001). *Fundamentals of Pumping*, CRC Press, Boca Raton, FL.

Tata, P., Patel, K., Soszynski, S., Lue-Hing, C., Carns, K., and Perkins, D. (2000). Potential for the use of on-line respirometry for the control of aeration, in *Proceedings of 2000 WEFTEC Annual Conference*, Water Environment Federation, Alexandria, VA.

Thumann, A. and Dunning, S. (2008). *Plant Engineers and Managers Guide to Energy Conservation*, 9th ed., Fairmont Press, New York.

Trillo, I., Jenkins, T., Redmon, D., Hilgart, T., and Trillo, J. (2004). Implementation of feedforward aeration control using on-line offgas analysis: the Grafton WWTP experience, in *Proceedings of 2004 WEFTEC Annual Conference*, Water Environment Federation, Alexandria, VA.

USDOE. (1996). *Fact Sheet: Replacing an Oversized and Underloaded Electric Motor*, DOE/GO-10096-287, U.S. Department of Energy, Washington, DC.

USEPA. (1999a). *Wastewater Technology Fact Sheet: Fine Bubble Aeration*, EPA 832-F-99-065, U.S. Environmental Protection Agency, Washington, DC (water.epa.gov/scitech/wastetech/upload/2002_06_28_mtb_fine.pdf).

USEPA. (1999b). *Wastewater Technology Fact Sheet: Ultraviolet Disinfection*, EPA 832-F-99-064, U.S. Environmental Protection Agency, Washington, DC (water.epa.gov/scitech/wastetech/upload/2002_06_28_mtb_uv.pdf).

USEPA. (2003). *Technology Transfer Network Support Center for Regulatory Air Models*, U.S. Environmental Protection Agency, Washington, DC (http://www.epa.gov/ttn/scram/).

USEPA. (2008). *Ensuring a Sustainable Future: An Energy Management Guidebook for Wastewater and Water Utilities*, EPA 832-R-08-002, U.S. Environmental Protection Agency, Washington, DC.

USEPA. (2010). *Evaluation of Energy Conservation Measures for Wastewater Treatment Facilities*, EPA 832-R-10-005, U.S. Environmental Protection Agency, Washington, DC, pp. 4-11, 6-4, 6-12.

Wagner, M. and von Hoessle, R. (2004). Biological coating of EPDM-membranes of fine bubble diffusers, *Water Science and Technology*, 50(7), 79–85.

Wallis-Lage, C.L. and Levesque, S.D. (2009). Cost Effective & Energy Efficient MBR Systems, paper presented at Singapore International Water Week, June 22–26 (bvwater.files.wordpress.com/2009/05/abstract_siw09_wallis-lage.pdf).

Weerapperuma, D. and de Silva, B. (2004). On-line analyzer applications for BNR control, in *Proceedings of 2004 WEFTEC Annual Conference*, Water Environment Federation, Alexandria, VA.

WEF. (2009). *Energy Conservation in Water and Wastewater Facilities*, Manual of Practice (MOP) No. 32, McGraw Hill, New York.

WEF/ASCE/EWRI. (2006). *Biological Nutrient Removal (BNR) Operation in Wastewater Treatment Plants*, Manual of Practice (MOP) No. 30, McGraw Hill, New York.

WEF/ASCE/EWRI. (2010). *Design of Municipal Wastewater Treatment Plants*, 5th ed., Manual of Practice (MOP) No. 8, ASCE Manuals and Reports on Engineering Practice No. 76, McGraw Hill, New York.

Section VI

Appendices

Innovative Energy Conservation

This section supplements technical information presented in this text on innovative energy conservation measures (ECMs) for wastewater treatment plants (WWTPs) with real-world experience and analyses of ECM implementations at nine plants. Each appendix includes a brief description of the ECMs installed at the facility, capital costs, energy savings, and payback period, along with a short discussion and analysis.

Appendix A. Magnetic Bearing Turbo Blowers at the Green Bay Metropolitan Sewerage District De Pere Wastewater Treatment Facility

The Green Bay, Wisconsin, Metropolitan Sewerage District's De Pere Wastewater Treatment Facility is an 8.0-MGD (average daily flow), two-stage activated sludge plant with biological phosphorus removal and tertiary effluent filtration. The plant had been running five 450-hp Roots multistage centrifugal blowers for the first-stage aeration process. The blowers were approaching the end of their service life and required either extensive rebuild or replacement. An evaluation conducted by the plant operations manager and their engineer determined that replacement of the multi-stage centrifugal blowers with magnetic bearing turbo blowers would reduce the aeration system's energy consumption and associated costs and reduce the level of maintenance required for the aeration system.

Six ABS, Inc., HST 330-hp magnetic bearing turbo blowers were installed as replacements for the five 450-hp multistage centrifugal blowers in first-stage aeration for a capital cost of $850,000. The turbo blower project reduced the electrical energy consumption by approximately 2,144,000 kWh/yr (a 50% reduction) compared to the costs to operate the multi-stage centrifugal blowers ($170,000 per year), representing an estimated $63,758 per year reduction (38% savings) in electrical energy costs and resulting in a payback estimated at 13.3 years. The blower project also contributed to the facility's objective of reducing the wastewater treatment plant's electrical energy consumption to enable a maximum percentage of the required electrical energy being provided by onsite generation using digester gas-fueled microturbines.

A.1 Evaluation of Energy Conservation Measures

A.1.1 Plant Site/Process Description

The De Pere Wastewater Treatment Facility (WWTF) operated by the Green Bay, Wisconsin, Metropolitan Sewerage District (GBMSD), serves the City of De Pere, portions of the village of Ashwaubenon, and portions of the Towns of Lawrence, Belleview, and Hobart. GBMSD acquired ownership of the De Pere WWTF from the City of De Pere on January 1, 2008. The original circa mid-1930s plant (a primary treatment facility with biosolids digestion) was upgraded in 1964 to an activated biosolids process, with chlorination for disinfection. In the late 1970s, a major upgrade to the facility (representing the current operational scheme) included a two-stage activated biosolids process with biological phosphorus removal, tertiary filtration (gravity sand filters), solids dewatering with incineration, and liquid chlorine disinfection. Influent data for the De Pere WWTF are presented in Table A.1.

TABLE A.1

Profile of De Pere WWTF Influent Data (2009)

Parameter	Daily Average
Flow (MGD)	8
BOD (lb/day)	29,070
TSS (lb/day)	18,587
Ammonia-N (lb/day)	Not monitored
Phosphorus (lb/day)	307.5

In 1997, additional upgrades to the facility were initiated, beginning with ultraviolet (UV) disinfection replacing the liquid chlorine system. The chlorine disinfection system is currently maintained as a backup system. Several other major upgrades included replacement of the coarse influent screen with fine screens (1998–1999), renovation of the multimedia tertiary filtration system to a single-media U.S. Filter Multiwash air scour system (1999–2000), and a solids handling upgrade that included installation of two gravity belt thickeners (replacing dissolved air flotation) and the addition of two filter presses (2001–2002).

Figure A.1 presents the process flow diagram for the De Pere WWTF, a two-stage activated biosolids treatment plant (online 1978 to present) with tertiary filtration and design flows as follows:

- Average dry, 9.5 MGD
- Design flow, 14.2 MGD
- Maximum hourly dry, 23.8 MGD
- Maximum hourly wet, 30.0 MGD

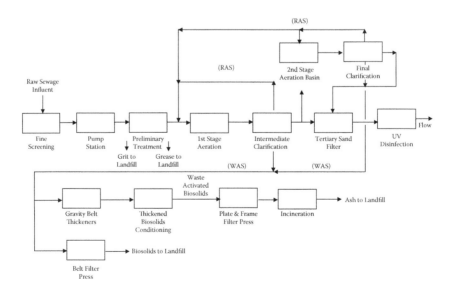

FIGURE A.1

De Pere WWTF process flow diagram (biological treatment). (From USEPA, *Evaluation of Energy Conservation Measures for Wastewater Treatment Facilities*, EPA 832-R-10-005, U.S. Environmental Protection Agency, Washington, DC, 2010.)

Influent to the plant undergoes fine screening and is subsequently pumped to pre-liminary treatment (grit removal followed by grease removal, utilizing two 50 ft × 50 ft clarifiers with grease/scum collection). The influent pump station consists of four 150-hp, 10-MGD pumps. Screenings are disposed of in a landfill. Grit, oil, and grease removed in preliminary treatment units are also disposed of in a landfill.

Biological treatment is conducted in two serial stages, each with a 1.1-MGD anaerobic zone (for phosphorus removal) followed by a 2.2-MGD aeration zone. Approximately 100% of the mixed liquor suspended solids from the aeration zone is recycled to the anoxic zone. Aeration is provided by six 6000-scfm (standard cubic feet per minute), 330-hp turbo blow-ers for the first-stage aeration process and three 4000-scfm, 250-hp multi-stage centrifugal blowers for the second-stage aeration process.

The first-stage biological treatment is followed by clarification—two each, 100-ft-diame-ter, 13.7-ft-side-water-depth clarifiers (one online for each aeration basin). Clarifier effluent from the first-stage biological treatment process can be further treated in the second-stage treatment process; however, all wastewater is currently treated only in the first-stage bio-logical process. The second stage of biological treatment is not currently utilized, as it is not required to achieve discharge compliance. Biological treatment is followed by 125-ft-diameter, 10.9-ft-side-water-depth clarifiers. Clarifier effluent is polished by tertiary sand filtration and disinfected using UV prior to discharge. During periods of high flow, UV disinfection is supplemented by disinfection with liquid chlorine.

Clarifier underflow (waste activated sludge, WAS) from biological treatment undergoes one of two dewatering processes. Approximately 75% of the WAS undergoes thickening (two 2-m gravity belt thickeners), chemical conditioning (lime and ferric chloride), dewa-tering (two 1.5 m × 2 m plate and frame filter press), and incineration (18.75-ft-diameter, 7-hearth, 7500-lb/hr, multiple-hearth incinerator). The incinerator ash is disposed of in a landfill. The balance of the WAS is chemically conditioned with polymer and dewatered in two 2-m belt filter presses. The dewatered sludge is disposed of in a landfill. Filtrate from sludge thickening and dewatering operations is returned to first-stage biological treatment.

The most recent upgrade (2003–2004) replaced the facility's first-stage treatment cen-trifugal blowers with high-speed, magnetic turbo blowers, which was the first installation of this new energy-efficient technology in the country. Because the second-stage aeration process is currently not utilized, only the first-stage process blowers were replaced under the ECM project.

A.1.2 Description of Energy Conservation Measures (ECMs), Drivers, and Issues

In October 2004, the City of De Pere commissioned six new high-speed, magnetic bearing turbo blowers (330-hp HRST Model S90000-1-H-5 manufactured by ABS, Inc.) for the facil-ity's first-stage aeration process. Prior to the blower replacement ECM project, the facility's existing blowers (five 450-hp multi-stage centrifugal blowers) had reached the end of their service life, requiring an extensive and expensive rebuild or replacement. An evaluation by the WWTP operations manager and their engineer determined that replacement of the existing multi-stage centrifugal blowers with magnetic bearing turbo blowers would meet the wastewater utility's mission of providing the highest quality of wastewater treatment for their service area customers at the lowest cost. Six new 330-hp magnetic bearing turbo blowers were commissioned on October 18, 2004, replacing five 450-hp centrifugal blow-ers. The new blowers would help meet this goal by

TABLE A.2

Monthly Averages, 2003 and 2009

		Concentration Range (mg/L)	
Parameter		2003	2009
BOD	Influent	453.10–704.07	380.09–499.43
	Effluent	0.62.–5.69	2.31–6.10
	Permit limit	9	9
TSS	Influent	267.43	229.36–352.46
	Effluent	0.66–3.33	1.39–3.35
	Permit limit	10	10
NH_3	Influent	Not required to be reported	Not required to be reported
	Effluent	0.14–2.44	0.16–4.14
	Permit limit	24 (summer), 34 (winter)	24 (summer), 34 (winter)
P	Influent	5.43–8.24	4.05–6.12
	Effluent	0.15–0.33	0.1–0.24
	Permit limit	1	1

- Increasing the efficiency of the aeration system and reducing energy costs
- Reducing or eliminating normal routine maintenance requirements of the aeration system
- Reducing operating costs through better matching of blower capacity to aeration demand (operating two to three 330-hp turbo blowers vs. two to three 450-hp centrifugal blowers)
- Maintaining high efficiency over a wide range of operation, while offering a high degree of turndown

Implementation of the blower replacement project was initiated by the city of De Pere in response to the need to address the situation in which the aeration system blowers had reached the end of their design operating life. Other drivers for the project included requirements for increasing the plant's capacity, reducing the amount of staff time for maintenance and control of the aeration system, maintaining high-quality effluent, and achieving energy savings. The WWTF management and their engineer had determined that replacement of the blowers was necessary to continue to provide their customers with high-quality wastewater treatment services at the lowest price. Table A.2 provides a comparison of the De Pere WWTF performance prior to the implementation of aeration system improvements (2003) and 2009 (post-ECM implementation) performance.

A.2 Results

A.2.1 ECM Implementation Cost

Table A.3 summarizes the implementation costs for the De Pere turbo blower ECM project. In the table, capital cost included replacement of the plant's medium power supply voltage system (2400 volts) to a lower voltage system (480 volt) to provide compatibility with the new turbo blowers.

TABLE A.3

ECM Implementation Cost

Cost Category	Cost (in 2004$)
Capital cost	$850,000
Installation cost	Not available[a]
Total cost	Not available[a]

[a] The blower replacement project was implemented as part of a larger ($2,000,000) plant infrastructure improvement project. Installation costs exclusively associated with the blower upgrade are not available, as these costs are site specific and equipment specific; however, for estimating purposes, installation costs often run 10 to 15% of capital costs.

A.2.2 Energy

Table A.4 summarizes the De Pere WWTF electrical energy consumption and savings for the aeration system improvement project.

A.2.3 Payback Analysis and Benefits

The blower replacement project did not focus solely on investment payback, but weighed heavily on energy conservation, full automation of the aeration process while maintaining high-quality effluent, ensuring protection of the environment, and gaining equipment dependability for the plant operation. The old aeration system experienced frequent blower surging when multiple blowers were in operation, requiring that the aeration system be operated in the manual mode. A simple payback analysis is determined by dividing the available blower project cost ($850,000 from Table A.3) by the projected annual electricity cost savings following the project implementation. The electrical energy costs savings provided by the turbo blower implementation resulted in a project payback in 13.3 years.

TABLE A.4

Electricity Use and Estimated Savings

Year	Annual Electrical Energy Use	Electrical Energy Cost	
		Rate	Annual Cost
Prior to ECM Implementation			
2003	4,325,700 kWh	$0.0393/kWh	$170,000
Following ECM Implementation			
2005	2,181,725 kWh	$0.0487/kWh	$106,250
Savings			
	2,143,975 kWh (50% reduction)		$63,758 (38% reduction)

A.3 Conclusions

A.3.1 Factors Leading to Successful ECM Implementation/Operation

This project was unique in that the blower equipment to be installed was the first of its kind in this country. This retrofit was intriguing and vitally important to the De Pere WWTP operation, creating an ownership stake in the project. All stakeholders involved in this project (city officials, management, operations staff, and the utility's engineering firm) worked as a team.

A.3.2 Impact on Other Operating Costs Resulting from ECM Implementation

Energy conservation (see Table A.4) was the most important factor, along with the importance of replacing outdated and failing equipment. In addition to energy savings resulting from this project, the following benefits were also realized.

A.3.2.1 Labor

Maintaining the old blower equipment was labor intensive. Maintenance needs for the old system included greasing the centrifugal blower motors, monitoring and filling oil levels in the blower bearing lubrication reservoirs, changing out the inlet filters, seal replacement, and vibration analysis on a weekly basis. Operationally, the old system was designed for coarse air bubble aeration; therefore, when fine-bubble diffusers were installed for energy conservation, blower surging (especially when multiple blowers were in operation) became a major problem when trying to operate in the automatic mode. As a result, operation of the aeration system required manual control of the blowers. System demand variations required staff to make changes to blower output many times throughout the day. The new high-speed turbo blower system is SCADA controlled, making the operation fully automatic and reducing staff surveillance of the aeration process.

A.3.2.2 Maintenance

Routine maintenance to the turbo blowers is minimal. Changing of the inlet filters is done on an as-needed basis, normally once per year, unless outside conditions become very dusty. The location of the new blowers (away from other plant process areas) contributed to reducing air inlet filter maintenance. Additionally, without the presence of any wear parts (e.g., bearings, gears), maintenance on the turbo blowers is virtually eliminated and vibration analysis is not required for the turbo blowers.

A.4 Lessons Learned

As a result of this project and facility upgrade projects implemented subsequent to the blower replacement, De Pere WWTF management learned that developing and executing a carefully planned program for replacing outdated and failing equipment can be accomplished without increasing treatment costs and customer user fees. Additionally, the blower replacement project provided other benefits and lessons that included the following:

- The magnetic bearing turbo blowers operated at significantly lower noise levels (75 dBA compared to 100 dBA) and are vibration free compared to the multi-stage centrifugal blowers, providing a comfort benefit to wastewater treatment staff working in the blower room. With the multi-stage centrifugal blowers, spending any extended time in the blower room was difficult to tolerate and unsafe without hearing protection.
- Heat from the turbo blower's cooling air exhaust is recirculated to plant buildings, reducing the demand for auxiliary heat.
- The De Pere WWTF was the first in the United States to install magnetic bearing turbo blowers. Following a thorough investigation of the magnetic bearing turbo blower technology, the stakeholder/project team concluded that the benefits offered by this new cutting-edge technology far outweighed the risk.

Reference

Shumaker, G. (2007). High-speed technology brings low costs, *Water and Wastes Digest*, August 13.

Appendix B. Turblex® Blowers and Air Flow Control Valves on the Sheboygan Regional Wastewater Treatment Plant

The Sheboygan Regional Wastewater Treatment Plant in Sheboygan, Wisconsin, is an 18-MGD (average daily flow) activated biosolids plant with biological phosphorus removal. The plant has been running four 250-hp positive-displacement blowers for the aeration basins. Typically, two blowers were required in the summer to provide sufficient aeration, with the remaining two blowers available as standby units. In 2005, one of the blowers failed and inspection of the remaining three indicated they would all require complete rebuilds. Instead of rebuilding the existing blowers, plant managers decide to replace the four positive-displacement blowers with two 300-hp, high-efficiency-motor Turblex® centrifugal blowers equipped with inlet guide vanes and variable diffuser vanes. One of the centrifugal blowers was sufficient to provide the necessary aeration year-round with the second unit being operated as a standby unit. Following commissioning of the Turblex blowers, the plant operators experienced difficulty controlling dissolved oxygen (DO) in the individual aeration basins. DO levels reached as high as 6 mg/L during evening hours and during the winter months, wasting blower output and energy. To correct this problem, air flow control valves were installed on the headers to each aeration basin. PLC programming was also upgraded to provide improved control of the DO levels through automatic operation of the air flow valves and blowers.

The capital cost of the Turblex blowers was $504,000 with a total project cost (capital plus installation) of $790,000. The plant received a $17,000 energy efficiency grant that reduced project costs to $773,000. The control valves had an installed cost of $128,000.

The centrifugal blowers saved the plant an estimated 6.2% in annual electricity bills, amounting to an average yearly energy savings of $25,644 for the first 3 years of operation from 2005 to 2008. The addition of the control valves more than doubled the energy savings, resulting in total savings of $63,889 in 2009 and a payback period for both projects of 14 years. The reduction in energy consumption averaged approximately 358,000 kWh/yr for the blower replacement project. The installation of the air control valves resulted in an additional reduction in electrical energy consumption of 459,000 kWh/yr. Although the plant had delayed installing the air control valves because of cost issues, they found that poor air control limited their ability to realize the full potential energy savings of the new blowers.

B.1 Evaluation of Energy Conservation Measures

B.1.1 Plant Site/Process Description

The Sheboygan Regional Wastewater Treatment Plant (WWTP) serves approximately 68,000 residential customers in the cities of Sheboygan and Sheboygan Falls, the Village of Kohler, and the towns of Sheboygan, Sheboygan Falls, and Wilson. The plant was originally constructed in 1982 as a conventional activated biosolids plant using turbine aerators with

TABLE B.1

Profile of Sheboygan WWTP Influent Data (2009)

Parameter	Daily Average
Flow (MGD)	11.78
BOD (mg/L)	175
TSS (mg/L)	203
Ammonia-N (mg/L)	Not monitored
Phosphorus (mg/L)	5.7

sparger rings. In 1990, the plant was upgraded to include a fine-bubble diffused air system with positive-displacement blowers. From 1997 through 1999, additional improvements were made to the facility to implement biological nutrient removal and to the bar screens, grit removal facilities, biosolids storage tanks, and the primary and secondary clarifiers. The plant currently operates as an 18.4-MGD biological nutrient removal plant with fine screens, grit removal, primary clarification, biological nutrient removal, activated biosolids thickening, and liquid (6% solids) biosolids storage. Table B.1 provides average daily influent data for the plant. Figure B.1 provides a process flow diagram of the plant treatment scheme.

Influent to the plant goes through two automatic self-cleaning fine screens. A 20-ft-diameter cyclone separator removes grit before the wastewater enters primary clarification. Primary clarification is provided by four primary clarifiers. Secondary biological treatment is conducted in six basins. The first two basins are anaerobic to provide phosphorus removal. They are configured with baffles in an "N" pattern. The remaining four basins are currently aerated using two Turblex blowers. Following aeration, secondary clarification is provided by four clarifiers. Return activated sludge (RAS) from the clarifiers is sent to the anaerobic zone. A portion of the RAS is conveyed upstream of the primary clarifier. Plant effluent is disinfected with chlorine and is then dechlorinated before discharge into Lake Michigan.

The combined primary and secondary biosolids underflow from the primary clarifier (waste sludge, or biosolids) is sent to three primary anaerobic digesters. From the primary digesters the biosolids flow to a single secondary anaerobic digester. Methane from the digesters is used to provide heat to the digesters as well as fuel for ten 30-kW microturbines that provide electricity to the plant. Two belt thickeners (2 m and 3 m) increase the solids content of the digested biosolids from 2.5% to 6% solids. Digested, thickened biosolids are held in two storage tanks before being land applied.

B.1.2 Description of Energy Conservation Measures (ECMs)

The Sheboygan plant implemented improvements to its aeration system in order to improve dissolved oxygen (DO) control and to replace failing blower equipment. The upgrade consisted of

- Replacing the four 250-hp positive-displacement blowers with two single-stage centrifugal blowers with inlet guide vanes and variable outlet vanes (Turblex model KAS SV-GL210 blowers equipped with 350-hp premium efficiency motors)
- Replacing the DO blower controls
- Upgrading the SCADA system

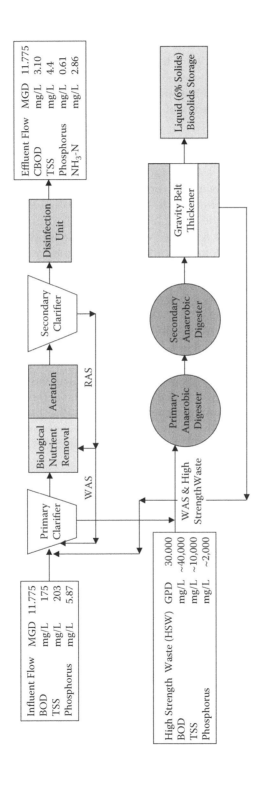

FIGURE B.1

Sheboygan wastewater treatment plant process flow diagram. (From USEPA, *Evaluation of Energy Conservation Measures for Wastewater Treatment Facilities*, EPA 832-R-10-005, U.S. Environmental Protection Agency, Washington, DC, 2010.)

- Installing air control valves on the headers to the individual aeration basin (AUMA valve actuators and DeZURIK butterfly valves were used on the headers of the 6 aeration basins)
- Upgrading the PLC programming for the blowers

The Turblex blowers were installed in the summer of 2005, and the aeration basin header control valves were installed in the spring of 2009.

B.1.3 Description of ECM Project Drivers and Issues

Prior to 2005, the Sheboygan plant was operating its aeration system using four 250-hp Gardner Denver positive-displacement blowers (installed in 1990). During the summer months, two blowers were required to provide sufficient aeration, with the remaining two serving as standby units. In 2005, one of the blowers failed. This prompted an investigation into the remaining blowers, the results of which were that all three of the remaining operational blowers would require total rebuild. The plant managers determined that two new Turblex blowers with larger, energy-efficient motors would allow a single blower to meet year-round aeration needs while saving energy and allow the second blower to be used for standby (i.e., the four 250-hp positive-displacement blowers could be replaced with two 350-hp centrifugal blowers).

Following commissioning of the Turblex blowers, the plant operators experienced difficulty controlling DO in the individual aeration basins. DO levels could reach as high as 6 mg/L during evening hours and during the winter months, wasting blower output and energy. In the spring of 2009 the plant installed air flow control valves on the headers to each aeration basin to control DO levels in the basins to match aeration requirements and to decrease wasted blower output and energy consumption. PLC programming was also upgraded to provide improved control of the DO levels through automatic operation of the air flow valves and blowers. Tables B.2 and B.3 show the influent and effluent qualities both before and after implementation of the ECMs.

B.2 Results

B.2.1 ECM Implementation Cost

Table B.4 summarizes the implementation costs for the new blowers and motors. Table B.5 summarizes the costs for the air control valves. The system received a Focus on Energy™ grant that offset a portion of the costs of the blowers. Focus on Energy is a quasi-governmental agency in Wisconsin that distributes grants for electricity conservation. The source of this funding is a tax levied on electric utility bills.

The installation costs listed in Tables B.4 and B.5 for the blower and control valve ECM projects include modifications to the existing blower building to pull in outside air, installing new electrical conductors and large diameter wire to accommodate the increased blower motor horsepower, installing new soft start controls and cabinets, installing valve actuators on the discharge valves required for the blower start-up sequence, and modifications done to the aeration basin air piping to install individual basin air flow control valves and the air blow-off channel air.

TABLE B.2

Monthly Averages, 2003 and 2009

		Concentration Range (mg/L)	
Parameter		2003	2009
BOD/CBOD	Influent	246.00	175.00
	Effluent	12.00	3.10
	Permit limit	30.00	25.00
TSS	Influent	244.00	203.00
	Effluent	6.10	4.40
	Permit limit	30.00	30.00
NH_3	Influent	Not measured	Not measured
	Effluent	2.40	2.86
	Permit limit	N/A	23.00
P	Influent	5.96	4.70
	Effluent	0.75	0.60
	Permit limit	1.00	1.00

TABLE B.3

Daily Maximums, 2003 and 2009

		Concentration Range (mg/L)	
Parameter		2003	2009
BOD/CBOD	Influent	420.0	397.0
	Effluent	64.0	12.0
TSS	Influent	1650.0	872.0
	Effluent	63.0	12.0
NH_3	Influent	Not measured	Not measured
	Effluent	19.0	16.4
P	Influent	11.8	9.8
	Effluent	4.5	2.2

B.2.2 Energy

Table B.6 summarizes the electricity savings from the two components of the aeration system upgrade ECM project. The utility estimated an average reduction in annual energy costs of $25,644 following commissioning of the Turblex blowers (for the years 2006 through 2008), representing an average reduction in annual electrical energy consumption

TABLE B.4

Blower ECM Implementation Costs

Cost Category	Cost (in 2005$)
Capital cost	504,000
Installation costs	286,000
Focus on Energy grant	–17,000
Total cost	773,000

TABLE B.5

Control Valve ECM Implementation Costs

Capital Category	Cost (in 2009$)
Capital cost	60,000
Installation cost	68,000
Total cost	128,000

TABLE B.6

ECM Implementation Electrical Energy Savings

	Energy Consumption and Savings		Energy Costs and Savings	
Year	kWh Used	kWh Annual Reduction	Rate ($0.00/kWh)	Annual Savings
	Before ECM Implementation			
2004	2,760,000	Baseline year	$0.0538	Baseline year cost ($148,888)
	After ECM Implementation			
2006[a]	2,402,000	358,000 (13%)	$0.0665	$23,807
2007[a]	2,402,000	358,000 (13%)	$0.0720	$25,776
2008[a]	2,402,000	358,000 (13%)	$0.0764	$27,350
2009[b]	1,943,000	817,000 (30%)	$0.0782	$63,889

[a] Electrical energy savings, from blower upgrade only, estimated by utility; blower electrical energy consumption is not submetered.

[b] Electrical energy savings from blower upgrade and air control valve combined, estimated by utility; blower electrical energy consumption is not submetered.

of 358,000 kWh over this period. Based on the three previous years' average annual energy cost savings, implementation of the air control valves resulted in an additional energy cost savings of $38,245 for 2009, representing an additional reduction in annual energy consumption of 459,000 kWh.

B.2.3 Payback Analysis/Benefits

The ECMs presented in this case study were part of a larger plan implemented by the Sheboygan WWTP management to become energy self-sufficient. The facility installed 30-kW microturbines in 2006 that allow it to burn biogas from the anaerobic digesters to provide electricity and heat to the plant. In conjunction with the microturbine implementation, the plant has undertaken various energy conservation measures to reduce energy consumption and increase the percentage of the plant's electrical power that can be supplied by the microturbines.

The plant superintendent estimated that the Turblex blowers will have a payback period of 14 years and included the avoided cost of rebuilding the positive-displacement blowers in the payback analysis. Installing the control valves resulted in a shorter payback period compared to the blowers. Although there was less than a year's worth of operating data for the air flow control valves, the initial 2009 data showed an additional $38,245 decrease in electricity costs (from the average savings of the previous 3 years from the blower replacement). Extrapolating this figure gives a payback period of less than 4 years for the air control valves. If the costs and energy savings of both the air control valves and blowers are taken together and the 2009 energy cost savings are taken as typical, the payback period is 14 years.

Although the payback period for this ECM project is longer than 10 years, the project was a significant component in the utility's objective of reducing the facility's electrical power demand toward achieving the status of meeting total plant electrical demand by onsite generation using digester gas-fueled microturbines.

B.3 Conclusions

B.3.1 Factors Leading to Successful ECM Implementation/Operation

The facility staff took a proactive approach to saving energy and reducing its dependence on electrical utility purchased power. Staff evaluated and continues to evaluate the energy efficiency of all projects instead of implementing the least capital cost fix for addressing failing equipment. The control valves significantly improved both the resulting energy cost savings and efficient operation of the new blowers. Controlling the air flow and eliminating wasted blower output were essential to realizing the full potential of the new blower equipment.

B.3.2 Impact on Other Operating Costs Resulting from ECM Implementation

In addition to energy cost savings, the ECM project provided the following additional benefits.

B.3.2.1 Labor

Using automatic air flow control, plant operators no longer need to make seasonal adjustments to the aeration system valves for the individual aeration basins to control DO concentration. Plant operators were required to make manual adjustments to the aeration system drop leg valves as needed to maintain the appropriate dissolved oxygen level in the aeration basins. In addition, twice a year, at the beginning of winter and the beginning of summer when influent water temperature changes, the plant operators spent additional time adjusting the drop leg air valves. When the operating blower was rotated once a year, the operators again adjusted the drop leg air valves to maintain the proper DO level in the aeration basins. The adjustments to the drop leg air valve required approximately 90 man-hours annually (~$2250/year).

B.3.2.2 Maintenance

Less maintenance is required on the air piping system with the new blowers in place. The former positive displacement blowers caused a hammering effect on the air piping system, which created the need for frequent maintenance. The repairs to the leaking air header system resulting from the hammering effect required 30 man-hours annually (~$750/year).

B.4 Lessons Learned

The plant initially postponed the installation of the control valves because of cost. Without the control valves, DO concentrations in the individual aeration basins could not be properly controlled, leading to wasting both blower output and energy. Installation of the control valves not only improved process performance (by properly controlling DO concentrations in the aeration basins) but also resulted in greater energy cost savings than might have been achieved by blower replacement alone.

Appendix C. Upgrade from Mechanical Aeration to Air-Bearing Turbo Blowers and Fine-Bubble Diffusers at the Big Gulch Wastewater Treatment Plant

The Big Gulch Wastewater Treatment Plant (WWTP), owned and operated by the Mukilteo Water and Wastewater District in Washington, is a 2.6-MGD (average daily flow) oxidation ditch plant operating two parallel oxidation ditches. Ditch A treats approximately 40% of the plant flow, and Ditch B treats approximately 60% of the flow. To address increases in biochemical oxygen demand (BOD) and total suspended solids (TSS) loadings, the oxidation ditch aeration system has been upgraded as follows:

- The existing mechanical brush (rotor) aeration systems in the oxidation ditches were replaced with Sanitaire® fine-bubble diffusers and three air bearing turbo blowers (K-turbo, TB 50-0.65).

- Dissolved oxygen (DO) probes with a PLC-based control system were installed to automate blower operation.

- A dNOx anoxic control system was installed to detect the nitrate knee, which is the point in the nitrogen conversion process when complete denitrification has occurred—that is, NO_3 has been converted to N_2. With the dNOx control system, the blowers are allowed to go idle as loading increases and automatically switch on when the nitrate knee is detected.

The aeration systems were replaced sequentially, with Ditch A being upgraded in 2008 and Ditch B in 2009. The total cost of the Ditch A upgrade (including the removal of one aeration rotor and installation of one turbo blower, diffusers, probes, and controls) was $487,066. The Big Gulch WWTP received a $39,191 grant from its electric utility to offset the cost of the project, bringing the total costs down to $447,875. The total cost of the Ditch B upgrade (including the removal of four aeration rotors and the installation of two turbo blower, diffusers, probes, and controls) was $1,045,023. An additional utility company grant of $46,594 reduced the Ditch B project cost to $998,449.

Observed energy savings following the aeration system upgrade were 148,900 kWH for 2010 (average energy use of 1,405,540 kWH for 2005 through 2008 minus energy use of 1,256,640 kWh in 2010). At a current electricity rate of $0.072/kWH, this translates to an electricity cost savings of $10,721. Based solely on energy cost savings, this project shows a 135-year payback. The Big Gulch project serves as an example of energy savings derived as a collateral benefit from a major plant upgrade and expansion. Although a greater reduction in energy consumption would be expected from this type of aeration system upgrade, it should be noted that the plant experienced a 40% increase in influent organic loading during the construction and commissioning period, compared to the period prior to the aeration system upgrade.

If an alternative payback analysis is considered that accounts for the 40% increase in the plant's organic loading during the construction and commissioning of the Ditch A and Ditch B upgrades, the project shows a significantly lower payback of 33 years. In the year following commissioning of the total project (2010), the plant removed approximately 34% more carbonaceous biological oxygen demand (CBOD) compared to the period 2004 through 2008 (prior to implementation of the oxidation ditch upgrades) while consuming less electrical power (an average of 1.59 kWh/lb CBOD during the period 2004 through 2008 compared to an estimated 1.06 kWh/lb CBOD in 2010). This translates to a savings in electricity cost of $0.037 per pound of CBOD removed and an estimated $43,756 for 2010 (a 33-year payback for the total project cost of $1,446,304).

In addition to improved treatment at lower electric consumption, theBig Gulch WWTP experienced other benefits from the aeration system improvements; for example, labor and maintenance costs have been reduced, as the turbo blowers do not require the level of maintenance required for the mechanical brush system. Also, chlorine use (for bulking biosolids control) has decreased due to improved settling.

C.1 Evaluation of Energy Conservation Measures

C.1.1 Site/Facility/Process Description

The Big Gulch WWTP provides wastewater treatment service for 22,455 people residing in portions of the City of Mukilteo and Snohomish County in Washington. Originally constructed in 1970, the WWTP consisted of a coarse bar screen and single oxidation ditch using brush rotor aerators, followed by a secondary clarifier and chlorine disinfection. Between 1989 and 1991, the Big Gulch WWTP underwent significant upgrades including the following:

- New headworks with a grit removal channel
- Influent screw pumps
- Selector tank
- Second oxidation ditch
- Third secondary clarifier
- Aerobic biosolids holding tanks within a rotary drum thickener
- Biosolids return piping
- Scum and waste active biosolids pumps
- Biosolids pumps
- Biosolids dewatering belt filter press
- Chlorine contact chamber

Subsequent to the 1991 facility upgrade, the following upgrades to the treatment plant were implemented:

- Influent screening (perforated-plate fine screens)
- Submersible mixers (in the oxidation ditches)
- Ultraviolet (UV) disinfection (replacing chlorine disinfection)

TABLE C.1

Profile of Big Gulch WWTP Influent Data (2004 to 2010)

Parameter	Average	Minimum	Maximum
Flow (MGD)	1.68	1.21	2.40
CBOD (mg/L)	217	116	462
TSS (mg/L)	255	131	398

To address a need for additional oxidation ditch aeration capacity to handle intermittent increases in BOD loading, the aeration system in both ditches was upgraded with fine-bubble diffusers and automatically controlled turbo blowers. As a result of upgrading the aeration system from mechanical rotors to automatically controlled turbo blowers and installing the diffused air system, the Big Gulch WWTP realized an incremental savings in energy cost, as detailed later in Table C.6. Influent data for the Big Gulch WWTP are presented in Table C.1. Figure C.1 presents a process flow diagram for the Big Gulch WWTP.

Influent to the plant passes through a perforated-plate mechanical fine screen (rate capacity of 6.5 MGD) into a gravity grit channel. Effluent from the grit removal system is returned to the headworks, and grit is sent to the dumpsters. Degritted influent, combined with return active sludge (RAS) from the secondary clarifiers and filtrate from the sludge dewatering belt filter press, is lifted to the selector mixing basin using the two influent screw lift pumps (3.83-MGD capacity each). Selector mixing basin effluent is conveyed to the oxidation ditches via overflow channels equipped with adjustable weir gates to distribute the flow to the ditches (40% to Ditch A and 60% to Ditch B). The two oxidation ditches, operating in parallel and providing a combined 18-hour hydraulic residence time, are followed by three secondary clarifiers.

Effluent from the secondary clarifiers is conveyed to the UV disinfection system. The UV system consists of 96 lamps and provides 35 mJ/cm^2 (energy) at a peak flow of 8.7 MGD (based on 60% UV transmittance). The UV disinfection system produces an effluent with fecal coliform counts below the facility's permit limit of 200 colonies/100 mL (monthly average).

Waste activated biosolids and scum from the secondary clarifiers are transferred via a rotary lobe pump to a pair of two-cell aerobic biosolids holding tanks for aerobic digestion, producing Class B biosolids. In 2006, the aerobic biosolids digestion system was upgraded with fine-bubble air diffusers and positive-displacement blowers. The biosolids

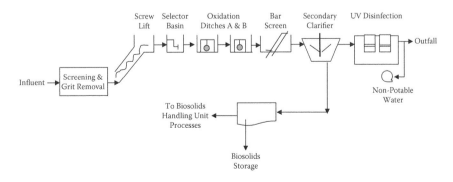

FIGURE C.1
Simplified schematic flow diagram of Big Gulch WWTP, an activated biosolids treatment plant with UV disinfection.

TABLE C.2

Big Gulch Activated Biosolids System Overview

Component	Description
Oxidation Ditch A	Constructed in 1970 640,000 gallons Receives/treats approximately 40% of plant flow
Oxidation Ditch B	Constructed in 1991 1,070,000 gallons Receives/treats approximately 60% of plant flow
Fine-bubble diffuser aeration system	Installed in 2008 (Ditch A) Installed in 2009 (Ditch B) Aeration provided by three, high-speed turbo blowers Horizontal momentum maintained by submersible pump
Monitoring and control system	Dissolved oxygen (DO) and oxidation reduction Potential (ORP) probes located in Ditch A and Ditch B Controls blower air flow to oxidation ditches (DO/PLC) Provides anoxic phase zone DO control for nitrate reduction (ORP/PLC)
Secondary clarifier 1	Constructed in 1970 Diameter = 58 ft, side-water depth = 9 ft Center feed, peripheral withdrawal (incorporated in 1990s)
Secondary clarifier 2	Constructed in 1980s Diameter = 54 ft, side-water depth = 9 ft Peripheral feed, center weir withdrawal
Secondary clarifier 3	Constructed in 1991 Diameter = 54 ft, side-water depth = 13 ft Center feed, peripheral withdrawal

is thickened through either settling in the aerobic biosolids holding tanks or rotary drum thickening. In 2007, the rotary drum biosolids thickener was installed to increase digestion capacity. Digested biosolids are dewatered using a gravity belt dewatering press and the dewatered biosolids are transported for land application. Table C.2 provides additional detail on the active biosolids process.

C.1.2 Description of Energy Conservation Measures (ECMs), Drivers, and Issues

To address regulatory agency compliance notifications (and associated corrective action recommendations) regarding the operation of the treatment plant (the treatment plant was operating at design capacity and exceeding BOD_5 and TSS influent loading rate limitations contained in the permit), the Mukilteo Water and Wastewater District implemented improvements to the aeration system at the Big Gulch WWTP to increase the oxidation ditch aeration capacity and provide additional treatment for the intermittent peak BOD and TSS loadings. These improvements were based on recommendations provided by the Washington State Department of Ecology and a 2008 wastewater treatment plant capacity study and engineering report by the District's engineer, Gray & Osborne. The aeration system improvements implemented by the Mukilteo Water and Wastewater District at the Big Gulch WWTP consisted of the following:

- Replace the existing mechanical brush aeration systems in Oxidation Ditch A and Oxidation Ditch B with a Sanitaire fine-bubble diffuser aeration system and turbo blowers. Each oxidation ditch has one 50-HP, high-speed turbo blower (K-Turbo, TB 50-0.65) equipped with a variable-frequency drive. A third 50-HP blower serves as a spare. A portion of the mechanical and electrical power infrastructure for Ditch B was included in the cost reported for the Ditch A project (although this work could not be defined as a separate line item by the utility).

- Install DO probes and a PLC-based control system that provide automated monitoring of the oxidation ditch DO concentration and automatic response control of the aeration blowers.

- Implement an automated ORP-based control system (dNOx anoxic control system) to detect the occurrence of the nitrate knee in the anoxic zone (i.e., the ORP vs. time inflection point indicating when denitrification is complete). The dNOx system uses ORP readings to detect nitrate levels during the anoxic phase. As loading to the activated biosolids system increases, the blowers are allowed to idle, and at the time of the nitrate knee the blowers are automatically turned back on.

The aeration system upgrades (fine-bubble diffusers, turbo blowers and controls, and blower building) were commissioned in Oxidation Ditches A and B in 2008 and 2010, respectively. These system improvements at the Big Gulch WWTP were implemented to accommodate an increase in BOD and TSS loadings at the facility. Primary drivers for the project included the regulatory requirements for increasing the plant's capacity and improving effluent quality. Reducing energy consumption was a secondary consideration. Prior to the design phase of the aeration system improvements project, the local publicly owned electric utility (Snohomish County PUD) was actively seeking ECM projects to grant funding assistance. The Mukilteo Water and Wastewater District had already received public and regulatory agency approval to implement the aeration system ECM project as the Big Gulch WWTP but the project had not yet been started. The aeration system improvements ECM project that was being planned for the Big Gulch WWTP satisfied the electric utility's project criteria, and the facility was provided a financial incentive of $39,191 (Oxidation Ditch A) and $46,594 (Oxidation Ditch B) to help offset project costs (see Tables C.4 and C.5).

As a result of the aeration system improvements ECM project implementation, the Big Gulch WWTP has decreased its energy usage, benefited from reduced maintenance requirements, reduced noise levels, and decreased the amount of aerosols emitted from the oxidation ditches. Many of these improvements and their impact on operations are listed in Table C.3, which provides a comparison of the WWTP's performance prior to implementation of the aeration system improvements (2008) and current performance, after the ECM implementation. According to the 2008 capacity study conducted by Gray & Osborne, the intermittent spikes in the influent loading that led to the loading exceedances were probably caused by non-residential (i.e., commercial or industrial) loadings. Permit level exceedances are highlighted in Table C.3.

C.2 Results

C.2.1 ECM Implementation Costs

The Ditch A project involved removing one aeration rotor and installing one turbo blower, air piping, and air diffusers. A portion of the mechanical and electrical power infrastructure for Ditch B was included in the cost reported for the Ditch A project (although this work could not be defined as a separate line item by the utility). The total cost of the Ditch A

TABLE C.3

Big Gulch WWTP Influent and Effluent Data

		CBOD[a]		TSS[b]	
Year	Average Flow (MGD)	Influent Loading (lb/day)	Effluent Loading (lb/day)	Influent Loading (lb/day)	Effluent Loading (lb/day)
Before ECM Implementation					
2004	1.80	2178	47	2771	103
2005	1.74	2052	45	2603	102
2006	1.81	2751	64	2805	78
2007	1.82	2611	53	4340	118
2008	1.56	2809	72	4087	117
After ECM Implementation					
2009 (Ditch A)	1.45	3579	117	3910	258
2010 (Ditches A and B)	1.48	3327	87	4063	194

Source: USEPA, *Evaluation of Energy Conservation Measures for Wastewater Treatment Facilities*, EPA 832-R-10-005, U.S. Environmental Protection Agency, Washington, DC, 2010.

Note: Average $CBOD_5$:BOD_5 concentration ratio of 0.88 based on January 2004 to June 2007 data measurements. In 2006, the BOD_5 loading limitation was exceeded for 3 months.

[a] NPDES permit limit, 3953 lb/day influence; 544 lb/day effluent. Maximum capacity based on capacity analysis (5813 lb/day). Request for increase in loading limit: 6039 lb/day.

[b] NPDES permit limit, 3605 lb/day influence; 653 lb/day effluent. Maximum capacity based on capacity analysis (6082 lb/day). Request for increase in loading limit: 6082 lb/day.

aeration system upgrade (including the turbo blowers, diffusers, probes, and control, as well as capital and installation) was $487,066. Construction costs were $389,653, and the remainder (or 25% of the total costs) were engineering and construction administration costs of $97,413. The Big Gulch WWTP received a $39,191 grant from its electrical utility (Snohomish County PUD), which was used to offset the construction costs incurred for the ECM implementation. Implementation costs for Oxidation Ditch A are presented in Table C.4.

Ditch B (whose treatment capacity is 1.5 times the capacity of Ditch A) involved removing four aeration rotors; installing two blowers, air piping, and diffusers; and construction of a blower building to house all three blowers. The total cost of the Ditch B aeration system upgrade (which included blower housing construction costs) was $1,045,022. Construction costs were $836,018, and the remainder (or 25% of the total costs) were engineering and

TABLE C.4

Implementation Cost Estimates for Oxidation Ditch A

Cost Category	Cost (2007$)
Construction cost	$389,653
Engineering and construction administration (25%)	$97,413
Subtotal	$487,066
Incentives[a]	–$39,191
Total	$447,875

[a] Incentive offered by Snohomish County PUD was used to offset construction costs incurred for the ECM implementation.

TABLE C.5

Implementation Cost Estimates for Oxidation Ditch B

Cost Category	Cost (2007$)
Construction cost	$836,018
Engineering and construction administration (25%)	$209,005
Subtotal	$1,045,023
Incentives[a]	–$46,594
Total	$998,429

Note: Oxidation Ditch B upgrades included construction of build-ing to house the three blowers for Oxidation Ditches A and B.

[a] Incentive offered by Snohomish County PUD was used to offset construction costs incurred for the ECM implementation.

construction administration costs of $209,005. The Big Gulch WWTP received a $46,594 grant from its electric utility (Snohomish County PUD), which was used to offset the con-struction costs incurred for the ECM implementation. Implementation cost estimates for Oxidation Ditch B are presented in Table C.5.

C.2.2 Energy

Table C.6 summarizes the Big Gulch WWTP electrical energy consumption and costs prior to and following implementation of the aeration system improvements to Oxidation Ditch A (2009) and Oxidation Ditches A and B (2010). The last column presents an estimate of the electrical cost savings per year at the plant. Observed energy savings following the Oxidation Ditch A and Ditch B aeration system upgrades was 148,900 kWh for 2010, a reduction of nearly 11% (compared to the average electricity consumption to the years 2005 through 2008 before the ECM project implementation). At a current electricity rate of $0.072/kWh, this translates to a total cost savings of $10,721 in 2010.

TABLE C.6

Electricity Use and Estimated Savings Based on ECM Implementation in Oxidation Ditches A and B

Year (A)	Total Electricity Use (kWh) (B)	Average Flow (MGD) (C)	Electricity Use/Flow (kWh/MGD) (D = B/C)	Average Electricity Rate ($/kWh) (E)	Estimated Electricity Cost Savings ($) $(B_{pre\text{-}ECM} - B_{post\text{-}ECM}) \times E$
		Before ECM Implementation			
2005	1,358,720	1.74	779,380	0.068	—
2006	1,355,440	1.81	750,590	0.068	—
2007	1,353,200	1.82	743,857	0.069	—
2008	1,554,800	1.56	997,199	—	—
		After ECM Implementation			
2009 (Ditch A)	1,261,600	1.45	867,576	0.070	10,076
2010 (Ditches A and B)	1,256,640	1.48	849,081	0.072	10,721

Note: A full year of data for 2010 was not available at the time of the study report, so the average electricity use from January 2010 through August 2010 was used for a monthly projection of electricity use from September 2010 through December 2010. 2010 estimated electricity cost savings are based on average elec-tricity usage prior to any ECM implementation (2005 through 2008, before upgrades were implemented in Oxidation Ditch A)

C.2.3 Payback Analysis/Benefits

Dividing the total project cost minus the incentive by the 2010 observed energy savings of 148,900 kWh and rate of $0.072/kWh, the simple payback for this ECM project is 135 years. From strictly an energy savings perspective, this project does not represent an economic benefit. The primary driver for this project was the utility's need to replace aging and undersized aeration equipment in Oxidation Ditches A and B as a means of addressing chronic wastewater discharge compliance issues. The utility took this opportunity to also consider incorporating energy-efficient technologies into the treatment facility upgrade. Reducing energy consumption was not the primary motivation for implementing the aeration system upgrades, but rather a consideration that was prioritized by the utility when planning the required wastewater treatment facility upgrade.

The Big Gulch ECM project case study is a good example of energy savings derived as a collateral benefit from a major plant upgrade and expansion. As noted earlier, the primary objective of this project was to increase the plant's capacity to accommodate increases in BOD and TSS loadings to improve effluent quality to comply with permit discharge limits and restore permit compliance. Reducing energy consumption was a secondary consideration. In fact, providing additional oxygen to satisfy the increased organic loading was expected to increase the total energy used by the facility. Note that the average CBOD influent loading following the upgrade of the oxidation ditches' aeration systems (2009 and 2010) was 3453 lb/day, an increase of 40% compared to the average CBOD loading of the 5 years prior to the aeration system upgrade project (2480.2 lb/day for the period 2004 to 2008). However, the replacement of existing mechanical brush aerators in the oxidation basins with new fine-bubble diffusers and turbo blowers did result in actually lowering the overall energy used by the Big Gulch Facility. While this project resulted in a long payback period based solely on energy savings, the benefits of increased plant capacity and returning the plant to compliance should not be overlooked.

Using electricity consumption and cost per pound of CBOD removed, an alternative payback analysis was conducted. At the average annual electricity cost and pounds CBOD removed for the period 2005 through 2008 or $96,692 per year and 884,760 lb CBOD removed per year (respectively), the electricity cost per pound of CBOD removed for this period is $0.109. For 2010, the annual cost of electricity (based on plant operating data for the period January 20101 through August 2010) and pounds CBOD removed are $90,478.08 and 1,182,600 lb (resulting in a cost of $0.072 per pounds CBOD removed). Comparing the pre-ECM and post-ECM electricity cost per pound of CBOD removed results in a cost savings $0.037 per pound of CBOD removed through implementation of the ECM project. The energy cost savings in 2010 associated with the ECM project implementation (to remove 1,182,600 pounds of CBOD) is $43,756. With a total project cost for the Ditch A and B modifications and upgrades of $1,446,304, the simple payback for the project using this alternative analysis is 33 years.

C.3 Conclusions

C.3.1 Factors Leading to Successful ECM Implementation/Operation

The WWTP manager gained the support of the engineer and design group for replacing the surface aerators with a fine-bubble diffuser system by presenting the benefits of reduced plant maintenance that would result from implementation of the ECM project.

Additionally, Big Gulch WWTP contacted equipment vendors and made site visits to learn how to best employ their ECM equipment. Finally, the WWTP contacted their local publicly owned electric utility (Snohomish County PUD) to obtain an energy audit of their existing equipment and subsequently entered into an agreement with the electric utility that provided incentive payments of $39,191 to help offset project costs for Oxidation Ditch A and $46,594 for Oxidation Ditch B.

C.3.2 Impact on Other Operating Costs Resulting from ECM Implementation

In addition to energy saving resulting from this project, the following benefits were also realized.

C.3.2.1 Labor

Big Gulch WWTP staff used to lubricate the bearings of the surface aerators two times per week and change out the transmission oil twice per year. Additionally, the drive belts for the surface aerators had to be regularly maintained or replaced. Since the turbo blowers do not require the same level of maintenance as the rotor aerators, the WWTP no longer has to expend staff resources for aeration system maintenance at the same level and does not have to purchase, store, or dispose of aerator lubrication oil (which needs to follow hazardous waste guidelines for disposal).

C.3.2.2 Chemicals

With the aeration system automated, mixed liquor settling has improved, and chlorine usage to control filamentous bacteria has been reduced to an as-needed basis.

C.3.2.3 Maintenance

The surface aerators used to produce a mist of aerosol that would coat the steel columns, handrails, and grating in the area contiguous to the oxidation ditches. These areas required cleaning on a regular basis. As a result of replacing the surface aerators with fine-bubble diffusers (which sit on the floor of the basin and gently produce bubbles), the Big Gulch WWTP staff no longer has to clean up such aerosol deposits. This upgrade has resulted in a cleaner, quieter, and safer WWTP. Additionally, the turbo blowers in the oxidation ditches require minimal maintenance to replace the air filters when they become dirty.

C.4 Lessons Learned

The project and the operational issues that provided the drivers for the aeration system improvements have emphasized to Big Gulch management and operations staff the message of this text and, according to plant manager Thomas Bridges, the importance of "constantly investigating new and innovative technologies. It's an exciting time for the wastewater industry, we're able to make improvements to our wastewater treatment system and realize significant energy savings as a result."

Reference

Gray & Osborne, Inc. (2008). *Wastewater Treatment Plant Capacity Study and Engineering Report*, Seattle, WA.

Appendix D. Optical DO Sensor Technology and Aerator Rotor VFD Control at the City of Bartlett, Tennessee, Wastewater Treatment Plant

The City of Bartlett, Tennessee, Wastewater Treatment Plant (WWTP) is a 1.0-MGD (average daily flow) secondary facility utilizing two mechanically aerated oxidation ditches to provide secondary treatment. Each of the aeration basins is equipped with three rotor aerators. Prior to implementing the aeration system modification ECM project, each basin was operated using one aeration rotor running continuously and a second rotor activated daily (and run at full speed) during periods of peak flow.

Under the Demonstration of Energy Efficiency Development Research Program funded by the Tennessee Valley Authority (TVA) and the American Public Power Association (APPA), the City of Bartlett Wastewater Division implemented optical dissolved oxygen (DO) sensor technology integrated with variable-frequency drive (VFD) speed control of the oxidation ditch rotor aerator. The objective of the TVA/APPA research/demonstration project was to advance the use of optical DO sensor technology integrated with VFD motor speed control to achieve energy savings at small to medium sized wastewater treatment plants (<10 MGD) within the TVA service area. Under the demonstration program, the DO control setpoint was established in each basin at 1.2 mg/L, and the rotor speed was controlled based on the DO readings in the oxidation ditches, relative to setpoint. During the demonstration program, one rotor in each basin reached full speed for only 20 to 30 minutes each day during the peak flow period. The second rotor was not, and has not ever been, required to maintain the oxidation ditch DO setpoint concentration. The optical DO sensor technology and aeration rotor VFD controls were installed and commissioned in 2007 for $13,500. Following implementation of the aeration system modifications, first-year (2008) energy consumption was reduced by nearly 72,000 kWh (13% reduction), and peak demand was reduced by 51 kW (a 39% reduction). The resulting energy cost savings were $9176/year (a 22% savings). The project resulted in a payback of 1.5 years.

D.1 Evaluation of Energy Conservation Measures

D.1.1 Site/Facility/Process Description

The City of Bartlett's WWTP #1, located in West Tennessee near Memphis, serves approximately 24,000 residential customers and one school. One hundred percent of the plant influent is domestic wastewater. The facility was originally commissioned in 1994 as a 0.5-MGD aerated lagoon and has undergone three major expansions (in 1999, 2003, and 2005) to meet the city's growing population. In 1993, the facility was upgraded to a secondary treatment facility (one oxidation ditch and secondary clarification). In 2003, the facility was upgraded with solids handling (aerobic digester and belt filter press). In 2005, a second oxidation ditch was added. Influent data for the City of Bartlett WWTP #1 is presented in Table D.1. Figure D.1 is a process flow diagram depicting the current configuration of the treatment plant.

TABLE D.1

City of Bartlett WWTP #1 Influent Data (2009)

Parameter	Daily Average
Flow (mg/L)	1.0
BOD (mg/L)	130
TSS (mg/L)	180
Ammonia-N (mg/L)	Not monitored
TKN (mg/L)	41
Phosphorus (mg/L)	6

Plant influent undergoes mechanical screening followed by biological treatment in two mechanically aerated oxidation ditches. Each oxidation ditch is equipped with three 60-hp rotor aerators. Oxidation ditch effluent undergoes secondary clarification followed by ultraviolet (UV) disinfection prior to discharge to the Loosahatchie River. Waste biosolids from the secondary clarifiers undergoes aerobic digestion. Digested biosolids are dewatered in a belt filter press and then are land applied as an agricultural soil amendment and fertilizer.

D.1.2 Description of Energy Conservation Measures (ECMs)

The City of Bartlett Wastewater Division implemented energy efficiency improvements to its aeration system as a result of a successful demonstration project conducted at WWTP #1 by the utility, Tennessee Valley Authority, and two technology vendors. The aeration system improvements consisted of the following:

- Installing InsitelG optical DO sensor technology (Model 10) to provide reliable DO monitoring in the oxidation ditches
- Integrating optical DO monitoring instrumentation output (4- to 20-mA signal) with VFD control (ABB variable-speed motor drives) of the aeration rotor speed

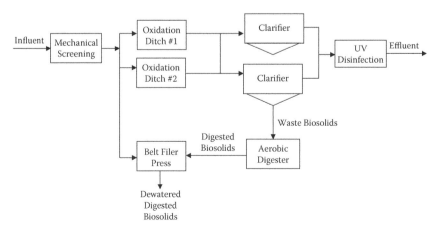

FIGURE D.1
City of Bartlett WWTP #1 process flow diagram.

Prior to implementing the aeration system's optical DO sensor technology with integrated VFD control of the oxidation ditch aeration rotor speed, the City of Bartlett WWTP #1 was operating each of the two oxidation ditches with two of the available three 60-hp aeration rotors (in each oxidation ditch). A single aerator was run continuously (at full speed, 60 Hz) and the second rotor was run (at full speed, 60 Hz) during periods of peak flow (activated by a timer/clock). The third rotor was installed but never operated in the original plant configuration; it is not currently required but is available as a backup spare.

The objective of the demonstration project was to advance the use of optical dissolved oxygen sensor technology coupled with variable-speed drive motor controllers to achieve energy savings at small to medium sized (<10 MGD) wastewater treatment plants. Funding and technical support for the demonstration project was provided by the American Public Power Association's Demonstration of Energy-Efficient Development research program and the Tennessee Valley Authority. Additional in-kind support was provided by the technology manufacturers (InsitelG and ABB), which consisted of personnel to assist in the installation and commissioning of the ECM equipment and providing DO instrumentation and VFD equipment (at no cost) during the initial demonstration trial period.

Currently and during the demonstration program (with a DO control setpoint of 1.2 mg/L in each of the oxidation ditches), one rotor in each of the plant's two oxidation ditches operates continuously at full speed and the second rotor's speed varies, depending on the DO reading in the oxidation ditch. The VFD controls the second rotor's input electrical power frequency between 60 Hz (at full speed) and 30 Hz (at minimum speed) and rotational speed depending on the input DO reading. The second rotor reaches full speed for only 30 to 45 minutes each day during the peak flow period. The frequency input to the second rotor from the VFD is 30 Hz for much of the day, and occasionally the VFD controlled rotor is turned off when the setpoint DO concentration can be maintained exclusively by the primary single, full-speed rotor. Tables D.2 and D.3 show the influent and effluent qualities both before and after implementation of the ECMs.

TABLE D.2

Monthly Averages, 2006 and 2009

Parameter		Concentration (mg/L)	
		2006	2009
CBOD	Influent	160	130
	Effluent	5	5
	Permit limit	20	20
TSS	Influent	279	280
	Effluent	12	12
	Permit limit	30	30
NH_3	Influent	Not measured	Not measured
	Effluent	0.11	0.15
	Permit limit	5	5
TKN	Influent	Not measured	41
	Effluent	Not measured	10
	Permit limit	Not measured	Monitor only
P	Influent	Not measured	6
	Effluent	Not measured	4
	Permit limit	Not measured	Monitor only

TABLE D.3

Daily Averages, 2006 and 2009

		Concentration (mg/L)	
Parameter		2006	2009
CBOD	Influent	212	200
	Effluent	5	5
	Permit limit	30	30
TSS	Influent	500	580
	Effluent	27	35
	Permit limit	45	45
NH$_3$	Influent	Not measured	Not measured
	Effluent	0.40	0.20
	Permit limit	10	10
TKN	Influent	Not measured	42
	Effluent	Not measured	15
	Permit limit	Not measured	Monitor only
P	Influent	Not measured	6.5
	Effluent	Not measured	7
	Permit limit	Not measured	Monitor only

D.2 Results

D.2.1 ECM Implementation Cost

Table D.4 summarizes the implementation costs for the optical DO sensor technology and the aeration rotor VFD motor speed controls.

D.2.2 Energy

Table D.5 summarizes the electricity energy consumption and costs prior to and following implementation of the aeration system ECMs and energy savings. Implementation of the aeration system control ECMs realized reductions of 13% in kWh per year consumed, 39% in peak demand, and 22% in annual electrical energy cost.

D.2.3 Payback Analysis/Benefits

At an annual energy savings of $9176 per year and ECM implementation cost of $13,500, the ECM project at the Bartlett WWTP #1 realized a payback in less than 1.5 years.

TABLE D.4

ECM Implementation Cost

Cost Category	Cost (2007$)
Capital and installation costs for optical DO sensor technology	$3500
Capital and installation costs for VFD	$10,000
Total installed cost	$13,500

TABLE D.5

Electrical Energy Cost and Savings

	Electrical Energy Consumption and Costs	
	2006 (Rotor Controls w/Timers)	2008 (Optical DO/VFD Rotor Controls)
Energy consumption		
kWh/day	1553	1356
kWh/yr	566,845	494,940
Peak demand (kW)	130	79
Total annual energy savings	—	71,905 kWh/yr (13%)
Energy costs		
At $0.05/kWh	$28,342/yr	$24,747/yr
Peak demand charge	$14,277/yr	$8,646/yr
Total energy cost	$42,569/yr	$33,393/yr
Energy cost savings	—	$9,176/yr (22%)

Source: USEPA, *Evaluation of Energy Conservation Measures for Wastewater Treatment Facilities*, EPA 832-R-10-005, U.S. Environmental Protection Agency, Washington, DC, 2010.

D.3 Conclusions

D.3.1 Factors Leading to Successful ECM Implementation/Operation

The collaborative effort by the project team and stakeholder interest in demonstration of this technology/ECM resulted in a successful demonstration of energy savings leading to full-scale operation with continuing energy savings results.

D.3.2 Impact on Other Operating Costs Resulting from ECM Implementation

In addition to the energy costs savings, the ECM project provided the following additional benefits.

D.3.2.1 Labor

Prior to implementing the aeration system ECM, manual monitoring of the DO concentration in the oxidation ditches required an operator's attention for approximately 1 hour/day (260 hours per year) at an associated labor cost of $4680 per year. This requirement has been eliminated by automatic DO monitoring.

D.3.2.2 Maintenance

The InsitelG DO sensor requires no periodic maintenance other than monthly inspection and rinsing with a garden hose and annual calibration. The annual cost associated with this maintenance activity is approximately $200/year.

D.4 Lessons Learned

Variable-frequency drives provide a soft start to the aeration rotor motors which should extend the operating life of the motors. Additionally, plant personnel learned that small changes in process control can lead to large savings in energy costs to the City.

Appendix E. Advanced Aeration Control for the Oxnard, California, Wastewater Treatment Plant

The Oxnard, California, Wastewater Treatment Plant (WWTP) serves a population of approximately 200,000 people and treats an average daily flow of 22.4 MGD. The trickling filter-activated biosolids treatment plant uses Turblex® blowers and associated proprietary pressure-based control software to automate the activated biosolids aeration process. To address aeration basin foaming and clarifier biosolids bulking problems, the plant implemented activated biosolids process optimization and automation utilizing the following integrated components:

- Replaced the aeration blowers' pressure based control software with DOmaster™ (Ekster and Associates, Inc.) control software. DOmaster is proprietary biological process control software that utilizes algorithms based on biological treatment process modeling and process data mining to effect dissolved oxygen (DO) control.

- Installed InsitelG optical DO sensor technology, replacing outdated membrane probes.

- Installed two total suspended solids monitors, one in the mixed liquor channel and one in the return activated sludge wet well.

- Installed SRTmaster™ (Ekster and Associates, Inc.) proprietary software to provide real-time control of the active biosolids process solids retention time (SRT). The software utilizes a biological process modeling-based control algorithm to maintain minimum variability of wasted solids (over the course of a day), resulting in significant improvements in solids settling and thickening.

- Installed OPTImaster™ (Ekster and Associates, Inc.) proprietary software that optimizes the process control setpoints for SRT and DO in each of the plant's aeration basins.

The effect of these modifications was improved biological process stability and discharge permit compliance and reduced sludge volume index (SVI): 20% for average SVI and 50% for maximum SVI. Since the implementation of this ECM, foaming in the aeration basin has not occurred.

In addition to improved stability of the biological treatment process, improved solids settling and thickening, and elimination of foaming, blower energy consumption was reduced by 306,600 kWh/yr (a 20% reduction). This reduction in energy consumption represented a nearly $27,000/year savings in electrical energy costs. Polymer dosage for thickening was reduced as a result of the improved settleability of the biological solids, resulting in a reduction in chemical costs of $7500/year. Additionally, the improved automation of the aeration process reduced labor costs by $$18,500/year.

The total project implementation cost was approximately $135,000. The payback considering only energy savings was approximately 5 years. Including the chemical cost savings and labor savings in the payback analysis reduced the payback period to approximately 2.5 years.

TABLE E.1

Oxnard WWTP Influent Data

Parameter	Average	Daily Maximum
Flow (MGD)	22.4	26.9
BOD (mg/L)	328	369
TSS (mg/L)	265	788

E.1 Evaluation of Energy Conservation Measures

E.1.1 Site/Facility/Process Description

The Oxnard WWTP serves approximately 200,000 people from the city of Oxnard, California. In the early 1970s, the WWTP was originally commissioned as an advanced primary plant. In 1977, trickling filters were installed, and in 1989 the plant was upgraded to a trickling filter-activated biosolids system with an increase in capacity from 24.5 MGD to 31.7 MGD. Current influent data for the Oxnard WWTP are presented in Table E.1. Figure E.1 presents a process flow diagram for the Oxnard WWTP.

Influent undergoes screening followed by primary clarification. Primary clarified effluent is distributed to two trickling filters filled with plastic media. One trickling filter is 40 ft in diameter and 25 ft deep, and the other is 100 ft in diameter and 26 ft deep. Under normal conditions, only the large trickling filter is used. From the trickling filter, flow enters the aeration process (two basins, each consisting of three compartments measuring 450 ft × 27 ft × 15 ft). The two aeration basins are of equal size, but only one basin is used at one time. Each compartment in an aeration basin has three individually controlled aeration

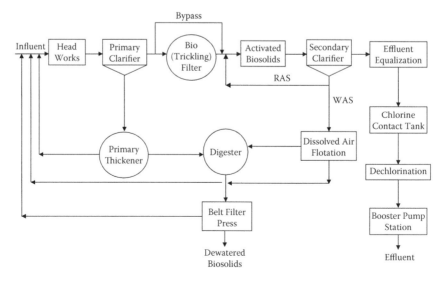

FIGURE E.1

Oxnard WWTP process flow diagram.

grids equipment with ceramic diffusers. Each aeration basin has nine dissolved oxygen meters (one per each grid), nine air flow meters (FCI, Inc.), and correspondingly nine valves that are automatically controlled using Rotork® electrical actuators. In the original design, a proprietary (Turblex) automatic DO control system was installed in combination with five 350-hp Turblex blowers. Each blower is rated at maximum flow of 6950 actual cubic feet/minute (acfm) and 10 pounds per square inch gauge (psig) pressure. Aeration is followed by secondary clarification, flow equalization, and chlorination/dechlorination. Treated effluent is discharged to the Pacific Ocean.

Primary biosolids are thickened in a gravity thickener to 4.8% solids, while secondary sludge is thickened in the dissolved air flotation units to 6.2% solids. Both primary and secondary biosolids are mixed prior to entering two digesters. Anaerobic digesters operate under methophilic conditions (30 to 35°C). Gas from the digesters is used for electrical energy production by three gas-driven generators (500 kW each). Typically, only two generators are used simultaneously. Digested biosolids are dewatered in a belt filter press (to 20% solids concentration) and disposed of in a landfill.

E.1.2 Description of Energy Conservation Measures (ECMs), Drivers, and Issues

The optimization and automation of the activated biosolids system included the following measures:

- Install two online total suspended solids (TSS) meters (InsitelG, Inc). One meter was installed in the mixed liquor channel, another in the return activated sludge (RAS) wet well. The TSS instrumentation provides suspended solids concentrations of the aeration process mixed liquor and of the RAS, which are required input to the process control and optimization algorithms implemented as part of the subject ECM.

- Replace outdated GLI dissolved oxygen meters with optical sensor technology (InsitelG, Inc.).

- Install SRTmaster software providing real-time control of solids retention time. The software utilizes a biological process model-based control algorithm and employees multilayer data filtration that guarantees that malfunctions of TSS or flow meters will not lead to erroneous control actions and process upset. The software alerts operators about both meter problems as well as changes in the patterns of process biological oxygen demand (BOD) loading or migration of solids to the clarifiers. Finally, the software maintains minimum variability of wasted solids over a day, resulting in significant improvements in biosolids thickening. SRTmaster receives instrumentation readings from the plant SCADA system and sends an optimized waste biosolids flow setpoint to the SCADA system electronically using industry-standard Open Productivity and Connectivity (OPC) drivers.

- Replace the Turblex blower pressure-based control software with DOmaster, which uses biological process model based algorithms instead of traditional PID algorithms (see end of Appendix) for DO control. The software also uses a data mining algorithm instead of pressure data as a DO control criterion, guaranteeing the lowest blower energy consumption. Utilization of these algorithms allows precise control of DO in each of the aeration basin compartments, minimizing energy used by blowers without aeration system oscillations. DOmaster uses multilayer data filtration to guarantee reliability of automatic control even if one of the control elements (meters or actuators) fails. The software communicates with the plant SCADA system in the same manner as SRTmaster, receiving instrumentation readings and returning airflow set points to the SCADA system for each aeration diffuser grid control valve opening and blower vane positioning.

- Use OPTImaster software to optimize setpoints for SRT and DO for each aeration compartment diffuser grid.

SRTmaster (in addition to the associated TSS and DO instrumentation) was implemented at the Oxnard WWTP in 2003. DOmaster for control of the blower was implemented in 2004, and OPTImaster was implemented in 2005. Activated biosolids automation and optimization ECMs were initiated to address foaming and intermittent bulking problems, to reduce operator workload, and to reduce energy and chemical usage. The data in Tables E.2 to E.4 compare the WWTP performance prior to implementation of the aeration system improvements (2002) and after (2009).

TABLE E.2

Monthly Averages, 2002 and 2009

Parameter		Concentration (mg/L)	
		2002	2009
BOD	Influent	262	328 (includes recycled flow)
	Effluent	17	17
	Monthly limit	30	30
TSS	Influent	221	265 (includes recycled flow)
	Effluent	5	5
	Monthly limit	30	30

TABLE E.3

Daily Maximums, 2002 and 2009

Parameter		Concentration (mg/L)	
		2002	2009
BOD	Influent	480	369 (includes recycled flow)
	Effluent	74	35
	Weekly limit	45	45
TSS	Influent	370	788 (includes recycled flow)
	Effluent	31	11
	Weekly limit	45	45

TABLE E.4

Sludge Volume Index (SVI)

Parameter	2002 (mL/g)	2009 (mL/g)
Average	165	130
Maximum	385	170

TABLE E.5

ECM Implementation Cost

Cost Category	Cost (2002$)
Software cost[a]	$100,000
Instrumentation costs	$30,000
Installation costs[b]	$5,000
Total cost	$135,000

[a] Due to the pilot nature of the project, the City contribution was $25,000 toward software purchase.
[b] Most of the ECM installation was done by plant personnel. Implementation of the control algorithms and associated instrumentation required approximately 2 to 3 days, while the design of the user interface required an additional 2 weeks.

E.2 Results

E.2.1 ECM Implementation Cost

Table E.5 summarizes the implementation costs for the Oxnard WWTP ECM project.

E.2.2 Energy

Following implementation of the ECM project, the average energy usage by blowers was reduced from 175 kW in 2002 to 140 kW in 2009. This resulted in a reduction of 306,600 kWh per year, or a 20% energy savings. The average electricity cost in 2009 was $0.088/kWh. The annual energy savings attributed to the ECM implementation is $26,980 (35 kW × 24 hr × 365 days × $0.088/kWh). Relatively small cost savings can be attributed to significant additional removal of BOD by the trickling filter.

E.2.3 Payback Analysis/Benefits

For this analysis, simple payback is determined by dividing the total project cost ($135,000 from Table E.5) by the resultant savings ($26,980 per year for energy only, $52,730 per year total). Based on these data, the project payback period is 5 years considering only the electrical energy savings. Including the chemical cost savings and labor savings in the payback analysis reduces the payback period to approximately 2.5 years.

E.3 Conclusions

E.3.1 Factors Leading to Successful ECM Implementation/Operation

The Oxnard WWTP operations manager was a champion of the innovations implemented under this ECM project and was also personally involved in integration of the Ekster software packages within the plantwide control system. Under his supervision, the operation staff embraced innovative ideas and worked directly with the vendors to speed up the implementation and commissioning of new optimization and automatic control methods. The fact that automatic control reduced operators' workload and improved NPDES compliance has helped adaptation of new technology at the Oxnard WWTP.

E.3.2 Impact on Other Operating Costs Resulting from ECM Implementation

E.3.2.1 Chemicals

Based on WWTP records, by reducing the polymer dosage used for biosolids thickening (a benefit of improved biosolids settleability), chemical costs were reduced by approximately $7500 a year.

E.3.2.2 Labor

Improved process monitoring and automation reduced the number of operator hours by at least 1 hour per day by eliminating sampling, frequent field measurements, and manual adjustments. The average labor cost savings resulting from this ECM are approximately $18,250 per year (1 hr/day × $50/hr × 365 days/year). The total energy savings for this ECM are $26,980 per year. Including chemical cost savings and reduced manpower requirements, the total savings resulting from this ECM are $52,730 per year.

E.3.3 Additional Benefits from ECM Implementation

In addition to the cost savings resulting from this project, the major benefit was improved process stability, reducing the sludge volume index (SVI) on average by 20% and maximum SVI by 50% (see Table E.4). As a result, in 2009 effluent water quality never exceeded NPDES limits (see Table E.3). In addition, foam observed periodically before this ECM implementation has not been seen since the project was implemented.

E.4 Lessons Learned

- Operation management leadership was a key factor in the project's success.
- ECMs need to provide multiple benefits to reduce payback period and to speed up adaption of new technology. Benefits related to improving reliability of operation are especially valuable.
- Reliable operation of ECMs provides the confidence required by operating staff in a new technology.
- When it comes to automation, the control algorithms need to take into account potential failures of control elements and ensure that these failures will not have negative impacts on operations.
- Third-party specialized automatic control software packages can be easily integrated with SCADA control systems using modern communication protocols. Ready-to-use automation software reduces algorithm design and programming costs and provides better reliability of automated control.

References

CEC. (2004). *Development of Software for Automatic Control of Dissolved Oxygen Concentration*, California Energy Commission, Sacramento.

Ekster, A. (2004). Golden age, *Water Environmental Technology*, 16(6), 62–66.

Ekster, A. and Wang, J. (2005). Effective DO control is available, *Water Environment Technology*, 17(2), 40–43.

Moise, M. and Ekster, A. (2007). Operations of a solids contact tank at low dissolved oxygen and low total suspended solids concentrations, in *Proceedings of WEFTEC 2007*, Water Environment Federation, Washington, DC.

Moise, M. and Norris, M. (2005). Process optimization and automation improves reliability and cost efficiency of Oxnard WWTP, in *Proceedings of WEFTEC 2005*, Water Environment Federation, Washington, DC.

A Sign of Things to Come[*]

During a 1989 Board of Commissioners monthly meeting held at a large southeastern Virginia sanitation district, one of the nine commissioners, a licensed Professional Engineer (PE) with a degree in computer science, proposed a question to the group of district plant managers, plant operators, and support staff members who were in attendance. To several of the 48 attendees (including the author) the commissioner's statement seemed more like a suggestion than a question. Whatever it was, we attendees were more shocked by the commissioner's words than calmed by their common-sense implications. I remember thinking that the commissioner's suggestion or question seemed more like a fishing net thrown into the seating area to net as many guppies as possible. What did the commissioner say? Clearly and succinctly, he stated: "In this new age of computers and fax machines, why don't we look at automating our wastewater treatment plants? That is, should we let the computers do much of the work?"

Well, somewhat in keeping with that old proverbial saying, you could say that the silence was deafening the instant after the commissioner's statement; it was so total that one could hear the dust motes striking against each other. The implications? Loud, very loud, to say the least. I remember sitting there in the momentary silence that seemed endless at the time and thinking, computers? You have to understand that few of the war baby (i.e., World War II) attendees, along with many of the others in attendance that day, had any extensive or even cursory knowledge of that huge, mysterious and mind-boggling concept of *computers*. Remember, it was 1989 and we were trained to rely on administrative clerks and their trusty typewriters to do the administrative work. Computers? Computerized operations? Automated plants? What? Why?

While we all sat there in silence, the commissioner who had asked why not automate the plants stood up and walked over to the large wall-sized chalk board to the right side of where we were sitting. After picking up a cucumber-sized hunk of chalk, he stated: "Consider the following." He then chalked a bulleted list and explained each entry:

---

[*] Adapted from Olsson, G., *Instrumentation, Monitoring, Control and Automation in Water and Wastewater Operations*, Lund University, Sweden, 2010; Araki, M., *Control Systems Robotics, and Automation*. Vol II. *PID Control*, Kyoto University, Japan, 2010.

- *Process engineering*—"We must focus any computerized automation makeover we make on ensuring that the correct process information is included to properly operate the process. We can use dynamic modeling and simulation to ensure this happens."

- *Instrumentation*—"Computerized automated operations will allow the plant to measure the states and parameters of the process, and this is a reasonable expectation because today online instrumentation for some key components (such as ammonia, nitrite/nitrite, and phosphate sensors) are available or are a work in progress and are getting more reliable and profitable to use with every passing day."

- *Signal processing*—"The sensors and instruments we use will remove inadequate data and extract essential information."

- *Monitoring*—"To make the operation reliable a monitoring and detection function based on measurement information must be included."

- *Operator interface*—"My proposal is not to replace the operator [a collective audible sigh of relief pervaded the room] but instead to augment his or her abilities to operate the plant and meet permit by providing him or her proper guidance based on measurements and monitoring the computer systems. Another good thing is that treatment plant operations simulators are being developed for training operators."

- *Actuator technology*—"I envision automated operation of pumps, compressors, and valves using computer-controlled actuators. Consider, for example, variable-speed technology for electric motors. This is a proven technology. What this really means is operators will have flexibly in operations of both pumps and compressors, which can have a significant impact on the quality of operations."

- *Advanced planning*—"An automated computer-operated treatment system can afford planners the time necessary to properly plan for future disturbances, such as rain events, seasonal fluctuations like temperature changes, and external changes."

- *Automatic control*—"Finally, a computerized system makes the decisions on how to change the control variables such as liquid and air flow rates, reactions, dosages, and so forth. Moreover, keep in mind that distributed computer systems, including programmable logical controllers, or PLCs, are available in modules and can be adjusted to the size of the plant. This technology, though in the developmental stage, is proving to be quite sufficient for most uses and is often packaged in real-time computer systems."

Putting the chalk down and looking out over his attentive audience, the commissioner continued: "We all know that our biggest headache in plant operations and in maintaining permit compliance is control of plant disturbances. These disturbances, again, as you know, include variations in flow rate, variation in influent concentration, and variations in influent composition. Under these varying load conditions, strict operator vigilance and instigation of proper operational controls are required. Keep in mind we already use SCADA to alert us to alarm conditions at pump stations and other locations, so why not expand our coverage of operational parameters controlling our treatment unit processes with automatic computer controls? What I am suggesting here is that we incorporate simple PID controllers into our systems to serve as a control loop feedback mechanism or tuner or controller."

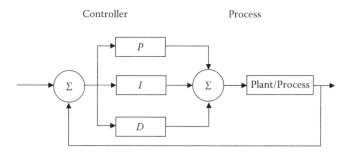

FIGURE E.2
Block diagram of a PID controller.

At this point the commissioner picked up the chalk again and drew Figure E.2. Upon completion of the simple drawing, the commissioner explained: "Now, this drawing may seem complicated to many of you [a collective nod of heads], but it shows a simple PID controller, which is actually based on a mathematical algorithm involving three separate constant parameters: the proportion (P), the integral (I), and derivative (D) values. Based on experience in problem solving, learning, and discovery, these values can be interpreted in terms of time: P depends on the present error, I on the accumulation of past errors, and D is a prediction of future errors, based on current rate of change. My point is that the weighted sum of these three actions is used to adjust the unit process via a control element such as the position of a control value, or the power supplied to a pump."

The commissioner put the chalk down again and smiled at all of us confused "students" in the room. Then he said, "If we think of controlling the treatment plant as being analogous to a helmsman steering a ship at sea, then I think you can see and understand that subtle feedback operation is the key to steering a steady course at sea and to treating wastewater to meet permit compliance. Think about how much smoother the operation of a ship at sea or the operation of a wastewater treatment can be when operating in autopilot. Thank you for your attention and patience."

Well, today as I sit here composing this account, I clearly remember not only the commissioner's suggestions about shifting the sanitation district's standard plant operation routine to computer-automated operations—a significant change for the bunch of computer-illiterate dinosaurs in the room—but also the impression those remarks made on all 48 of us in attendance at the meeting that day. Upon leaving the meeting room, the common refrain I heard was, "Auto-pilot? Is he suggesting that we operate the entire plant in auto using one of them new-fangled space cadet computers? No way, José." The funny thing is that within a week or so most of us doubting Thomases had a change of heart on employing computer-assisted operations; the new refrain became, "Why not!" Why the change of heart? We realized that a time of change was upon us, and we either had to board the digital 1 + 0 train or get run over by it. The train ride has been enlightening and expanding and ongoing, to say the least.

Appendix F. DO Optimization Using Floating Pressure Blower Control in a Most Open Valve Strategy at the Narragansett Bay Commission Bucklin Point WTTP, Rhode Island

The Narragansett Bay Commission's (NBC) Bucklin Point Wastewater Treatment Plant (WWTP) in East Providence, Rhode Island, is a 23.7-MGD (average daily flow) activated biosolids nitrification/denitrification facility employing the Modified Ludzak–Ettinger (MLE) process. Commissioned in 2006, the MLE process aeration system utilized a conventional aeration/blower control system consisting of individual proportion–integral–derivative (PID) loops to control dissolved oxygen (DO) in and air flow to each of the MLE process aerobic zones. This control system was based on most open valve (MOV) logic. The control system modulated blower air flow based on discharge pressure. Each of the air distribution system's 16 drop legs was modulated independently based on DO measurements in the aeration basins. The MOV logic was programmed to use the positions of the 16 valves to increase or decrease the pressure setpoint.

Following commissioning of the MLE process, the plant experienced difficulty attaining consistent nitrogen removal because the constant-pressure-based PID aeration control system was unable to adjust to changing process conditions and maintain proper DO concentration in the MLE process aerobic zones, resulting in insufficient conversion of ammonia nitrogen to nitrate and inhibition of denitrification due to high DO concentration in the return activated sludge (RAS) flow to the MLE process anoxic zones. During normal influent loading conditions, the system was unable to maintain DO levels in the aeration process closer than 1.0 ppm from setpoint. The DO control problem was exacerbated during wet weather events. An analysis of the process concluded that the interaction between the PID control loops was causing instability in the control of aeration process DO.

The following modifications to the aeration process control system were implemented:

- A specialized proprietary DO/blower control algorithm (Dresser Roots Wastewater Solutions Group) replaced the PID control loops.
- Direct air flow control was substituted for the pressure control logic, basing the MOV strategy on air flow vs. pressure.

The results of these modifications were

- The RAS flow DO control is consistent and no longer inhibits biological denitrification.
- Aeration system energy consumption and costs are below projections.
- DO excursions are less than 0.5 ppm from setpoint.
- MOV logic is effective in minimizing blower energy consumption.
- Operator intervention (manual control of blowers and drop leg valves) is no longer required to effect discharge compliance.

The annual electrical energy consumption reductions achieved through the implementation of this ECM project were 1,068,700 kWh for 2007 (10.3% reduction), 1,464,800 kWh for 2008 (14% reduction), and 1,207,600 kWh for 2009 (11.6% reduction). The electrical energy savings for the first 3 years following implementation of the aeration control system were $115,881 (2007), $155,457 (2008), and $136,022 (2009). The cost of implementing the aeration control system modification was $200,000. The payback for this project was achieved in 1.5 years.

F.1 Evaluation of Energy Conservation Measures

F.1.1 Site/Facility/Process Description

The Narragansett Bay Commission's (NBC) Bucklin Point Wastewater Treatment Plant (WWTP) serves a population of approximately 130,000 residing in the cities of Central Falls, Cumberland, East Providence, Lincoln, Pawtucket, and Smithfield in the Blackstone River Valley and east Providence, Rhode Island, area. The WWTP is operated for NBC under a management contract with United Water.

The Bucklin Point WWTP was originally commissioned in 1950 and has since undergone four major upgrades. The last comprehensive upgrade of the Bucklin Point WWTP was completed in 2006, a reconfiguration of the conventional activated biosolids aeration process to a Modified Ludzak–Ettinger process to effect nitrogen removal. The MLE biological treatment process consists of an anoxic basin upstream of an aerobic zone. An internal recycle carries nitrates created during the nitrification process in the aerobic zone along with mixed liquor to the anoxic zone of nitrification. RAS is mixed with the influent to the anoxic zone. The extent of denitrification is tied to the mixed liquor recycle flow; higher recycle rates increase denitrification. Because only recycled nitrate has the opportunity to be denitrified, the MLE alone cannot achieve extremely low final nitrogen concentrations. The maximum denitrification potential is approximately 82% at a 500% recycle rate (WEF/ASCE/EWRI). Total nitrogen (TN) effluent concentrations typically range from 5 to 8 mg/L (Barnard, 2006). Actual denitrification might be limited by other factors, such as carbon source availability, process kinetics, and anoxic or aerobic zone sizes. Furthermore, oxygen recycled from the aerobic zone can negatively affect the denitrification rate in the anoxic zone (WEF/ASCE/EWRI, 2006). Performance factors include limitations due to the single anoxic zone and the internal recycle rate that returns nitrates to the anoxic zone. Selection factors include the possibility of constructing walls in existing basins to create an anoxic zone; additional pumping, piping, and electricity to accommodate the internal recycle; and the possible need for an additional carbon source to promote denitrification. The Modified Ludzack–Ettinger process is illustrated in Figure F.1. Aeration for the aerobic stage of the MLE process is provided by three 600-hp Dresser Roots single-stage centrifugal blowers, each capable of delivering 12,100 cubic feet per minute (cfm).

Prior to the 2006 upgrade, conventional biological treatment was accomplished in the same aeration tanks (and tank volumes) currently being used in the MLE process. At that time, the four aeration trains each consisted of four aerobic zones only, for a total of 16 aerobic zones. The MLE process utilized the same configuration but provided for 12 anoxic zones (three sequential tanks at the head end of the aeration train) followed by 16 aerobic zones (four in each train following the three sequential anoxic zones). The 2006 upgrade

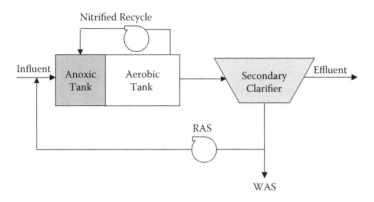

FIGURE F.1
Modified Ludzack–Ettinger process.

provided for a fine-bubble diffusion system in place of the mechanical aerators. The additional equipment at the aeration basin included optical DO probes and 16 electric-motor-operated butterfly valves for modulating air flow. Each of the 16 control zones also had a flow tube and flow transmitter to provide measurement of actual cubic feet per minute (acfm) to each zone.

The 2006 plant upgrade was designed with conventional aeration/blower control utilizing individual proportional–integral–derivative (PID) loops for controlling dissolved oxygen and air flow in each of the MLE process aerobic zones. The DO/blower control system was based on constant discharge pressure using most open valve logic to minimize blower energy consumption by manipulating the air delivery system's pressure setpoint (through opening and closing the air distribution system's drop leg valves to each of the aeration system's 16 aerobic zones) in response to DO readings in the aerobic zones. Influent data for the Bucklin Point WWTP is presented in Table F.1. Figure F.2 is a process flow diagram depicting the current configuration of the treatment plant.

Influent (dry and wet weather flow) from the plant's two main sewer interceptors (Blackstone Valley and East Providence) are collected in the facility's influent pump station and conveyed (using three 100-hp, 38.7-MGD screw pumps) to preliminary treatment for screening and grit removal (four 40-MGD screens with 0.75-in. openings followed by four 40-MGD 19-ft-diameter grit vortex units). Grit and screenings are disposed in a landfill. Primary treatment for dry weather flow (up to 46 MGD) follows using three circular

TABLE F.1

Profile of the NBC Bucklin Point
Influent Data (2009)

Parameter	Daily Average
Flow (MGD)	23.7
BOD (mg/L)	155
TSS (mg/L)	147
Ammonia-N (mg/L)	15.37
TKN (mg/L)	15.70
Phosphorus (mg/L)	4.17

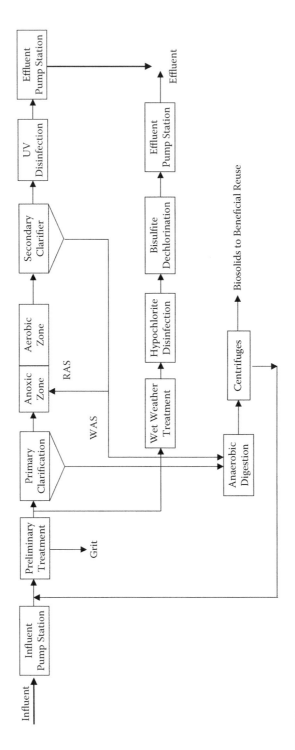

FIGURE F.2
Bucklin Point WWTP process flow diagram.

clarifiers (each 102 ft in diameter and 14 ft deep). Wet weather flow (i.e., influent exceeding 46 MGD) is collected, following preliminary treatment, in two 2.5-million gallon (total volume) holding tanks. The contents of the wet weather holding tanks are returned to the treatment plant to undergo primary and secondary treatment once the wet weather event flow ceases. During a wet weather event, any flow that exceeds the capacity of the holding tanks undergoes chlorination followed by dechlorination and is discharged to the Seekonk River.

Primary effluent undergoes biological treatment in a four-train MLE process. Each train consists of three sequential anoxic zones (0.59 million gallons each) followed by four sequential aerobic zones (2.28 million gallons each). Anoxic/aerobic treatment is followed by six secondary clarifiers (four at 111 ft in diameter and 11 ft deep each and two at 110 ft in diameter and 12 ft deep). An internal mixed liquor recycle carries nitrates from the aerobic zone to the anaerobic zone for denitrification. A portion of the settled biosolids from the secondary clarifiers is returned and mixed with the influent to the anoxic zones. Secondary clarifier effluent is disinfected using ultraviolet radiation prior to discharge to the Seekonk River.

Primary biosolids underflow from the primary clarification process and waste activated sludge (WAS) underflow from the secondary clarifiers are anaerobically digested and then dewatered in centrifuges to produce biosolids that are recycled as compost for use in non-agricultural land application.

F.1.2 Description of Energy Conservation Measures (ECMs), Drivers, and Issues

The ECM implemented at the Bucklin Point WWTP is a DO/blower control system utilizing proprietary control algorithms (developed by ESCOR, Inc., a subsidiary of Dresser Roots) in lieu of PID loop control, allowing DO/blower control based on air flow vs. pressure. Unlike the original pressure-based system, the Dresser Roots system employs direct flow control of the blowers. As the DO varies from setpoint, the required incremental changes in air flow are used to modify both aeration drop leg air flow and blower flow. The most open valve logic directly manipulates basin air flow control valve positions to ensure that at least one valve is always at maximum position, thereby minimizing system pressure without using a pressure setpoint. The reduced complexity makes the control more robust and more accurate. Elimination of the pressure control loop also minimizes tuning. In August 2006, an integrated air flow control system was implemented to provide stable control of the aeration system blowers. This air flow control-based technology replaced the facility's pressure based aeration control system.

Following commissioning of the Bucklin Point WWTP's conversion of the active biosolids process to an MLE process in 2006, the plant experienced difficulty attaining consistent nitrogen removal because the constant-pressure-based aeration/blower control system was unable to adjust to changing process conditions and maintain proper DO concentration in the aeration basins, resulting in insufficient conversion of ammonia nitrogen to nitrate. The constant-pressure-based control system was unable to maintain DO levels in the aeration tanks closer than 1.0 ppm compared to setpoint. This problem was exacerbated during wet weather events. The MOV control was unable to consistently minimize the system discharge pressure, resulting in wasted energy. An analysis conducted by ESCOR determined that the interaction between the PID control loops was causing instability in the control of aeration basin DO. The DO/blower system control instability in turn caused the following problems:

- Biological nitrogen removal (denitrification) was being inhibited by high DO in the internal mixed-liquor recycle (IMLR) flow to the MLE process anoxic zones.
- Energy consumption and costs exceeded expectations.
- A utility rebate was being jeopardized because the constant-pressure MOV logic was ineffective.
- Plant operations staff were forced to manually intervene in the operation of the blowers/aeration system to maintain performance and compliance.

The drivers for implementing an improved DO/blower control system were primarily to enable the WWTP to maintain effluent total nitrogen levels below 85 mg/L (monthly average) during the permitted seasonal compliance period (May through October) and to provide consistent nitrification and denitrification during varying flows from wet weather events. The ECM significantly revised the aeration system control strategy while maintaining existing control devices by doing the following:

- Substituting specialized proprietary DO/blower control algorithms for PID control loops
- Eliminating pressure control in lieu of direct air flow control
- Basing MOV logic on zone air flow control

The results of the ECM implementation included the following:

- The IMLR flow DO control is consistent and no longer inhibits denitrification in the anoxic zone.
- Aeration system energy consumption and costs are below original projections (and the utility rebate was secured).
- DO excursions are less than 0.50 ppm from setpoint.
- MOV logic is effective in minimizing blower discharge pressure (and associated energy consumption).
- Operator intervention (manual control) with the aeration system is no longer required to effect discharge compliance.

Table F.2 presents the monthly average influent and effluent quantities both before and after implementation of the ECMs.

F.2 Results

F.2.1 ECM Implementation Costs

Table F.3 provides the installed cost (as estimated by the utility and operations management contractor) of the implementation of the proprietary DO/blower control system.

F.2.2 Energy

Table F.4 summarizes the electricity energy consumption and costs prior to and following implementation of the aeration system ECM and resultant energy savings.

TABLE F.2

Monthly Averages, 2004 and 2009

		Concentration (mg/L)	
Parameter		2004	2009
BOD	Influent	232	155
	Effluent	14	4
	Permit limit	30	30
TSS	Influent	143	147
	Effluent	15	7
	Permit limit	30	30
NH_3	Influent	14.819	15.37
	Effluent	11.526	0.69
	Permit limit	No applicable limit	15[a]
TKN	Influent	23.647	25.7
	Effluent	14.375	2.1
	Permit limit	No applicable limit	No applicable limit
TN	Effluent	15.614	7.95
	Permit limit	No applicable limit	8.5[a]
P	Influent	4.995	4.17
	Effluent	1.884	2.01
	Permit limit	No applicable limit	No applicable limit

[a] May to October.

TABLE F.3

ECM Implementation Cost

Cost Category	Cost (2007$)
Proprietary DO control/blower supply management system	$170,000
Installation/commissioning	$30,000
Total installed cost	$200,000

TABLE F.4

Electrical Energy Cost and Savings

Year (A)	Monthly Electricity Use (kWh) (B)	Average Daily Flow (MGD)	Annual Energy Use Reduction (kWh)	Average Electricity Rate ($/kWh) (C)	Annual Electricity Cost Savings (D)[a]
		Before ECM Implementation			
2006	864,612	—	—	0.0990	
		After ECM Implementation			
2007	775,553	20.33	1,068,700 (10.3%)	0.10843	$115,880 (11%)
2008	742,547	21.95	1,464,800 (14.0%)	0.10613	$155,457 (15%)
2009	763,980	21.66	1,207,600 (11.6%)	0.11264	$136,022 (13%)

[a] $D = (B_{evaluation\ year} - B_{2006} \times C \times 12.$

F.2.3 Payback Analysis/Benefits

For this analysis, simple payback is determined by dividing the ECM project cost ($200,000) from Table F.3 by the electricity cost savings following project implementation. The reduced electrical energy consumption and electricity cost savings provided by the aeration system control ECM resulted in a payback in the seventh month (July) of the second year following commissioning of the ECM (2008), a 1.5-year payback.

F.3 Conclusions

F.3.1 Factors Leading to Successful ECM Implementation/Operation

Careful evaluation and documentation of operating conditions led to an identified need to address dissolved oxygen control after eliminating other possible sources. Collaboration by all stakeholders (internal and external to the NBC and Bucklin Point WWTP) resulted in a successful resolution to the problem.

F.3.2 Impact on Other Operating Costs Resulting from ECM Implementation

In addition to the energy costs savings, the ECM project provided the following additional benefits.

F.3.2.1 Labor

Implementation of the blower control system ECM eliminated field sampling/testing for aeration basin DO readings and manual manipulation of the aeration basin's drop leg valves.

F.3.2.2 Chemicals

The more stabilized operation resulting from the implementation of the ECM reduced sodium bicarbonate additions, as a more consistent alkalinity in the effluent is achieved.

F.4 Lessons Learned

In larger facilities, every component has a significant impact when it is not operating efficiently. Constant diligence to review and improve operational procedures is critical in a biological nutrient removal process, especially during wet weather events when the process is adversely impacted. Biological nutrient removal processes must operate within narrow parameters (those controllable by operations staff) to achieve the best possible steady-state conditions under significant variable conditions as compared to conventional wastewater treatment facilities. Monitoring data for the operating parameters must be timely, accurate, and repeatable to ensure operational integrity during each shift. Professionals involved in the design and operation of wastewater treatment plants should be aware of the need to operate biological systems closely and respond to changes in process conditions within short periods of time. DO control systems and strategies must be properly designed and integrated to maintain reliability under frequent operational changes (e.g., for blowers, inlet and outlet guide vanes, and drop leg valves which are designed for frequent changes of varying nature).

References

Barnard, J. (2006). Biological nutrient removal: where we have been, where we are going, in *Proceedings of WEF 79th Annual Technical and Educational Conference*, Dallas, TX.

WEF/ASCE/EWRI. (2006). *Biological Nutrient Removal (BNR) Operation in Wastewater Treatment Plants*, Manual of Practice (MOP) No. 30, McGraw Hill, New York.

Appendix G. Capacity and Fuel Efficiency Improvements at Washington Suburban Sanitary Commission Western Branch WWTP, Prince Georges County, Maryland

The Washington Suburban Sanitary Commission's (WSSC) Western Branch Wastewater Treatment Plant (WWTP) is a 21.6-MGD (average daily flow) denitrified activated biosolids (DNAB) biological nutrient removal facility utilizing methanol supplementation for biological denitrification and chemical precipitation for supplementing biological phosphorus removal. Biosolids are thickened and dewatered using dissolved air flotation and centrifuges, respectively, and are subsequently incinerated in two natural gas-fired multiple hearth furnaces (MHFs) prior to landfill disposal. The furnaces were commissioned in 1974 and 1977 and were originally designed to process 26 dry tons per day (DTPD) of heat treated biosolids. WSSC decommissioned the biosolids heat treating process at the Western Branch WWTP soon after commissioning the second MHF in 1977.

The MHF consists of a circular steel shell surrounding a number of hearths. Scrapers (rabble arms) are connected to a central rotating shaft. Units range from 4.5 to 21.5 ft in diameter and have from 4 to 11 hearths. Dewatered sludge solids are placed on the outer edge of the top hearth. The rotating rabble arms move them slowly to the center of the hearth. At the center of the hearth, the solids fall through ports to the second level. The process is repeated in the opposite direction. Hot gases generated by burning on lower hearth dry solids. The dry solids pass to the lower hearths. The high temperature on the lower hearths ignites the solids. Burning continues to completion. Ash materials discharge to lower cooling hearths where they are discharged for disposal. Air flowing inside center column and rabble arms continuously cools internal equipment.

Fuel consumption, in many cases, is high due to the design of the furnace. Cold biosolids are fed to the top (hearth) of the furnace which is also the exhaust point for the furnace combustion gases. The cold biosolids, contacting the hot furnace, release volatile hydrocarbons that do not have adequate residence time in the furnace of exposure to adequately high temperature to be completely burned. The result is a high-hydrocarbon, smoky, and odorous emission that is usually addressed by increasing the operating temperature of the upper hearths of the furnace or adding an afterburner section to increase the final temperature of the furnace's exhaust gas stream. Both of these options increase furnace fuel consumption. Additionally, operating the furnace at a higher temperature may also cause slagging (melting) of the residual ash inside the furnace, resulting in higher maintenance costs. In 2001, to comply with promulgation of Title V and the "zero visible emissions" requirements of the Clean Air Act, the capacity of the MHFs at WSSC was reduced to 12 DTPD, and external afterburners were installed, which increased natural gas consumption.

Many furnaces that are now employed to incinerate biosolids are equipped with flue gas recirculation and waste heat recovery systems integral to the design of the furnace. Flue gas recirculation controls air emissions while allowing the furnace to be operated at a lower temperature. Recovered heat energy from the furnace exhaust can be utilized to preheat the combustion air fed to the furnace. Because combustion air represents a significant heat load (fuel consumption requirement) to the furnace, preheating combustion air with waste heat reduces the amount of fuel required by the furnace to achieve biosolids incineration. Older MHFs that are not equipped with such energy-saving features can be modified and upgraded to incorporate waste heat recovery/combustion air preheating, not only resulting in fuel savings but also increasing furnace capacity and reducing maintenance.

The Western Branch MHFs were retrofitted in 2009 and 2010 with flue gas recirculation (FGR) systems that take exhaust flow from the top hearth of the furnace and reinject it into the one of the lower hearths. Recirculation flue gas mixes with the higher temperature combustion gases, providing ample time and temperature to completely oxidize the volatile hydrocarbons released from the biosolids feed entering the top hearth of the MHF. This allows the furnace to run at a lower temperature (or without exhaust gas afterburner), optimizing fuel consumption and eliminating ash slagging. The MHFs were also retrofitted with air-to-air waste heat recovery heat exchangers that recover and utilize the heat contained in the furnace exhaust steams (exhaust combustion air and center shaft cooling air) to preheat the furnace combustion air, reducing the heat load to and fuel consumption of the furnace. Circle slot jets were added to convey the preheated combustion air into the hearths to increase turbulence and air–fuel mixing. The improved convection and turbulence increase the drying and combustion rates of the furnaces, resulting in an increase in their capacity. The modifications to the Western Branch WWTP MHFs have resulted in the following benefits:

- Increased the throughput capacity of the furnace to 17 to 19 dptd from 12 DTPD (a 42 to 58% increase), postponing the need to build new furnaces to meet growing demand.
- Reduced NO_x emissions from the MHFs to meet best available control technology.
- Reduced natural gas consumption and cost by 320,000 therms per year (a 76% reduction) and $400,000 per year, respectively. Energy and maintenance savings will achieve a payback on the $4.5 million in approximately 11 years.

G.1 Combustion Theory

Because this case study is designed to provide basic information on how one facility installed various improvements and modifications to its multiple hearth furnaces to ensure operational sustainability, to increase operator efficiency, and to conserve energy, it is important for the reader to have some basic knowledge of combustion theory. Accordingly, a very fundamental presentation of the aspects related to combustion of biosolids is presented in the following section.

G.1.1 Principles of Combustion

When dried, biosolids rank alongside peat and lignite coal in heat content. Biosolids do, however, contain higher levels of volatile matter with a lower fixed carbon content. Biosolids combustion, therefore, produces more flaming than coal as volatiles are oxidized. The point is that, although the composition of biosolids differs considerably from that of other fuels, the principles of combustion are basically the same.

Combustion, in contrast to incineration, which is an engineered process using controlled flame combustion to thermally degrade waste substances, is the rapid chemical combination of oxygen with carbon, hydrogen, and sulfur; this phenomenon is usually referred to as burning. It should be pointed out that the sulfur content of sewage biosolids is low and does not significantly contribute to the overall combustion process (USEPA, 1978).

To achieve complete combustion under ideal conditions (efficient combustion), the three-T rule of combustion must be followed: *temperature, time,* and *turbulence.* Temperature is important because, in order for the feedstock to be combusted, it must be introduced to the bed at an operating temperature that exceeds the ignition temperature of the waste constituents. Time is required to volatize the feed material. The actual time required to volatize the feedstock depends on the manner in which the feed is introduced; for example, a 2.5-in.-diameter plug discharged to the bed requires approximately 25 seconds to volatize, while feedstock discharged at the top of the system may require about 2.5 seconds to be gasified and oxidized. Turbulence works to distribute (mix) the feedstock with optimal contact with oxygen in the air and volatized organics, which leads to more efficient combustion. There is a general rule of thumb with regard to the three-T rule of combustion. This rule basically states, assuming sufficient fuel and air, that as reaction time is increased temperature is raised, and as turbulence is maximized excess air can be decreased (WEF, 1992). The preceding explanation assumes ideal combustion. Because an ideal combustion or incineration system is only possible on paper, excess air greater than the ideal stoichiometric combustion requirement is provided to ensure sufficient oxygen when complete combustion is required (USEPA, 1978). Depending on the method of operation and type of furnace used, a typical sewage biosolids incinerator may utilize excess air quantities varying from about 40% to over 100%. It is desirable to maintain the excess air at a minimum to reduce stack losses.

G.1.2 Biosolids Incineration

Dewatering biosolids prior to incineration is very important, because even though the heat value of biosolids is relatively high (averaging about 7000 Btu/lb dry solids), the water content of most biosolids requires the addition of auxiliary fuel to maintain combustion in the incinerator (USEPA, 1978). Obviously, the costs involved with providing this auxiliary fuel can impact the economics of efficient incinerator operation. The key factor involved with reducing this cost is the solids content of the biosolids. For this reason, many facilities have opted for centrifugation of biosolids to increase the solids content. This increase in solids content directly affects the net heat value of the feedstock. Moreover, with increased heat value or heat content of the biosolids feedstock, autogenous (self-sustained) combustion is possible. Biosolids incineration can be described as follows (USEPA, 1978):

1. Temperature of the biosolids is raised to 212°F.
2. Water is evaporated from the biosolids.
3. Water vapor temperature and air temperature are increased.
4. The temperature of the biosolids is elevated to the ignition point of the volatiles.

For the biosolids incinerator operator and maintenance technician, the important points discussed in this section can be summed up as follows: (1) for successful biosolids incineration, it is important to ensure that there is a proper mixing of combustion gases; and (2) the fuel mixture and the volatile solids in the biosolids are critical.

G.2 Evaluation of Energy Conservation Measures

G.2.1 Site/Facility/Process Description

The Washington Suburban Sanitary Commission's Western Branch WWTP is located in Upper Marlboro, Maryland, on the Western Branch of the Patuxent River. The facility was originally commissioned in 1966 as a 5-MGD primary/secondary plant utilizing anaerobic digestion and vacuum filters to process biosolids for land application. The current 30-MGD plant serves residential, commercial, and industrial users in the east-central portion of Maryland's Prince George's County. Since its original commissioning in 1966, the Western Branch WWTP has undergone several significant upgrades to accommodate the service area's growing population and to implement treatment process enhancements required to meet increasingly stringent effluent quality requirements promulgated by the State of Maryland and U.S. Environmental Protection Agency (USEPA). In 1974, a 15-MGD nitrification plant (Phase I) was constructed as well as biosolids incineration facilities. In 1977, a mirror-image 15-MGD nitrification plant (Phase II) was commissioned. The Phase I and Phase II nitrification plants utilized a two-biosolids process with final filtration. In 1989, a denitrification activated sludge (DNAS) process was implemented to comply with seasonal (summer) permitted nitrogen removal effluent limits (1.5 ppm NH_3 and 3.0 ppm TKN). The DNAS process uses methanol as a carbon source for denitrification. Figure G.1 provides a process flow diagram of the Western Branch WWTP. Additional enhancements to the Western Branch WWTP are currently being designed to optimize the facility's nitrogen removal performance (to achieve an annual average effluent nitrogen concentration of 3 mg/L). These enhancements are scheduled to be commissioned in 2013 and include the following:

- High-rate activated sludge (HRAS) process upgrades
  - Centralized HRAS return activated sludge (RAS)/waste activated sludge (WAS) pumping station
  - Scrubber blow-down recycle isolation to HRAS process
  - Anaerobic zone baffle walls and mixers or plug flow reactor staging
  - HRAS surface wasting
  - JRAS enhanced nitrogen removal (ENR) monitoring and control systems
- Nitrification activated sludge (NAS) process upgrades
  - Centralized NAS RAS/WAS pumping system
  - Anoxic zone baffles walls and mixers
  - NAS plug flow baffle walls
  - NAS ENR monitoring and control systems
- Denitrification activated sludge (DNAS) process upgrades
  - DNAS ENR monitoring and control systems
- Solids handling process upgrades
  - Dissolved air flotation thickening improvements
  - Increased thickened sludge storage capacity
  - Dual centrifuge operation capability
- HRAS bypass with grit removal

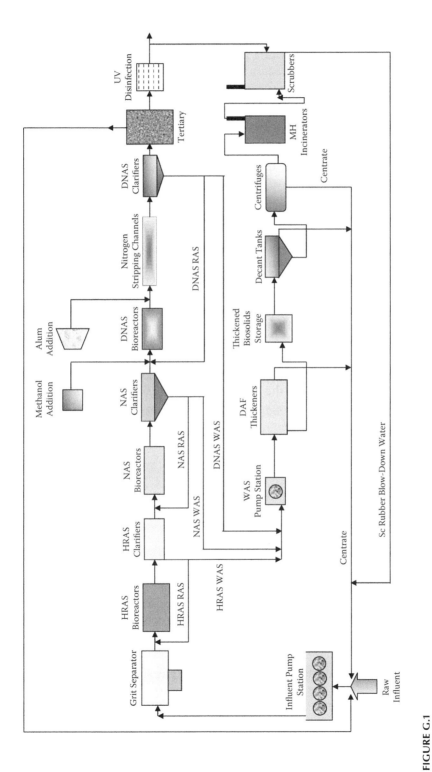

FIGURE G.1
WSSC Western Branch WWTP process flow diagram.

G.2.2 Description of Energy Conservation Measures (ECMs), Drivers, and Issues

The ECM implemented at the Western Branch WWTP involved upgrades to the facility's biosolids incineration multiple hearth furnaces (MHFs). Commissioned in 1974, the MHFs were designed to process (burn) 26 DTPD of heat-treated biosolids. WSSC decommissioned the biosolids heat-treating process soon after commissioning the MHFs. In 1996, WSSC replaced the centrifuges to increase the capacity of the MHGs, but in 2001 Title V and "zero visible emissions" requirements required reducing the capacity of the MHFs to 12 DTPD. To mitigate visible emissions, WSSC installed external afterburners, which increased the consumption of natural gas.

In the MHF process, the biosolids were introduced into the top of the furnace. The top hearth was also the exhaust point for the combustion gases. The cold biosolids feed, coming into contact with the hot furnace, released volatile hydrocarbons that did not have sufficient residence time in the furnace nor adequately high temperature to be oxidized (burned) completely before being emitted. The result was a high hydrocarbon content, smoky, and odorous emission. This emission exhaust stream was controlled by increasing the operating temperature at the top of the furnace and utilizing the external afterburner to increase the exhaust temperature and residence time. This operating mode resulted in high natural gas consumption. The higher operating temperature of the furnace, in turn, created slagging (i.e., melting) of the ash inside the furnace, increasing maintenance and associated costs (to remove slag).

Exacerbating the increase in fuel consumption, the incoming furnace combustion air entered the furnace at room temperature. This cooled the furnace and added significant heat load (and increased fuel consumption) to maintain furnace temperature. The hot MHF exhaust stream was cooled and cleaned in a wet scrubber before being discharged to that atmosphere, and the heat in the exhaust stream was lost (i.e., waste heat was not utilized). The following energy-savings modifications to the MHFs were implemented by WSSC at the Western Branch WWTP.

G.2.2.1 Flue Gas Recirculation

A flue gas recirculation (FGR) system was installed to collect exhaust flow from the top hearth of the furnace and reinject it into the lower hearths. This recirculated flue gas accomplishes the following:

- Unburned vapors and gases from hearth 1 are redirected through the burn zone in the furnace, providing sufficient contact time and temperature to complete the hydrocarbon oxidation process before exhausting.

- The additional air flow through the furnace tends to cool the hot hearths (reducing slagging) and helps to heat the cooler drying hearths, thus stabilizing the furnace operations.

- The high water vapor content of the recirculated gas stream (entrained from the drying zones) reduces the production of nitrogen oxides (NO_x) in the burning hearths. Stable temperatures in an MHF due to the addition of an FGR system are also known to reduce the production of thermal NO_x.

G.2.2.2 Exhaust Waste Heat Recovery

An air-to-air heat exchanger was installed in the exhaust stream of the furnace, upstream of the quench and wet scrubber, allowing recovery of the waste heat from the furnace exhaust. The recovered heat is utilized to preheat the combustion air entering the furnace, reducing the consumption of natural gas. Additionally, the center shaft cooling air exhaust (heated air) is returned to the furnace as preheated combustion air.

G.2.2.3 Circle Slot Jets

Circle slot jets, a ring of small air jets located near the top of each hearth, concentric with the center shaft of the furnace and about one-half the diameter of the furnace, were installed in the MHFs. Preheated combustion air is injected downward into the hearths through the circle slot jets, creating an impingement region and dual set of donut-shaped vortices in each hearth. This increases turbulence and air–fuel mixing. Simultaneously, a small portion of the required supply air is introduced (at room temperature) into the bottom hearth to cool the ash as it exits the furnace. The result of this modification is improved convection and turbulence, which increases drying rates in the drying zone and combustion rates in the burn zones. In 2009, one of the facility's multiple hearth furnaces was retrofitted with flue gas recirculation, exhaust waste heat recovery, and circle slot jets to improve fuel efficiency and capacity. In 2010, the facility's second multiple hearth furnace was similarly modified. Submetering of the fuel to the furnaces (single meter) was implemented as part of the ECM project. The drivers for this ECM were to

- Significantly reduce the amount of natural gas required to burn the biosolids produced from the Western Branch WWTP.
- Meet air emission requirements.
- Pay 100% of the capital cost of upgrade through energy savings.
- Increase the throughput capacity of the furnaces.
- Reduce NO_x emissions to meet best available control technology regulatory requirements.

G.3 Results

G.3.1 ECM Implementation Costs

Table G.1 provides the installed cost for implementation of the MHF modifications.

TABLE G.1

ECM Implementation Cost

Cost Category	Cost (2008$)
Total installed cost	$4,500,000

TABLE G.2

Natural Gas Cost and Savings

Year	Natural Gas Consumption (therms/yr)	Gas Rate ($/therm)	Energy Cost ($/yr)
Before ECM Implementation			
2005	420,000	$1.25	$525,000
After ECM Implementation			
2009	100,000	$1.25	$125,000
Savings	320,000		$400,000

G.3.2 Energy

Table G.2 summarizes the energy consumption and costs (nature gas fuel) prior to and following implementation of the aeration system ECM and the resultant energy savings provided by the MHF modifications. Based on the first 6 months of operation of the first of two MHFs to be modified, an annual reduction of 320,000 therms of natural gas were projected for the biosolids incineration operation at the Western Branch WWTP (a 76% reduction), resulting in an annual fuel expenditure savings of $400,000 per year.

G.3.2.1 Payback Analysis/Benefits

For this analysis, simple payback is determined by dividing the total project cost ($4,500,000 from Table G.1) by the natural gas fuel cost savings projection for the year following the project implementation ($400,000 per year, per Table G.2). The fuel costs savings provide by the MHF modifications resulted in a project payback period of 11.3 years following commission in 2009. This payback period does not include the avoided cost benefit (of delaying the construction of additional incineration capacity) provided by the increased capacity of the existing MHFs resulting from the ECM modifications. The new upgraded MHF operates at a continual throughput of 17 to 19 DTPD (a 42 to 58% capacity increase).

G.4 Conclusions

G.4.1 Factors Leading to Successful ECM Implementation/Operation

- Project planning should consider the condition of existing equipment, as well as the impacts of future process, operations and maintenance (O&M), and energy improvements.
- Hands-on training of multiple operators is vital to the success of newly installed equipment.
- To obtain commitment to a new system, plant staff must be shown that upgraded equipment improves operations and reliability.

G.4.2 Impact on Other Operating Costs Resulting from ECM Implementation

G.4.2.1 Emergency Sludge Hauling

MHF modifications increase the MHF capacity, reducing the need for emergency hauling of biosolids that are not incinerated. The resulting savings are estimated at $100,000 to $200,000/yr. These savings, if included in the analysis of payback, result in a payback period of between 7.5 and 9 years.

G.5 Lessons Learned

- As seasons change, the characteristics of the biosolids change and can range anywhere from 21% solids to 29% solids. At high solids content, the MHF is almost autogenous, and in this mode the furnace operation would be better if there was an ability to add ambient air separately from hot air on different hearths. The circle slot jets, as installed currently, do not allow this operational mode.

- The existing condition of the MHFs should be closely evaluated when considering an ECM implementation, as repair costs can increase the overall capital cost of the project by 10 to 15%. If the existing furnace requires rehabilitation, it should take place before or during an ECM implementation. Rehabilitation becomes more difficult once circle slot jets and heat exchanger ductwork is installed.

References

Brown, C. (1983). Seabees in service, in *Proceedings of Coastal Structures '83: A Specialty Conference on the Design, Construction, Maintenance, and Performance of Coastal Structures*, American Society of Civil Engineers, Arlington, VA.

Brunner, C.R. (1980). *Design of Sewage Incinerator Systems*, Noyes Data Corporation, Park Ridge, NJ.

Brunner, C.R. (1984). *Incinerator Systems: Selection and Design*, Van Nostrand Reinhold, New York.

Coker, C.S., Walden, R., Shea, T.G., and Brinker, M. (1991). Dewatering municipal wastewater sludges for incineration, *Water Environment & Technology*, 16, 63–67.

Haller, E.J. (1995). *Simplified Wastewater Treatment Plant Operations*, Technomic Publishers, Lancaster, PA.

Hardaway, C.S., Thomas, G.R., Unger, M.A, Greaves, J., and Rice, G. (1991). *Seabees Monitoring Project, James River Estuary, Virginia*, Report for the Center of Innovative Technology by Virginia Institute of Marine Science, College of William and Mary, Virginia Institute of Marine Science, Gloucester Point.

Lester, F.N. (1992). Sewage and sewage sludge treatment, in Harrison, R., Ed., *Pollution: Causes, Effects, and Control*, Royal Society of Chemistry, London, pp. 33–62.

Lewis, F.M., Haug, R.T., and Lundberg, L.A. (1988). Control of Organics, Particulates and Acid Gas Emissions from Multiple Hearth and Fluidized Bed Sludge Incinerators, paper presented at 61st Annual Conference and Exposition of the Water Pollution Control Federation in Dallas, TX, October 2–6.

Metcalf & Eddy. (2002). *Wastewater Engineering: Treatment, Disposal and Reuse*, 4th ed., McGraw-Hill, New York.

Outwater, A.B. (1994). *Reuse of Sludge and Minor Wastewater Residuals*, CRC Press, Boca Raton, FL.

Peavy, S., Rowe, D.R., and Tchobanglous, G. (1985). *Environmental Engineering*, McGraw-Hill, New York.

USEPA. (1978). *Operations Manual: Sludge Handling and Conditioning*, EPA 430/9-78-002, U.S. Environmental Protection Agency, Washington, DC.

USEPA. (1982). *Dewatering Municipal Wastewater Sludges*, EPA 625/1-82-014, Center for Environment Research Information, U.S. Environmental Protection Agency, Cincinnati, OH.

USEPA. (1990). *Analytical Methods for the National Sewage Sludge Survey*, EPA 440/1-90-023, U.S. Environmental Protection Agency, Washington, DC.

USEPA. (1993). *Preparing Sewage Sludge for Land Application or Surface Disposal: A Guide for Preparers of Sewage Sludge on the Monitoring, Recordkeeping, and Reporting Requirements of the Federal Standards for the Use or Disposal of Sewage Sludge, 40 CFR Part 503*, EPA 831/B-93-002a, U.S. Environmental Protection Agency, Washington, DC.

Vesilind, P.A (1980). *Treatment and Disposal of Wastewater Sludges*, Ann Arbor Science, Ann Arbor, MI.

Appendix H. Permit-Safe and Energy-Smart Greening of Wastewater Treatment Plant Operations at the San Jose/Santa Clara, California, Water Pollution Control Plant

The San Jose/Santa Clara Water Pollution Control Plant (SJ/SC WPCP), one of the largest advanced wastewater treatment facilities in California, serves a population of 1,500,000 people in a 300-square-mile area encompassing San Jose, Santa Clara, Milpitas, Campbell, Cupertino, Los Gatos, Saratoga, and Monte Sereno. The plant, treating an average daily flow of 107 MGD, was last upgraded from a two-stage nitrification process to a step-feed biological nutrient removal (BNR) process in 1995.

Recent experience has shown that BNR systems are reliable and effective in removing nitrogen and phosphorus. The process is based on the principle that, under specific conditions, microorganisms will remove more phosphorus and nitrogen than is required for biological activity; thus, treatment can be accomplished without the use of chemicals. Not having to use and therefore having to purchase chemicals to remove nitrogen and phosphorus potentially has numerous cost–benefit implications. In addition, because chemicals are not required to be used, chemical waste products are not produced, thus reducing the need to handle and dispose of waste. Several patented processes are available for this purpose. Performance depends on the biological activity and the process employed.

In 2008, the SJ/SC WPCP implemented the following energy conservation measures (ECMs) projects with financial assistance from the California Wastewater Process Optimization Program (CalPOP):

- Optimization of three plant pumping systems (post screening, post primary settling and post clarification)
- Implementation of pulsed air mixing of the WPCP's BNR process anaerobic and anoxic reactors
- Optimization of the biosolids thickening dissolved air flotation (DAF) pressurization pumps

All three ECM projects involved the development and implementation of proprietary control system algorithms.

The post screening, post primary settling, and post clarification pumping system optimization project resulted in electrical energy use reductions of 13.33 kW/million gallons (MG) (22% reduction), 19.9 kW/MG (23.5% reduction), and 21.6 kW/MG (17% reduction), respectively. Pulsed air mixing of the BNR process anaerobic and anoxic reactors reduced natural gas consumption by 1.2×10^{11} BTU/yr (a 38% reduction) and electrical energy consumption by 4.8×106 kWh/yr, a 23% reduction (aeration for one of the two BNR trains at SJ/SC WPCP is provided by internal combustion engine-driven blowers; the blowers for the second BNR train are electric motor driven). Optimization of the operation of the DAF pressurization pumps reduced electrical energy consumption by 1,603,030 kWh/yr (64% reduction).

Testing and verification of these ECM projects by the SJ/SC WPCP's energy utility (Pacific Gas & Electric) reported a total energy savings of $1,178,811 per year. With a total implementation cost for these three ECM projects of $269,569, the payback for the combined ECM project was less than 3 months.

H.1 Evaluation of Energy Conservation Measures

H.1.1 Site/Facility/Process Description

The San Jose/Santa Clara Water Pollution Control Plant first began operations in 1956 as a primary treatment facility. The plant was upgraded in 1964 to secondary treatment and again in 1979 with the addition of a two-stage nitrification and filtration process. A step-feed biological nutrient removal (BNR) process was implemented in 1995. The BNR process commonly consists of alternating anoxic and aerobic states; however, influent flow is split to several feed locations, and the recycle biosolids stream is sent to the beginning of the process. BNR is used to remove total nitrogen. Implementation of BNR led to reductions in aeration energy consumption and costs, enhanced bulking control, and increased plant capacity. The single-stage BNR process has the advantage of operating two activated biosolids plants in parallel rather than in series (as was the case prior to the 1995 upgrade).

The plant has the capacity to treat 167 million gallons per day (MGD) and currently receives an average influent of 107 MGD. Approximately 10% of the plant effluent is reused as recycled water for irrigation and makeup water for cooling towers. The plant influent data are presented in Table H.1. Figure H.1 presents the process flow diagram for SJ/SC WPCP. Raw sewage entering the plant undergoes several stages of treatment. First, the influent passes through screening, grit removal, and primary settling. The flow then splits into two parallel BNR plants (BNR1 and BNR2). The BNR plants consist of multiple treatment zones with multiple aerators and clarifiers. The first compartment in each aeration basin is operated under anaerobic conditions, and the second and fourth compartments are operated under aerobic conditions. The third compartment is operated under anoxic conditions. Approximately 60% of the influent flow and 100% of the returned activated sludge (RAS) are fed to the first (anaerobic) compartment. Approximately 40% of the influent flow is fed to the third compartment, which is operated under anoxic conditions. To maintain solids in suspension in the anoxic/anaerobic compartments, approximately 1000 scfm of air are pumped into each of these compartments. To minimize capital costs when the plant was upgraded from the two-stage nitrification configuration to the BNR plant, existing coarse bubble diffusers were utilized for mixing the anaerobic/anoxic zones vs. installing new mechanical mixers. A schematic of the BNR process is depicted in Figure H.2.

TABLE H.1

SJ/SC WPCP Influent Data (2009)

Parameter	Average	Daily Maximum
Flow (MGD)	107	167
BOD (mg/L)	298	512
TSS (mg/L)	241	797
Ammonia-N (mg/L)	31	54

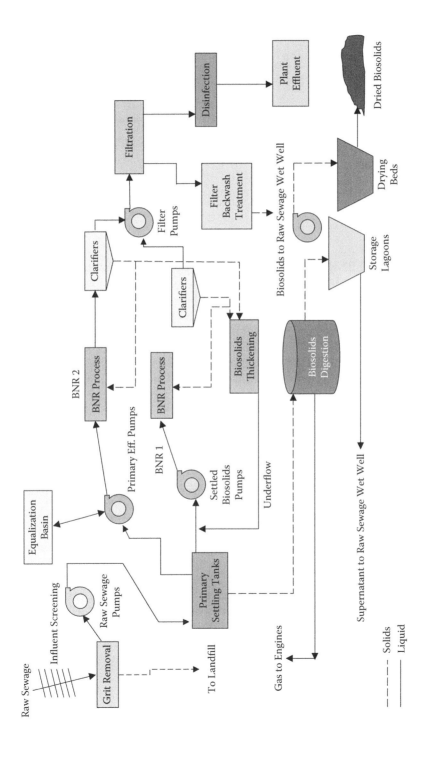

FIGURE H.1
SJ/SC WPCP treatment process flow diagram.

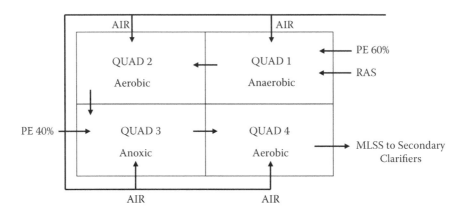

FIGURE H.2
SJ/SC WPCP BNR process flow diagram.

Air is supplied to the BNR1 plant aeration process by internal combustion engine drive blowers that utilize a mixture of digester, landfill, and natural gas for fuel. A byproduct of the operation of the internal combustion engines is hot water (spent cooling water), which is used for digester heating. Air is supplied to the BNR2 plant aeration process by electrical motor-driven blowers. BNR process effluent is filtered, disinfected with chlorine, and then dechlorinated prior to discharge to San Francisco Bay.

The solids wasted from the BNR processes are thickened in dissolved air flotation (DAF) tanks. The thickened biosolids are then fed to the plant's mesophilic digesters where they blend with the biosolids from the primary clarifiers. Digested solids are stored in biosolids stabilization lagoons for up to 3 years, and the dredge biosolids from these lagoons are dried in solar drying beds. The dried solids are then hauled to a nearby landfill and used as landfill cover.

H.1.2 Description of Energy Conservation Measures (ECMs), Drivers, and Issues

During the past several years, the plant has been quite active and successful in identifying and implementing energy saving projects without compromising effluent quality. The projects have produced significant sustained savings in operating costs. In addition to energy savings, financial incentives in the form of rebates from the local electric and natural gas utility, Pacific Gas & Electric (PG&E, San Francisco), helped to incentivize these projects. In many cases, the utility rebates covered the entire cost of the projects, resulting in a final implementation cost of zero. This case study describes several ECM projects completed in 2008, which were partially funded by the California Wastewater Process Optimization Program (CalPOP). The program was administered by QuEST, Inc. (Berkeley, CA). The ECM projects implemented at the SJ/SC plant are described below.

H.1.2.1 Pumping Systems Optimization

The first ECM project focused on reducing energy consumption by optimizing the operation of three of the plant's major pump stations. The information on the optimized pump stations is provided in Table H.2.

TABLE H.2

Optimized Pump Stations

Pump Station	Average Flow (MGD)	Total No. of Pumps	VFD-Equipped Motors	Power Use (kWh/day)
Post screening	113	7	3	282
Post primary settling	109	4	4	384
Post clarification	108	5	5	570

The plant implemented an optimization algorithm developed by Ekster & Associates, Inc. (Fremont, CA) to select the proper pump operating schedule and optimize energy consumption. This computer program utilizes field data such as pump station flows, pump discharge pressures, wet well levels, and power usage associated with pumps. The pertinent data were collected using a specially designed pump testing routine. Thereafter, the software program selects the combination of pumps and speed (for the existing variable speed motors) at each flow rate. To ensure that the global rather than local minimum power consumption is reached, the software program utilizes two optimization algorithms in tandem (generic and gradient reduction algorithms) rather than a single algorithm. This methodology guarantees that the selected pump and speed combination for each flow regime results in the consumption of less energy compared to any other possible combination. Plant staff programmed the selected schedule and pump speed for each flow range into the plant's distributed control system (DCS). In addition, discharge pressures and wet well levels were optimized by reassessing the minimum safety requirements. During the implementation phase of the project, the plant compared field data to the pump manufacturer's data. This included a comparison of the manufacturer's pump performance curves with the experimental curves. The study revealed that some of the pumps had lost 5 to 7% of their efficiency, probably due to age and wear. These findings reinforced the idea that pump curves generated using field data—not manufacturer's pump curves—should be used to develop optimized operating sequences. In addition, this effort resulted in identification and qualification of the pumping systems' information and data that will be used by the utility to plan for refurbishing and replacement of underperforming pumping system components.

H.1.2.2 BNR Improvements: Pulse Aeration of Anaerobic/ Anoxic Zones and Mixed Liquor Channels

A second ECM project implemented at the plant involved switching the mode of air mixing in the anoxic/anaerobic compartments of BNR plants from continuous to pulse (on/ off). This aeration method was also applied to the mixed liquor channels (MLCs) of both BNR plants. Prior to switching to the pulse air mixing mode, solids were maintained in suspension in the anaerobic/anoxic compartments and MLCs by continuous air flow. An Ekster patent-pending method was used to replace continuous air mixing with pulsed air mixing. The implementation of this ECM required significant modifications of the aeration system. These modifications included installation of new valves, actuators, pneumatic lines, electrical infrastructure, and special control system programming. These modifications were required to be completed within a 6-month period due to deadlines associated with the CalPOP program. To meet this unusually stringent schedule requirement, the plant staff performed the conceptual and detailed designs in-house and prepared all other pertinent documentations, including the bid packages. Plant staff also specified control valves, flow meters, actuators, and auxiliary control elements, such as air piping for actuators, input–output units for the plant's DCS, etc. Establishing the timing sequence of the activation and deactivation (i.e., on/off sequencing) of the pulsed air mixing control system is site specific, depending on the settleability of the mixed liquor suspended solids and the geometry of the anaerobic/anoxic compartment. For the SJ/SC WPCP, an air flow rate and timing sequence were established by trial and error that maintained solids in suspension while keeping the dissolved oxygen (DO) concentration low enough (0.2 mg/L or less) so as not to significantly hinder the anaerobic/anoxic process. The adequacy of the mixing provided by the pulsed air system was verified by confirming equal concentrations of solids on the bottom and at the surface of the anaerobic/anoxic compartments (at the end of the air "on" cycle). At the end of the "off" cycle, a bit of biosolids settling occurs. At the end of the "on" cycle, it is important to ensure that biosolids resuspension has occurred.

During implementation of the pulse air mixing control, engineers discovered that providing simultaneous pulsed air for multiple tanks could lead to oscillation of the blower output. A special programming routine was subsequently developed to avoid this control system oscillation. The new routine sequences the tanks rather than simultaneously providing air to all the tanks to resuspend solids.

H.1.2.3 Dissolved Air Flotation Process Optimization

Optimization of the DAF process was achieved by reducing the energy used by the pressurization pumps. This was made possible by utilizing proprietary algorithms developed by Ekster and Associates that provided the means to optimize the DAF control systems. Prior to optimization, each DAF tank was operating at a constant pressurized flow, which was significantly higher than required. The algorithm allowed automatic adjustment of the pressurized flow based on the number of DAF tanks in service and the incoming solids load to maintain the same air-to-solids ratio (A/S) under all operating and influent conditions. The new algorithm also provided a close approximation of equal solids loading throughout the day for each DAF unit. The minimum A/S was determined by trial and error. The criterion utilized to establish the minimum A/S was the equality of water and biosolids concentrations before and after A/S reduction. The current A/S stands at 0.005 and is one of the lowest ever reported in the literature. The control system algorithms for the in-plant systems, pulsed air mixing system, and DAF process were implemented and commissioned in 2008. Submetering of electrical power for the DAF process was also implemented as part of the ECM project. Electrical submetering for the in-plant pumping systems and BNR2 plant blowers was installed prior to the ECM project (with their original design) as was the gas metering (for the combined mixture of landfill gas, digester gas, and utility supplied natural gas) for the BNR1 plant internal combustion engine driven blowers. Tables H.3 and H.4 provide a comparison of the major water quality indicators associated with the SJ/SC WPCP influent and effluent in 2007 prior to implementation of improvements and performance after ECM implementation. Table H.5 provides a comparison of DAF performance before and after optimization.

H.2 Results

H.2.1 ECM Implementation Cost

Table H.6 summarizes the cost associated with the implementation of the ECM projects.

TABLE H.3

Monthly Averages, 2007 and 2009–2010

		Concentration (mg/L)	
Parameter		2007 Daily Average	November 2009 to February 2010
BOD	Influent	332.0	363.0
	Effluent	3.1	3.7
	Monthly permit limit	10.0	10.0
TSS	Influent	291.0	293.0
	Effluent	1.5	1.5
	Monthly permit limit	10.0	10.0
NH_3	Influent	27.9	31.0
	Effluent	0.5	0.6
	Monthly permit limit	3.0	3.0

TABLE H.4

Daily Maximums, 2007 and 2009–2010

		Concentration (mg/L)	
Parameter		2007 Daily Average	November 2009 to February 2010
BOD	Influent	438.0	516.0
	Effluent	5.0	6.0
	Permit limit	20.0	20.0
TSS	Influent	534.0	546.0
	Effluent	3.5	2.1
	Permit limit	20.0	20.0
NH_3	Influent	43.8	41.8
	Effluent	1.4	1.8
	Permit limit	8.0	8.0

TABLE H.5

DAF Performance Before and After ECM Implementation

Thickened Solids (%)		Underflow Total Suspended Solids (mg/L)	
Before Optimization	After Optimization	Before Optimization	After Optimization
3.8	3.8	92	87

TABLE H.6

ECM Implementation Cost

Project	Capital Cost	Implementation Cost	Total Cost
Liquid pumping optimization	$4,545	$39,223	$43,768
Pulse aeration	$62,822	$118,770	$181,592
DAF process optimization	$2,948	$41,261	$44,209
Total cost for all ECM			$269,569

TABLE H.7

Energy Use Before and After Pump Station Optimization

Pump Station	Before Optimization (kW/million gal)	After Optimization (kW/million gal)	Energy Use Reduction (kW/MG)	Percent Reduction
Post screening	59.58	46.25	–13.33	22.0%
Post primary settling	84.51	64.62	–19.89	23.5%
Post clarification	126.51	104.88	–21.63	17.1%

TABLE H.8

Energy Savings Achieved by Switching to Pulsed Aeration

	Annual Energy Consumption		Energy Savings	
BNR	Before Optimization	After Optimization	Net Annual Savings	Percent Reduction
No. 1	3.1×10^{11} BTU	1.9×10^{11} BTU	1.2×10^{11} BTU	38.0%
No. 2	6.2×10^{6} kWh	1.4×10^{6} kWh	4.8×10^{6} kWh	22.5%

TABLE H.9

Dissolved Air Floatation (DAF) Process Optimization: Energy Use and Savings

Annual Energy Consumption		Energy Savings	
Before Optimization	After Optimization	Net Annual Savings	Percent Reduction
2,496,600 kWh/yr	893,570 kWh/yr	1,603,030 kWh/yr	64%

H.2.2 Energy

H.2.2.1 Pump Station Optimization

Optimizing the pump stations led to energy reductions of between 17% and 23.5% (see Table H.7).

H.2.2.2 BNR Process Improvements

Table H.8 summarizes energy savings achieved by converting from continuous to pulsed aeration in the BNR process.

H.2.2.3 DAF Process Optimization

Table H.9 summarizes energy savings achieved through DAF optimization.

H.2.3 Payback Analysis/Benefits

Table H.10 summarizes the payback for the SJ/SC WPCP ECM projects.

TABLE H.10

Payback Analysis for the Energy Saving Projects at SJ/SC WPCP (2008)

Project	Total Cost	Annual Savings (@ $0.11/kWh)	Annual Savings ($1/Therm)	Payback Period (Months)
Liquid pumping optimization	$43,768	$244,858	N/A	2.1
BNR process improvements	$181,592	$176,339	$581,275	2.9
DAF process	$44,209	$176,339	N/A	3.0

Note: PG&E reimbursed the plant $269,569 for these three ECM projects.

H.3 Conclusions

H.3.1 Factors Leading to Successful ECM Implementation/Operation

- All ECMs were initially extensively tested on one unit before implementing the modification for the entire system. This provided the staff with the confidence that the ECMs would not have a negative effect on the performance of the treatment processes.
- Measures are quickly reversible and adaptable to changing operational situations. One example is the suspension of pulsed aeration for a few brief periods to accommodate operational corrective measures required to deal with foam observed from time to time on the surface of aeration basins.
- Rebates from electrical utilities provided additional incentives for prioritizing implementation of the ECMs.

H.3.2 Impact on Other Operating Costs Resulting from ECM Implementation

In addition to energy saving resulting from this project, it is expected that the service life of the pumps will increase as a result of operating the pumps closer to their best efficiency points (BEPs).

H.4 Lessons Learned

Significant savings can be achieved with minimum capital investment by optimizing operating procedures and process control setpoints; however, changes in operating protocols or parameters require significant testing to ensure that these measures do not jeopardize plant reliability and water quality.

References

Ekster, A. (2009). Optimization of pump station operation saves energy and reduces carbon footprint, in *Proceedings of WEFTEC 2009*, Water Environment Federation, Alexandria, VA.

Lemma, I.T., Colby, S., and Herrington, T. (2009). Pulse aeration of secondary aeration tanks holds energy saving potential without compromising effluent quality, in *Proceedings of WEFTEC 2009*, Water Environment Federation, Alexandria, VA.

Sinaki, M., Yerrapotu, B., Colby, S., and Lemma, I. (2009). Permit safe, energy smart—greening wastewater treatment plant operations, in *Proceedings of WEFTEC 2009*, Water Environment Federation, Alexandria, VA.

Appendix I. Diffuser Upgrades and DO Controlled Blowers at the Waco, Texas, Metropolitan Area Regional Sewer System Wastewater Treatment Facility

The Waco Metropolitan Area Regional Sewer System (WMARSS) Treatment Facility is a 22.8-MGD (average daily flow) single-stage nitrification plant with multi-stage centrifugal blowers. The facility was experiencing difficulty achieving single-stage nitrification because the existing aeration system was unable to deliver the required oxygen to complete the nitrification reaction. As a result, the air flow rate for the diffusers was set at a rate that exceeded the design rate of the diffusers, causing them to produce coarse bubbles vs. fine bubbles, further exacerbating the inadequate transfer of oxygen into the wastewater.

The WMARSS facility staff analyzed the problem and concluded that an upgrade to the aeration system was necessary. Under the upgrade, the number of fine-bubble diffusers in each basin was increased from 2800 to 3500. In addition, a dissolved oxygen (DO) probe was added to each of the aeration basins' three aeration zones. The control system was upgraded to provide automatic control based on the readings of the DO monitoring system.

The upgrades to the aeration system cost $397,708 (total installed cost). The installation was done by plant personnel so no outside installation costs were incurred by WMARSS. Electrical energy savings of $331,272 were realized in the first 2 years and $335,907 in the third year following the installation and commissioning of the ECM project, resulting in a payback of 2.4 years. Between 2003 and 2008, the system reduced electrical energy consumption by an average of 4,642,741 kWh/yr (33% reduction), representing over $2.5 million in energy cost savings over this period (an average annual energy cost savings of $423,226/yr). In addition to energy savings, labor costs have been reduced as operators no longer are required to perform routine DO readings and associated manual blower adjustments. Because the nitrification process is now working as designed, nitrate levels are lower in the effluent, reducing chlorine demand in the disinfection process, resulting in reduced chemical costs for chlorine.

I.1 Evaluation of Energy Conservation Measures

I.1.1 Site/Facility/Process Description

The Waco Metropolitan Area Regional Sewer System (WMARSS) treatment facility serves approximately 175,000 people from the cities of Bellmead, Hewitt, Lacy–Lakeview, Robinson, Waco, and Woodway, Texas. In the early 1970s, WMARSS was originally commissioned as a trickling filter plant and was upgraded to a 37.8-MGD activated biosolids plant from 1983 to 1985. In 1995, the plant was upgraded to perform single-stage

TABLE I.1

WMARSS Influent Quality, 2009

Parameter	Average	Daily Maximum	Peak
Flow (MGD)	22.8	—	83.2
BOD (mg/L)	251	608	—
TSS (mg/L)	300	2671	—
Ammonia-N (mg/L)	31.5	95.1	—

nitrification. WMARRSS facility influent data are presented in Table I.1. Figure I.1 presents the process flow diagram for the WMARSS wastewater treatment plant (WWTP), a single-stage nitrifying activated biosolids treatment plant.

Influent undergoes screening followed by primary clarification. Primary clarified effluent is distributed to five aeration basins, typically operated in a plug flow mode (in which primary settled wastewater and return activated sludge enter the end of the aeration tanks and air is distributed uniformly throughout the length of the aeration tank). During high influent flow conditions, the aeration process is operated in a step feed mode, a modification of plug flow in which the primary clarifier effluent is introduced at several points in the aeration tank and the secondary return activated sludge (RAS) is introduced in the anoxic zone, resulting in a reduced mixed liquor suspended solids (MLSS) loading entering the secondary clarifier. The first 50 ft of each aeration basin is maintained as an anoxic zone that receives RAS from the secondary clarifiers. Currently, RAS achieves oxygen savings in the aeration basins (as oxygen is stripped from the nitrate in the RAS stream in the anoxic zone). In the future, the anoxic zone and associated RAS will accommodate biological nutrient removal (phosphorus and nitrogen). Each aeration basin measures 50 ft wide by 251 ft long and has a side-water depth of 18 ft (total basin volume = 8.45 million gallons). Seven Hoffman multi-stage centrifugal blowers (five 250-hp blowers at 6000 standard cubic feet per minute [scfm] and two 650-hp blowers at 12,500 scfm) provide a total of 55,000 scfm through a fine-bubble diffuser system.

Aeration is followed by secondary clarification in four clarifiers. Variable portions of the secondary clarifier effluent are pumped to sand filters. Unfiltered secondary clarifier effluent is blended with filtered effluent, and the combined effluent stream undergoes chlorination and dechlorination prior to reuse or discharge to the Brazos River.

Primary biosolids undergoes two-stage thickening (gravity followed by rotary drum) prior to anaerobic digestion. Secondary waste activated sludge (WAS) is thickened in the rotary drum thickener (RDT) prior to digestion. The WAS is used as MLSS seed for sidestream treatment. Supernatant from the rotary drum thickener is combined with gravity thickener supernatant for sidestream treatment (existing trickling filters and aeration currently under construction followed by a final solids clarifier). Filtrate from biosolids dewatering may also undergo sidestream treatment or be returned to the head of the plant. Biosolids from the sidestream treatment final solids clarifier is thickened (by gravity or by rotary drum) prior to anaerobic digestion. A portion of the sidestream treatment final clarifier underflow is returned to the anoxic zone of the sidestream treatment process.

Overflow from the sidestream treatment clarifier is returned to the head end of the plant. The plant has four mesophilic anaerobic digesters that can be operated in series, parallel, complete mix, or as a combination of primary/secondary digesters. Industrial waste is received via tanker at the WMARSS facility in the form of blood, biosolids, and grease (animal source) from local food processing establishments (averaging 13,000

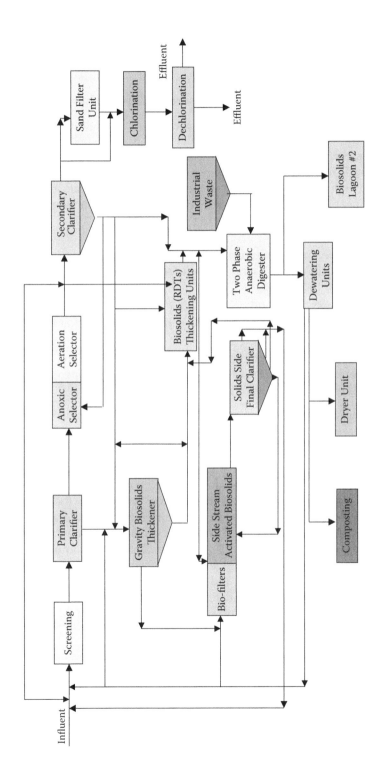

FIGURE I.1
WMARSS process flow diagram.

gallons per day). This industrial/commercial waste is treated with the primary and secondary biosolids in the anaerobic digesters. Digested biosolids are dewatered using a belt press or sent to surface disposal (biosolids lagoons), then dried and pelletized (or composted, in the future).

I.1.2 Description of Energy Conservation Measures (ECMs), Drivers, and Issues

The City of Waco, Texas, Utility Services Department implemented improvements to the aeration system at the WMARSS treatment facility to address deficiencies in the plant's nitrification process. Single-stage nitrification was not possible using the existing diffuser system because insufficient oxygen was being supplied to the aeration process. The existing diffusers were being operated in excess of their design air flow rate, producing coarse bubbles instead of fine bubbles, and the number of diffusers was inadequate. The aeration system improvements consisted of the following:

- Supplement the existing Sanitaire® fine-bubble membrane and ceramic disc diffusers with additional diffusers. The number of diffusers in each basin was increased from 2800 to 3500.
- Install Danfoss dissolved oxygen probes in each of the aeration basins' three aeration zones (midway in the first and second aeration zones and at the end of the third aeration zone).
- Implement automatic blower and aeration system control through the plant's PLC system using aeration basin DO readings. The plant's control system software provided the following output based on aeration basin DO readings:
  - On/off control of blowers
  - Blower inlet valve throttling
  - Aeration system drop leg throttling

The aeration basins' diffuser system was supplemented with additional diffusers in 2003. Additionally, DO probes were installed in the aeration basins, and the aeration control system was programmed to utilize DO reading in the aeration basin to control the operation of the blower inlet and basin drop leg throttling valves. Implementation of the aeration system improvements was initiated to address problems with completing the plant's nitrification cycle. Drivers for the project included requirements for increasing the plant's capacity and improving effluent quality and to effect energy savings. Tables I.2 and I.3 provide a comparison of the WMARSS facility performance prior to the implementation of aeration system improvements (2002) and performance after the ECM implementation.

I.2 Results

I.2.1 ECM Implementation Costs

Table I.4 summarizes the implementation costs for the WMARSS aeration system ECM project.

TABLE I.2

Monthly Averages, 2002 and 2009

Parameters		Concentration (mg/L) 2002	Concentration (mg/L) 2009
BOD	Influent	322.58	251.00
	Effluent	2.81	2.31
	Permit limit	10.00	10.00
TSS	Influent	419.56	300.00
	Effluent	3.06	1.20
	Permit limit	15.00	15.00
NH_3	Influent	15.78	31.50
	Effluent	1.45	0.33
	Permit limit	3.00	3.00

TABLE I.3

Daily Maximums, 2002 and 2009

Parameter		Concentration (mg/L) 2002	Concentration (mg/L) 2009
BOD	Influent	644.0	608.0
	Effluent	7.8	9.4
	Permit limit	25.0	25.0
TSS	Influent	1600.0	2671.0
	Effluent	11.7	5.5
	permit limit	40.0	40.0
NH_3	Influent	26.7	95.5
	Effluent	13.8	5.0
	Permit limit	10.0	10.0

TABLE I.4

ECM Implementation Costs

Cost Category	Cost (2002$)
Capital components	
Additional diffusers	$239,200
PLC automation	$24,906
DO instrumentation	$18,420
Air control valves	$66,692
Total capital cost	$349,218
Installation costs (performed by plant personnel)	
Aeration improvements	$18,390
Aeration improvements	$30,100
Total installation costs (est.)	$48,490
Total project cost	$397,708

TABLE I.5

Electricity Use and Estimated Savings

Year (A)	Total Annual Electricity Use (kWh) (B)	Annual Energy Consumption Reduction (kWh) (C)	Average Daily Flow (MGD) (D)	Electricity Use per Average Daily Flow (kWh/MGD) (E)	Average Electricity Rate ($/kWh) (F)	Electricity Cost Savings ($) (G)[a]
		Before ECM Implementation				
2002	14,076,530	—	26.4	$532,431	$0.0430	—
		After ECM Implementation				
2003	11,624,105	2,452,425 (17%)	24.3	477,996	$0.0537	$131,695.22
2004	11,006,112	3,070,418 (22%)	28.8	382,366	$0.0650	$199,577.17
2005	9,201,249	4,875,281 (35%)	24.6	373,701	$0.0689	$335,906.86
2006	7,969,924	6,106,606 (43%)	21.7	367,563	$0.0897	$547,762.56
2007	7,851,481	6,225,049 (44%)	27.6	284,400	$0.1150	$715,880.64
2008	8,949,861	5,126,669 (36%)	22.9	390,765	$0.1187	$608,535.61

[a] $G = (B_{\text{evaluation year}} - B_{2002}) \times (F)$.

I.2.2 Energy

Table I.5 summarizes the WMARSS facility electrical energy consumption and costs prior to and following implementation of the aeration system improvements. The last column presents the electrical cost savings per year at the facility.

I.2.3 Payback Analysis/Benefits

For this analysis, simple payback is determined by dividing the total project cost ($397,708 from Table I.4) by the electricity cost savings for each year following the project implementation. The electrical energy costs savings provided by the aeration system improvements resulted in a project payback in the first quarter (March) of the third year following commissioning (2005), a 2.4-year payback.

I.3 Conclusions

I.3.1 Factors Leading to Successful ECM Implementation/Operation

Facility staff was directly involved in identifying the treatment plant's operational problem and the process equipment and operational modifications to address the plant's operating problems. The staff was also involved in installation of the aeration system improvements. The aeration system modifications resulted in improved plant performance, reduction in energy consumption, and reduction in direct operator involvement required to maintain blower and aeration system performance. Plant personnel reported that this ownership stake in the project was a critical factor in the successful implementation and continued successful operation of treatment system.

I.3.2 Impact on Other Operating Costs Resulting from ECM Implementation

In addition to energy savings resulting from this project, the following benefits were also realized.

I.3.2.1 Labor

The standard operating procedure before automating the aeration process was to manually check the DO concentration in each of the basins' zones on an hourly basis and adjust the drop leg valves and blower demand in response to the DO readings. The utility estimates that automation of the aeration saves approximately 3 hours of operator labor per day (1095 hours per year) at a savings of $21,900 per year.

I.3.2.2 Chemicals

With the ammonia cycle stabilized (less nitrite/chlorine demand in the effluent), chlorine demand has been reduced and stabilized. Prior to implementation of the aeration system modifications, a daily maximum of approximately 6000 lb of chlorine was dosed per day (under normal flow conditions of approximately 25 MGD whenever the plant had high effluent nitrate levels). Currently (under average daily flow conditions of 22.5 MGD and complete nitrification), the average chlorine dosage ranges between 800 and 1200 lb/day. During implementation of the aeration system ECM, the effluent chlorination monitoring and control system was also upgraded. It is not possible to segregate the chlorine chemical consumption/cost reductions attributed exclusively to the ECM project, as both the upgraded chlorination monitoring and control system and the ECM were commissioned during the same period (2003/2004).

I.3.2.3 Maintenance

With the savings in energy cost and operator labor provided by automating the aeration process, WMARSS is able to maintain adequate staffing according to industry standards.

Glossary*

A²/O process: Modified Ludzack–Ettinger (MLE) process preceded by an initial anaerobic stage; used to remove both total nitrogen (TN) and total phosphorus (TP).

Absorber: In a photovoltaic device, the material that readily absorbs photons to generate charge carriers (free electrons or holes).

Access time: (1) The time it takes a computer to locate data or an instruction word in its storage section and transfer it to its arithmetic unit, where required computations are performed. (2) Time it takes to transfer information that has been operated on from the arithmetic unit to the location in storage where the information is stored.

Acid: A substance that dissolves in water and releases hydrogen ions.

Acid hydrolysis: A chemical process in which acid is used to convert cellulose or starch to sugar.

Activated sludge: Sludge particles produced in raw or settled wastewater (primary effluent) by the growth of organisms (including zoogleal bacteria) in aeration tanks in the presence of dissolved oxygen.

Actuator: A mechanism for translating a signal into the corresponding movement of control; typically the actuator moves a valve.

Aeration: The process of adding air to water. Air can be added to water by either passing air through water or passing water through air.

Aerobes: Bacteria that must have dissolved oxygen (DO) to survive.

Afterburner: A device that includes an auxiliary fuel burner and combustion chamber to incinerate combustible gas contaminants.

Agglomeration: The growing or coming together of small scattered particles into larger flocs or particles, which settle rapidly.

Air emission: For stationary sources, the release or discharge of a pollutant by an owner or operator into the ambient air either by means of a stack or as a fugitive dust, mist, or vapor inherent to the manufacturing or forming process.

Air pollutant: Dust, fumes, mist, smoke, other particulate matter, vapor, gas, aerosol, odorous substances, or any combination thereof.

Albedo: The ratio of light reflected by a surface to the light falling on it.

Alcohol: A general class of hydrocarbons that contains a hydroxyl group (OH). There are many types of alcohol (e.g., butanol, ethanol, methanol).

Alcohol fuels: Alcohol can be blended with gasoline for use as transportation fuel. It may be produced from a wide variety of organic feedstock. The common alcohol fuels are methanol and ethanol. Methanol may be produced from coal, natural gas, wood, and organic waste. Ethanol is commonly made from agricultural plants, primarily corn, containing sugar.

Algorithm: A prescribed set of well-designed rules or processes for the solution of a problem in a finite number of steps—for example, a full statement of an arithmetic procedure for evaluating $\sin(x)$ to a stated precision.

---

* Some definitions were retrieved from EERE, *Glossary of Energy-Related Terms*, Energy Efficiency and Renewable Energy, U.S. Department of Energy, Washington, DC, 2012 (http://www1.eere.energy.gov/site_administration/glossary.html).

Alkalinity: The capacity of water or wastewater to neutralize acids.

Alkaline fuel cell (AFC): A type of hydrogen/oxygen fuel cell in which the electrolyte is concentrated potassium hydroxide (KOH) and the hydroxide ions (OH⁻) are transported from the cathode to the anode.

Alphanumeric: Pertaining to a character set that contains both letters and digits and, usually, other characters such as punctuation marks.

Alternating current (AC): An electric current that reverses its direction at regularly recurring intervals, usually 50 or 60 times per second.

Alternator: A device that turns the rotation of a shaft into alternating current (AC).

Ambient: Natural condition of the environment at any given time.

Ampere (amp): A unit of electrical current can be thought of as like the rate of water flowing through a pipe (liters per minute).

Ampere-hour: A measure of the flow of current (in amperes) over 1 hour; used to measure energy production over time, as well as battery capacity.

Amphorous silicon: An alloy of silica and hydrogen, with a disordered, noncrystalline internal atomic arrangement, that can be deposited in thin layers (a few micrometers in thickness) by a number of a deposition methods to produce thin-film photovoltaic cells on glass, metal, or plastic substrates.

Amplifier: A device that enables an input signal to control power from a source independent of the signal and thus be capable of delivering an output that bears some relationship to, and is generally greater than, the input signal.

Anaerobes: Bacteria that do not need dissolved oxygen (DO) to survive.

Anaerobic digestion: A biochemical process by which organic matter is decomposed by bacteria in the absence of oxygen, producing methane and other byproducts.

Analog: Pertaining to representation of numerical quantities by means of continuously variable physical characteristic. Contrast with *digital*.

Analog control: Implementation of automatic control loops with analog (pneumatic or electronic) equipment.

Analog device: A mechanism that represents numbers by physical quantities—for example, by lengths, as in a slide rule, or by voltage or currents, as in a differential analyzer of a computer of the analog type.

Analog signal: A continuously variable representation of a physical quantity, property, or condition such as pressure, flow, temperature, etc. The signal may be transmitted as pneumatic mechanical or electrical energy.

Analog-to-digital (A/D): The conversion of analog data into digital data.

Analog-to-digital converter: Any unit or device used to convert analog information to approximate corresponding digital information.

Ancillary services: Operations provided by hydroelectric plants that ensure stable electricity delivery and optimize transmission system efficiency.

Anemometer: Wind speed measurement device used to send data to the controller; also used to conduct wind site surveys.

Angle of attack: In wind turbine operation, it is the angle of the airflow relative to the blade.

Angle of incidence: The angle that a ray of sun makes with a line perpendicular to the surface; for example, a surface that directly faces the sun has a solar angle of incidence of zero, but if the surface is parallel to the sun (such as sunrise striking a horizontal rooftop), the angle of incidence is 90°.

Anhydrous: Very dry; no water or dampness is present.

Anion: A negatively charged ion; an ion that is attracted to the anode.

Annual removals: The net volume of growing stock trees removed from the inventory during a specified year by harvesting, cultural operations such as timberland improvement, or land clearing.

Annualized growth rate: Calculated as $(x_n/x_1)1/n$, where x is the value under consideration and n is the number of periods.

Annunciator: A visual or audible signaling device and the associated circuits used for indication of alarm conditions.

Anode: The electrode at which oxidation (a loss of electrons) takes place. For fuel cells and other galvanic cells, the anode is the negative terminal; for electrolytic cells, the anode is the positive terminal.

Anoxic: A condition in which the aquatic environment does not contain dissolved oxygen (DO), referred to as an *oxygen-deficient condition*.

ANSI: American National Standards Institute, Inc.; formerly known as the USASI.

Aquifer: Water-bearing stratum of permeable sand, rock, or gravel.

ASCII: Abbreviation for American Standard Code for Information.

Asexual reproduction: The naturally occurring ability of some plant species to reproduce asexually through seeds, meaning the embryos develop without a male gamete. This ensures that the seeds will produce plants identical to the mother plant.

Ash: Inorganic residue or noncombustible mineral matter remaining after ignition.

Asynchronous: Pertaining to a lack of time coincidence in a set of repeated events; this term can be applied to computer operations to indicate that the execution of one operation is dependent on a signal that the previous operation is completed.

Autogenous/autothermic combustion: The burning of a wet organic material where the moisture content is at such a level that the heat of combustion of the organic material is sufficient to vaporize the water and maintain combustion. No auxiliary fuel is required except for start-up.

Automatic control system: A control system that operates without human intervention.

Availability factor: A percentage representing the number of hours a generating unit is available to produce power (regardless of the amount of power) in a given period, compared to the number of hours in the period.

Azimuth angle: The angle between true south and the point on the horizon directly below the sun.

Backflow: A reverse flow condition created by a difference in waster pressures that causes water to back into the distribution pipes of a potable water supply from any source or sources other than an intended source.

Bacteria: The simplest wholly contained life systems are *bacteria* or *prokaryotes*, which are the most diverse group of microorganisms. Among the most common microorganisms in water, they are primitive, unicellular (single-celled) organisms possessing no well-defined nucleus and that present a variety of shapes and nutritional needs. Bacteria contain about 85% water and 15% ash or mineral matter. The ash is largely composed of sulfur, potassium, sodium, calcium, and chlorides, with small amounts of iron, silicon, and magnesium. Bacteria reproduce by binary fission.

Note: Binary fission occurs when one organism splits or divides into two or more new organisms.

Bacteria, once called the smallest living organisms (it is now known that smaller forms of matter exhibit many of the characteristics of life), range in size from 0.5 to 2 microns in diameter and about 1 to 10 microns long.

Note: A *micron* is a metric unit of measurement equal to 1 thousandth of a millimeter. To visualize the size of bacteria, consider that about 1000 bacteria lying side by side would reach across the head of a straight pin.

Bacteria are categorized into three general groups based on their physical form or shape, although almost every variation has been found (see table below). The simplest form is the sphere. Spherical-shaped bacteria are called *cocci* ("berries"). They are not necessarily perfectly round, but may be somewhat elongated, flattened on one side, or oval. Rod-shaped bacteria are called *bacilli*. Spiral-shaped bacteria, called *spirilla*, have one or more twists and are never straight. Such formations are usually characteristic of a particular genus or species. Within these three groups are many different arrangements. Some exist as single cells; others as pairs, as packets of four or eight, as chains, and as clumps.

Forms of Bacteria

Form	Technical Name		Example
	Singular	Plural	
Sphere	Coccus	Cocci	*Streptococcus*
Rod	Bacillus	Bacilli	*Bacillus typhosis*
Curved or spiral	Spirillum	Spirilla	*Spirillum cholera*

Most bacteria require organic food to survive and multiply. Plant and animal material that gets into the water provides the food source for bacteria. Bacteria convert the food to energy and use the energy to make new cells. Some bacteria can use inorganics (e.g., minerals such as iron) as an energy source and exist and multiply even when organics (pollution) are not available.

Bacterial growth factors: Several factors affect the rate at which bacteria grow, including temperature, pH, and oxygen levels. The warmer the environment, the faster the rate of growth. Generally, for each increase of 10°C, the growth rate doubles. Heat can also be used to kill bacteria. Most bacteria grow best at neutral pH. Extremely acidic or basic conditions generally inhibit growth, although some bacteria may require acidic and some require alkaline conditions for growth. Bacteria are aerobic, anaerobic, or facultative. If *aerobic*, they require free oxygen in the aquatic environment. *Anaerobic* bacteria exist and multiply in environments that lack dissolved oxygen. *Facultative* bacteria (e.g., iron bacteria) can switch between aerobic and anaerobic environments. Under optimum conditions, bacteria grow and reproduce very rapidly. Bacteria reproduce by *binary fission*. An important point to consider in connection with bacterial reproduction is the rate at which the process can take place. The total time required for an organism to reproduce and the offspring to reach maturity is the *generation time*. Bacteria growing under optimal conditions can double their number about every 20 to 30 minutes. Obviously, this generation time is very short compared to that of higher plants and animals. Bacteria continue to grow at this rapid rate as long as nutrients hold out; even the smallest contamination can result in a sizable growth in a very short time.

Note: Even though wastewater can contain bacteria counts in the millions per milliliter, in wastewater treatment, under controlled conditions, bacteria can help to destroy and to identify pollutants. In such a process, bacteria stabilize organic matter (e.g., activated sludge processes) and thereby assist the treatment process in producing effluent that does not impose an excessive oxygen demand on the receiving body. Coliform bacteria can be used as an indicator of pollution by human or animal wastes.

Baffles: Deflection vanes, grids, grating, or similar devices constructed or placed in air or gas flow systems, flowing water, wastestreams, or slurry systems to achieve a more uniform distribution of velocities; to absorb energy; to divert, guide, or agitate the fluid; and to check eddies.

Bagasse: The fibrous material remaining after extraction of juice from sugarcane; often burned by sugar mills as a source of energy.

Bardenpho process (four-state): Continuous-flow, suspended-growth process with alternating anoxic, aerobic, anoxic, and aerobic stages; used to remove total nitrogen.

Baseload plants: Electricity-generating units that are operated to meet the constant or minimum load on the system. The cost of energy from such units is usually the lowest available to the system.

Batch processing: (1) Pertaining to the technique of executing a set of programs such that each is completed before the next program of the set is started. (2) Loosely, the serial execution of programs.

Battery: Device that stores electrical energy. Some renewable energy sources use batteries so that a system need provide only the average amount of power required, rather than having to provide for the largest anticipated loads.

Bias: (1) The departure from a reference value of the average of a set of values; thus, a measure of the amount of unbalance of a set of measurements or conditions. (2) The average DC voltage or current maintained between a control electrode and the common electrode in a transistor.

Binary coded decimal (BCD): Describing a decimal notation in which the individual decimal digits are represented by a group of binary bits; for example, in binary coded decimal notation that uses the weights 8-4-2-1, each decimal digit is represented by a group of four binary bits. The number 12 is represented as 0001 0010 for 1 and 2, respectively, whereas in binary notation it is represented by 1100.

Binary cycle: Binary geothermal systems use the extracted hot water or steam to heat a secondary fluid to drive the power turbine.

Biobased product: As defined by the Farm Security and Rural Investment Act (FSRIA), a product determined by the U.S. Secretary of Agriculture to be a commercial or industrial product (other than food or feed) that is composed, in whole or in significant part, of biological products or renewable domestic agriculture materials (including plant, animal, and marine materials) or forestry materials.

Biochemical conversion: The use of fermentation or anaerobic digestion to produce fuels and chemicals from organic sources.

Biodiesel: Fuel derived from vegetable oils or animals fats. It is produced when a vegetable oil or animal fat is chemically reacted with an alcohol.

Bioenergy: Useful, renewable energy produced from organic matter, which may either be used directly as a fuel or processed into liquids and gases.

Biofuels: Liquid fuels and blending components produced from biomass (plant) feedstocks; used primarily for transportation.

Biogas: A combustible gas derived from decomposing biological waste. Biogas normally consists of 50 to 60% methane.

Biomass: Any organic nonfossil material of biological origin constituting a renewable energy source. It is produced from organic matter that is available on a renewable or recurring basis, including agricultural crops and trees, wood and wood residues, plants (including aquatic plants), grasses, animal manure, municipal residues, and other residue materials. Biomass is generally produced in a sustainable manner from water and carbon dioxide by photosynthesis. The three main categories of biomass are primary, secondary, and tertiary.

Biomass gas (biogas): A medium-Btu gas containing methane and carbon dioxide, resulting from the action of microorganisms on organic materials such as a landfill.

Biomaterials: Products derived from organic (as opposed to petroleum-based) products.

Bio-oil: Intermediate fuel derived from fast pyrolysis.

Biopower: The use of biomass feedstock to produce electric power to heat through direct combustion of the feedstock, through gasification and then combustion of the resultant gas, or through other thermal conversion processes. Power is generated with engines, turbines, fuel cells, or other equipment.

Biorefinery: A facility that processes and converts biomass into value-added products. These products can include biomaterials, fuels such as ethanol, or important feedstocks for the production of chemicals and other materials. Biorefineries can be based on a number of processing platforms using mechanical, thermal, chemical, and biochemical processes.

Biosolids: A term that is not used in 40 CFR 503 (Standards for the Use or Disposal of Sewage Sludge), which was promulgated by the U.S. Environmental Protection Agency (USEPA), under the authority of the Clean Water Act as amended, to ensure that sewage sludge (biosolids) is managed or used in a way that protects both human health and the environment. Phase I of the risk-based regulation governs the final use or disposal of sewage sludge. The term *biosolids* is commonly used in the wastewater treatment industry as a replacement for the term *sewage sludge*. Biosolids are the solid, slime–solid, or liquid residue generated during the treatment of domestic sewage in a wastewater treatment facility. Biosolids include, but are not limited to, domestic sewage, scum, and solids removed during primary, secondary, or advanced treatment processes. The definition of biosolids also includes a material derived from biosolids.

Biosolids cake: The solids discharged from a dewatering apparatus.

Biosolids quality parameters: The USEPA determined that three main parameters should be used in gauging the quality of biosolids: (1) relative presence or absence of pathogenic organisms, (2) pollutant levels, and (3) degree of attractiveness of the biosolids to vectors (e.g., insects, rodents). There can be a number of possible biosolids qualities. To represent biosolids that meet the highest quality for all three biosolids quality parameters, the term *exceptional quality*, or EQ, has come into common use.

Bit: (1) An abbreviation of binary digit. (2) A single character in a binary number. (3) A single pulse in a group of pulses. (4) A unit of information capacity of a storage device. The capacity in bits is the logarithm to the base two of the number of possible states of the device.

Black liquor (pulping liquor): The alkaline spent liquor removed from the digesters in the process of chemically pulping wood. After evaporation, the liquor is burned as a fuel in a recovery furnace that permits the recovery of certain basic chemicals.

Blackbody: An ideal substance that absorbs all radiation falling on it and reflects nothing.

Bone dry: Having 0% moisture content.

Borehole breakout: Failure of a borehole wall due to stress in the rock surrounding the borehole. The breakout is generally located symmetrically in the wellbore perpendicular to the direction of greatest horizontal stress on a vertical wellbore.

Bottom ash: The solid material that remains on a hearth or falls off the grate after thermal processing is complete.

Breakpoint: The point at which most of the combined chlorine compounds have been destroyed and the free chlorine starts to form. The chlorine breakpoint of water can only be determined by experimentation. This simple experiment requires 20 1000-mL breakers and a solution of chlorine. Raw water is placed in the beakers and dosed with progressively larger amounts of chlorine; for example, one might start with no chlorine in the first beaker, then 0.5 mg/L, then 1.0 mg/L, and so on. After a period of time, say 20 minutes, each beaker is tested for total chlorine residual and the results plotted.

Breakpoint chlorination: To produce a free chlorine residual, enough chlorine must be added to the water to produce what is referred to as *breakpoint chlorination*, the point at which near complete oxidation of nitrogen compounds is reached. Any residual beyond breakpoint is mostly free chlorine. When chlorine is added to natural waters, the chlorine begins combining with and oxidizing the chemicals in the water before it begins disinfecting. Although residual chlorine will be detectable in the water, the chlorine will be in combined form with weak disinfecting power. Adding more chlorine to the water at this point actually decreases the chlorine residual, as the additional chlorine destroys the combined chlorine compounds. At this stage, water may have a strong swimming pool or medicinal taste and odor. Free chlorine has the highest disinfecting power.

Breakpoint chlorination curve: At the start of the curve, no residual exists, even though there was a dosage. This is called the *initial demand* and represents the microorganisms and interfering agents using the result of the chlorine. After the initial demand, the curve slopes upward. Chlorine combining to form chloramines produces this part of the curve. All of the residual measured on this part of the curve is combined residual. At some point, the curve begins to drop back toward zero. This portion of the curve results from a reduction in combined residual, which occurs because enough chlorine has been added to destroy (oxidize) the nitrogen compounds used to form combined residuals. The breakpoint is the point where the downward slope of the curve breaks upward. At this point, all of the nitrogen compounds that could be destroyed have been destroyed. After breakpoint, the curve starts upward again, usually at a 45° angle. Only on this part of the curve can free residuals be found. Notice that the breakpoint is not zero. The distance that the breakpoint is above zero is a measure of the remaining combined residual in the water. This combined residual exists because some of the nitrogen compound will not have been oxidized by chlorine. If irreducible combined residual is more than 15% of the total residual, chlorine odor and taste complaints will be high.

Breech: A passageway leading from a furnace to its emergency exhaust stack or air emissions control equipment.

Brine: A geothermal solution containing appreciable amounts of sodium chloride, or other salts.

Brinelling: Tiny indentations (dents) high on the shoulder of the bearing race or bearing; a type of bearing failure.

British thermal unit (Btu): A basic measure of thermal (heat) energy defined as the amount of energy (heat) required to increase the temperature of 1 pound of water by 1°F at normal atmospheric pressure. 1 Btu = 1055 joules.

Buffer: (1) An internal portion of a data processing system serving as intermediate storage between two storage or data handling systems with different access times or formats; usually to connect an input or output device with the main or internal high-speed storage. (2) An isolating component designed to eliminate the reaction of a driven circuit on the circuits driving it (e.g., a buffer amplifier).

Bulk density: Weight per unit of volume, usually specified in pounds per cubic foot.

Burner: A device that positions a flame in the desired location by delivering fuel and air to that location in such a manner that continuous ignition is accomplished.

Burning area: The horizontal projection of a grate, hearth, or both.

Burning hearth: A solid surface to support the solid fuel or solid waste in a furnace and upon which materials are placed for combustion.

Burning rate: The volume of solid waste incinerated or the amount of heat released during incineration.

Byte: (1) Generic term to indicate a measureable operation of one or more contiguous binary digits (e.g., 8-bit or 6-bit byte). (2) A group of binary digits usually operated upon as a unit.

Cap rocks: Rocks of low permeability that overlie a geothermal reservoir.

Capacity: The ability to store material or energy that acts as a buffer between the input and the output of a control loop element.

Capacity factor: The ratio of the electrical energy produced by a generating unit for the period of time compared to the electrical energy that could have been produced at continuous full-power operation during the same period.

Capacity, gross: The full-load continuous rating of a generator, prime mover, or other electric equipment under specified conditions as designated by the manufacturer. It is usually indicated on a nameplate attached to the equipment.

Capitol cost: The cost of field development and plant construction and the equipment required for the generation of electricity.

Carbon dioxide (CO_2): A product of combustion; the most common greenhouse gas.

Carbon monoxide (CO): A colorless, odorless gas produced by incomplete combustion. Carbon monoxide is poisonous when inhaled.

Carbon sequestration: The absorption and storage of carbon dioxide from the atmosphere by naturally occurring plants.

Carnot cycle: An ideal heat engine (conceived by Sadi Carnot) in which the sequence of operations forming the working cycle consists of isothermal expansion, adiabatic expansion, isothermal compression, and adiabatic compression back to the initial state.

Cascade control: The use of two conventional feedback controllers in series such that two loops are formed, one within the other. The output of the controller in the outer loop modifies the setpoint of the controller in the inner loop.

Cascading heat: A process that uses a stream of geothermal hot water or steam to perform successive tasks requiring lower and lower temperatures.

Casing: Pipe placed in a wellbore as a structural interface between the wellbore and the surrounding formation. It typically extends from the top of the well and is cemented in place to maintain the diameter of the wellbore and provide stability.

Cast silicon: Crystalline silicon obtained by pouring pure molten silicon into a vertical mold and adjusting the temperature gradient along the mold volume during cooling to obtain slow, vertically advancing crystallization of the silicon. The polycrystalline ingot thus formed is composed of large, relatively parallel, interlocking crystals. The cast ingots are sawed into wafers for further fabrication into photovoltaic cells. Cast-silicon wafers and ribbon-silicon sheets fabricated into cells are usually referred to as *polycrystalline photovoltaic cells*.

Cathode: The electrode at which reduction (a gain of electrons) occurs. For fuel cells and other galvanic cells, the cathode is the positive terminal; for electrolytic bells (where electrolysis occurs), the cathode is the negative terminal.

Cation: A positively charged ion.

Cavitation: Noise or vibration that can damage the turbine blades; it is a result of bubbles forming in the water as it passes through the turbine which causes a loss in capacity, head, and efficiency. The cavities, or bubbles, collapse when they pass into regions of higher pressure.

Cellulose: The main carbohydrate in living plants. Cellulose forms the skeletal structure of the plant cell wall.

Cellulosic ethanol: Ethanol derived for cellulosic and hemicellulosic parts of biomass.

Centrifuge: Apparatus that rotates at high speed and separates substances of different densities through centrifugal force.

Centrifugation: Centrifuges of various types have been used in dewatering operations for at least 30 years and appear to be gaining in popularity. Depending on the type of centrifuge used, in addition to centrifuge pumping equipment for solids feed and centrate removal, support systems for the removal of dewatered solids are required. Generally, in operation, the centrifuge spins at a very high speed. Chemically conditioned solids are pumped into the centrifuge, and the spinning action throws the solids to the outer wall of the centrifuge. The centrate (water) flows inside the unit to a discharge point. The solids held against the outer wall are scraped to a discharge point by an internal scroll moving slightly faster or slower than the centrifuge speed of rotation. In a continuous-feed, solid-bowl, conveyor type of centrifuge (the most common type currently used), solid/liquid separation occurs as a result of rotating the liquid at high speeds to cause separation by gravity. The solid bowl has a rotating unit with a bowl and a conveyor. The unit has a conical section at one end that acts as a drainage device. The conveyor screw pushes the sludge solids to outlet ports and the cake to a discharge hopper. The sludge slurry enters the rotating bowl through a feed pipe leading into the hollow shaft of the rotating screw conveyor. The sludge is distributed through ports into a pool inside the rotating bowl. As the liquid sludge flows through the hollow shaft toward the overflow device, the fine solids settle to the wall of the rotating bowl. The screw conveyor pushes the solids to the conical section, where the solids are forced out of the water and the water drains back in the pool. Expected percent solids for centrifuge dewatered sludge can vary from 10 to 15%. In most cases, chemical conditioning is required to achieve optimum concentrations. Centrifuge operators often find that the operation of centrifuges can be simple, clean, and efficient. However, they soon discover that centrifuges are noisemakers. Units run at very high speeds and produce high-level noise, which can cause loss of hearing with prolonged exposure; therefore, in areas where a centrifuge is in operation, special care must be taken to provide hearing protection. Actual operation of a centrifuge requires the operator to control and adjust chemical feed rates, to observe unit operation and performance, to control and monitor centrate returned to the treatment system, and to perform required maintenance as outlined in the manufacturer's technical manual.

Charge controller: Keeps batteries from overcharging.

Chips: Woody material cut into short, thin wafers. Chips are used as a raw material for pulping and fiberboard or as biomass fuel.

Chlorination: The addition of chlorine or chlorine compounds to water. Chlorination is considered to be the single most important process for preventing the spread of waterborne disease. Chlorine has many attractive features that contribute to its wide use in industry. Five of the key attributes of chlorine are as follows:

- It causes damage to the cell wall.
- It alters the permeability of the cell (the ability to pass water in and out through the cell wall).
- It alters the cell protoplasm.
- It inhibits the enzyme activity of the cell so it is unable to use its food to produce energy.
- It inhibits cell reproduction.

Some concerns regarding chlorine usage that may impact its use include the following:

- Chlorine reacts with many naturally occurring organic and inorganic compounds in water to produce undesirable DBPs.
- Hazards associated with using chlorine, specifically chlorine gas, require special treatment and response programs.
- High chlorine doses can cause taste and odor problems.

Chlorine is used in water treatment facilities primarily for disinfection. Because of chlorine's oxidizing powers, it has been found to serve other useful purposes in water treatment:[*]

- Taste and odor control
- Prevention of algal growths
- Maintenance of clear filter media
- Removal of iron and manganese
- Destruction of hydrogen sulfide
- Bleaching of certain organic colors
- Maintenance of distribution system water quality by controlling slime growth
- Restoration and preservation of pipeline capacity
- Restoration of well capacity, water main sterilization
- Improved coagulation

---

[*] White, G.C., *Handbook of Chlorination and Alternative Disinfectants*, Van Nostrand Reinhold, New York, 1992.

Chlorine is available in a number of different forms: (1) Pure elemental gaseous chlorine, a greenish-yellow gas possessing a pungent and irritating odor that is heavier than air, nonflammable, and nonexplosive; when released to the atmosphere, this form is toxic and corrosive. (2) Solid calcium hypochlorite, in tablets or granules. (3) Liquid sodium hypochlorite solution in various strengths. The selection of one form of chlorine over another for a given water system depends on the amount of water to be treated, configuration of the water system, local availability of the chemicals, and skill of the operator. A major advantage of using chlorine is the effective residual that it produces. A residual indicates that disinfection is completed and the system has an acceptable bacteriological quality. Maintaining a residual in the distribution system provides another line of defense against pathogenic organisms that could enter the distribution system and helps to prevent regrowth of those microorganisms that were injured but not killed during the initial disinfection stage. Often, new waterworks operators have difficulties understanding the terms used to describe the various reactions and processes used in chlorination. Common chlorination terms include the following:

– *Chlorine reaction*—Regardless of the form of chlorine used for disinfection, the reaction in water is basically the same. The same amount of disinfection can be expected, provided the same amount of available chlorine is added to the water. The standard term for the concentration of chlorine in water is milligrams per liter (mg/L) or parts per million (ppm); these terms indicate the same quantity.

– *Chlorine dose*—The amount of chlorine added to the system. It can be determined by adding the desired residual for the finished water to the chlorine demand of the untreated water. Dosage can be either milligrams per liter (mg/L) or pounds per day (lb/day). The most common usage is mg/L.

– *Chlorine demand*—The amount of chlorine used by iron, manganese, turbidity, algae, and microorganisms in the water. Because the reaction between chlorine and microorganisms is not instantaneous, demand is relative to time. For example, the demand 5 minutes after applying chlorine will be less than the demand after 20 minutes. Demand, like dosage, is expressed in mg/L. Chlorine demand is determined as follows:

$$Cl_2 \text{ demand} = Cl_2 \text{ dose} - Cl_2 \text{ residual}$$

– *Chlorine residual*—The amount of chlorine (determined by testing) that remains after the demand is satisfied. Residual, like demand, is based on time. The longer the time after dosage, the lower the residual will be, until all of the demand has been satisfied. Residual, like dosage and demand, is expressed in mg/L. The presence of a *free residual* of at least 0.2 to 0.4 ppm usually provides a high degree of assurance that the disinfection of the water is complete. *Combined residual* is the result of combining free chlorine with nitrogen compounds. Combined residuals are also referred to as chloramines. *Total chlorine residual* is the mathematical combination of free and combined residuals. Total residual can be determined directly with standard chlorine residual test kits.

– *Chorine contact time (CT)*—One of the key items in predicting the effectiveness of chlorine on microorganisms. It is the interval (usually only a few minutes) between the time when chlorine is added to the water and the time the water passes by the sampling point. The chlorine contact time is calculated based on the free chlorine residual prior to the first customer times the contact time in minutes:

$$CT = Concentration \times Contact\ time = mg/L \times minutes$$

A certain minimum time period is required for the disinfecting action to be completed. The contact time is usually a fixed condition determined by the rate of flow of the water and the distance from the chlorination point to the first consumer connection. Ideally, the contact time should not be less than 30 minutes, but even more time is needed at lower chlorine doses, in cold weather, or under other conditions. Pilot studies have shown that specific CT values are necessary for the inactivation of viruses and *Giardia*. The required CT value will vary depending on pH, temperature, and the organisms to be killed, and various state and federal charts and formulae are available to make such determinations. Charts in the USEPA *Surface Water Treatment Rule Guidance* manual list the required CT values for various filter systems. USEPA has set a CT value of 3-log ($CT_{99.9}$) inactivation to ensure that the water is free of *Giardia*. Filtration, in combination with disinfection, must provide 3-log removal/inactivation of *Giardia*. Under the 1996 Interim Enhanced Surface Water Treatment (IESWT) rules, the USEPA requires systems that filter to remove 99% (2 log) of *Cryptosporidium* oocysts. To be sure that the water is free of viruses, a combination of filtration and disinfection to provide 4-log removal of viruses has been judged the best for drinking water safety (99.99% removal). Viruses are inactivated (killed) more easily than cysts or oocysts.

Chlorine chemistry: The reactions of chlorine with water and the impurities that might be in the water are quite complex, but a basic understanding of these reactions can aid the operator in keeping the disinfection process operating at its highest efficiency. When dissolved in pure water, chlorine reacts with the H^+ ions and the OH^- radicals in the water. Two of the products of this reaction (the actual disinfecting agents) are *hypochlorous acid*, HOCl, and a *hypochlorite radical*, OCl^-. If microorganisms are present in the water, the HOCl and the OCl^- penetrate the microbe cells and react with certain enzymes. This reaction disrupts the organisms' metabolism and kills them. The chemical equation for hypochlorous acid is as follows:

$$Cl_2 \quad + \quad H_2O \quad \Leftrightarrow \quad HClO \quad + \quad HCl$$

(*chlorine*)   (*water*)   (*hypochlorous acid*)   (*hydrochloric acid*)

The symbol ⇔ indicates that the reactions are reversible. Hypochlorous acid is a weak acid, meaning it dissociates slightly into hydrogen and hypochlorite ions, but it is a strong oxidizing and germicidal agent. Hydrochloric acid is a strong acid that retains more of the properties of chlorine. HCl tends to lower the pH of the water, especially in swimming pools where the water is recirculated and continually chlorinated. The total hypochlorous acid and hypochlorite ions in water constitute the *free available chlorine*. Hypochlorites act in a manner similar to HCl when added to water, because hypochloric acid is formed. When chlorine is first added to water containing some impurities, the chlorine immediately reacts with the dissolved inorganic or organic substances and is then unavailable for disinfection. The amount of chlorine used in this initial reaction is the *chlorine demand* of the water. If dissolved ammonia (NH_3) is present in the water, the chlorine will react with it to form compounds called *chloramines*. Only after the chlorine demand is satisfied and the reaction with all the dissolved ammonia is complete is the chlorine actually available in the form of HOCl and OCl⁻. The equation for the reaction of hypochlorous acid (HOCl) and ammonia (NH_3) is as follows:

$$HOCl \quad + \quad NH_3 \quad \rightarrow \quad NH_2Cl \quad + \quad H_2O$$

(hypochlorous acid)   *(ammonia)*   *(monochloramine)*   *(water)*

The chlorine as hypochlorous acid and hypochlorite ions remaining in the water after the above reactions are complete is known as *free available chlorine*, which is a very active disinfectant.

Closed-loop: A signal path formed about a process by a feedback measurement signal (input to a controller) and the signal delivered to the final control element (controller output signal).

Closed-loop biomass: Crops grown in a sustainable manner for the purpose of optimizing their value for bioenergy and bioproduct uses. These include annual crops such as maize and wheat and perennial crops such as trees, shrubs, and grasses (e.g., switchgrass).

Coanda: A screening technology that can be self-cleaning.

Coagulation: Following screening and the other pretreatment processes, the next unit process in a conventional water treatment system is a mixer where the first chemicals are added in what is known as coagulation. The exception to this situation occurs in small systems using groundwater, when chlorine or other taste and odor control measures are introduced at the intake and are the extent of treatment. The term *coagulation* refers to the series of chemical and mechanical operations by which coagulants are applied and made effective. These operations are comprised of two distinct phases: (1) rapid mixing to disperse coagulant chemicals by violent agitation into the water being treated, and (2) flocculation to agglomerate small particles into well-defined floc by gentle agitation for a much longer time. The coagulant must be added to the raw water and perfectly distributed into the liquid; such uniformity of chemical treatment is reached through rapid agitation or mixing. Coagulation results from adding salts of iron or aluminum to the water. Common coagulants (salts) are as follows:

- Alum (aluminum sulfate)
- Sodium aluminate
- Ferric sulfate
- Ferrous sulfate
- Ferric chloride
- Polymers

Coarse materials: Wood residues suitable for chipping, such as slabs, edgings, and trimmings.

Co-firing: Practice of introducing biomass into the boilers of coal-fired power plants.

Combined cycle: An electricity-generating technology in which electricity is produced from otherwise lost waste heat exiting from one or more gas (combustion) turbines. The exiting heat is routed to a conventional boiler or to a heat recovery steam generator for utilization by a steam turbine in the production of electricity. Such designs increase the efficiency of the generating unit.

Combined heat and power (CHP) plant: A plant designed to produce both heat and electricity from a single heat source. This term is being used in place of the term *cogenerator*, which was used by the Energy Information Administration (EIA) in the past. CHP better describes the facilities because some of the plants included do not produce heat and power in a sequential fashion and, as a result, do not meet the legal definition of cogeneration specified in the Public Utility Regulatory Policies Act (PURPA).

Combustion air: The air used for burning a fuel.

Combustion efficiency: A result of time, turbulence, and temperature.

Commercial sector: An energy-consuming sector that consists of service-providing facilities and equipment of businesses; federal, state, and local governments; and other private and public organizations, such as religious, social, or fraternal groups. The commercial sector includes institutional living quarters. It also includes sewage treatment facilities. Common uses of energy associated with this sector include space heating, water heating, air conditioning, lighting, refrigeration, cooking, and running a wide variety of other equipment. This sector includes generators that produce electricity and useful thermal output primarily to support the activities of the above-mentioned commercial establishments.

Commercial species: Tree species suitable for industrial wood products.

Composite wastewater sample: A combination of individual samples of wastewater taken at selected intervals, generally hourly for some specified period, to minimize the effect of the variability of the individual samples.

Compressed natural gas (CNG): Mixtures of hydrocarbon gases and vapors, consisting principally of methane in gaseous form that has been compressed.

Concentrator: A reflective or refractive device that focuses incident insolation onto an area smaller than the reflective or refractive surface, resulting in increased insolation at the point of focus.

Condensate: Water formed by condensation of steam.

Condenser: Equipment that condenses turbine exhaust steam into condensate.

Conservation Reserve Program (CRP): Federal program that provides farm owners or operators with an annual per-acre rental payment and half the cost of establishing a permanent land cover in exchange for retiring environmentally sensitive cropland from production for 10 to 15 years. In 1996, Congress reauthorized CRP for an additional round of contracts, limiting enrollment to 36.4 million acres at any time. The 2002 Farm Act increased the enrollment limit to 39 million acres. Producers can offer land for competitive bidding based on an Environmental Benefits Index (EBI) during periodic signups, or they can automatically enroll more limited acreages in processes such as riparian buffers, field windbreaks, and grass strips on a continuous basis. CRP is funded through the Commodity Credit Corporation (CCC).

Conventional hydroelectric (hydropower) plant: A plant in which all the power is produced from natural streamflow as regulated by available storage.

Cooling tower: A structure in which heat is removed from hot condensate.

Core: A cylinder of rock recovered from the well by a special coring drill bit.

Crop failure: Consists mainly of the acreage on which crops failed because of weather, insects, and diseases, but includes some land not harvested due to lack of labor, lower market prices, or other factors. The acreage planted to cover and soil improvement crops not intended for harvest is excluded from crop failure and is considered idle.

Cropland: Total cropland includes five components: cropland harvested, crop failure, cultivated summer fallow, cropland used only for pasture, and idle cropland.

Cropland harvested: Includes row crops and closely sown crops; hay and silage crops; tree fruits, small fruits, berries, and tree nuts; vegetables and melons; and miscellaneous other minor crops. In recent years, farmers have double-cropped about 4% of the acreage.

Cropland pasture: Land used for long-term crop rotation; however, some cropland pasture is marginal for crop uses and may remain in pasture indefinitely. This category also includes land that was used for pasture before crops reached maturity and some land used for pasture that could have been cropped without additional improvement.

Cropland used for crops: Includes cropland harvested, crop failure, and cultivated summer fallow.

Crossflow: An impulse turbine that can be manufactured with limited technical means.

Crust: Earth's outer layer of rock; also called the *lithosphere*.

Cryogenic liquefaction: The process through which gases such as nitrogen, hydrogen, helium, and natural gas are liquefied under pressure at very low temperatures.

Cull tree: A live tree, 5.0 inches in diameter at breast height (dbh) or larger, that is not merchantable for saw logs now or prospectively because of rot, roughness, or species.

Cut-in speed: The speed at which a shaft must turn in order to generate electricity and send it over a wire.

Daylighting: The use of direct, diffuse, or reflected sunlight to provide supplemental lighting for building interiors; in streams, daylighting is the redirection of the stream into an above-ground channel.

dbh: The diameter of a tree measured at approximately breast height from the ground.

Denitrification: Process that removes nitrogen from the wastewater. When bacteria come in contact with a nitrified element in the absence of oxygen, they reduce the nitrates to nitrogen gas, which escapes the wastewater. The denitrification process can be carried out either in an anoxic activated sludge system (suspended growth) or in a column system (fixed growth). The denitrification process can remove up to 85% or more of nitrogen. After effective biological treatment, little oxygen-demanding material is left in the wastewater when it reaches the denitrification process. The denitrification reaction will occur only if an oxygen demand source exists when no dissolved oxygen is present in the wastewater. An oxygen demand source is usually added to reduce the nitrates quickly. The most common soluble biological oxygen demand (BOD) source is methanol. Approximately 3 mg/L of methanol is added for every 1 mg/L of nitrate-nitrogen. Suspended growth denitrification reactors are mixed mechanically, but only enough to keep the biomass from settling without adding unwanted oxygen. Submerged filters of different types of media may also be used to provide denitrification. A fine media downflow filter is sometimes used to provide both denitrification and effluent filtration. A fluidized sand bed where wastewater flows upward through a media of sand or activated carbon at a rate to fluidize the bed may also be used. Denitrification bacteria grow on the media.

Densification: A mechanical process to compress biomass (usually wood waste) into pellets, briquettes, cubes, or densified logs.

Depletion factor: Annual percentage of the depletion of the thermal resource.

Dewatering: A physical process that removes sufficient water from biosolids so that the physical form is changed from essentially that of a fluid to that of a damp solid.

Digester gas: Biogas that is produced using a digester, which is an airtight vessel or enclosure in which bacteria decompose biomass in water to produce the biogas.

Diode: A solid-state device that acts as a one-way valve for electricity.

Direct current (DC): An electric current that flows in a constant direction. The magnitude of the current does not vary or has only a slight variation.

Direct methanol fuel cell (DMFC): A type of fuel cell in which the fuel is methanol (CH_3OH) in gaseous or liquid form. The methanol is oxidized directly at the anode instead of first being reformed to produce hydrogen. The electrolyte is typically polymer electrolyte membrane (PEM).

Direct use: Use of geothermal heat without first converting it to electricity, such as for space heating and cooling, food preparation, or industrial processes.

Disinfection:[*] Disinfection is a unit process used in both water and wastewater treatment, although there are differences (mainly in the types of disinfectants used and applications) between the use of disinfection in water and wastewater treatment. The discussion here is limited to disinfection as it applies to water treatment. To comply with Safe Drinking Water Act (SDWA) regulations, the majority of public water systems (PWSs) use some form of water treatment. A Community Water System Survey reported that in the United States, in 1995, 99% of surface water systems provided some treatment to their water, with 99% percent of those treatment systems using disinfection/oxidation as part of the treatment process. Although 45% of groundwater systems provided no treatment, 92% of those groundwater plants that did provide some form of treatment included disinfection/oxidation as part of the treatment process. Why the public health concern with regard to groundwater supplies? According to USEPA's Bruce Macler, "There are legitimate concerns for public health from microbial contamination of groundwater systems. Microorganisms and other evidence of fecal contamination have been detected in a large number of wells tested, even those wells that had been previously judged not vulnerable to such contamination. The scientific community believes that microbial contamination of groundwater is real and widespread. Public health impacts from this contamination, while not well quantified, appear to be large. Disease outbreaks have occurred in many groundwater systems. Risk estimates suggest several million illnesses each year. Additional research is underway to better characterize the nature and magnitude of the public health problem."[†] The most commonly used disinfectants/oxidants (in no particular order) are chlorine, chlorine dioxide, chloramines, ozone, and potassium permanganate. As mentioned, the process used to control waterborne pathogenic organisms and prevent waterborne disease is called *disinfection*. The goal in proper disinfection in a water system is to destroy all disease-causing organisms. Disinfection should not be confused with sterilization. *Sterilization* is the complete killing of all living organisms. Waterworks operators disinfect by destroying organisms that might be dangerous; they do not attempt to sterilize water. Disinfectants are also used to achieve other specific objectives in drinking water treatment, including nuisance control (e.g., zebra mussels, Asiatic clams), oxidation of specific compounds (taste- and odor-causing compounds, such as iron and manganese), and as a coagulant and filtration aid.

---

[*] USEPA, *Alternative Disinfectants and Oxidants Guidance Manual*, U.S. Environmental Protection Agency, Washington, DC, 1999, chapters 1 and 2.

[†] Macler, B., What is the ground water disinfection rule?, *Hydrovisions Online*, 5(4), 1996 (http://www.grac.org/winter96/gwdr.htm).

Distributed generation (distributed energy resources): Refers to electricity provided by small, modular power generators (typically ranging in capacity from a few kilowatts to 50 megawatts) located at or near customer demand.

District heating: A type of direct use in which a utility system supplies multiple users with hot water or steam from a central plant or well field.

Downwind turbine: A turbine that does not face into the wind and whose direction is controlled directly by the wind.

Draft tube: A water conduit that can be straight or curved, depending on the turbine installation that maintains a column of water from the turbine outlet and the downsteam water level.

Drag bit: Drilling bit that drills by scraping or shearing the rock with fixed hard surfaces, or *cutters.*

Drilling: Boring into the Earth to access geothermal resources, usually with oil and gas drilling equipment that has been modified to meet geothermal requirements.

Dry steam: Very hot steam that does not occur with liquid.

Drying hearth: A solid surface in an incinerator upon which wet waste materials or waste matter that may turn to liquid before burning are placed to dry or to burn with the help of hot combustion gases.

Dump load: A device that allows excess energy to be safely disposed of.

E-10: A mixture of 10% ethanol and 90% gasoline based on volume.

E-85: A mixture of 85% ethanol and 15% gasoline based on volume.

Efficiency: The ratio of the useful energy output of a machine or other energy-converting plant to the energy input.

Electric power sector: Includes privately or publicly owned establishments that generate, transmit, distribute, or sell electricity, including combined heat and power (CHP) plants.

Electric utility: A corporation, person, agency, authority, or other legal entity or instrumentality aligned with distribution facilities for delivery of electric energy for use primarily by the public. Included are investor-owned electric utilities, municipal and state utilities, federal electric utilities, and rural electric cooperatives. A few entities that are tariff based and corporately aligned with companies that own distribution facilities are also included.

Emissions: Anthropogenic releases of gases to the atmosphere. In the context of global climate change, they consist of radiatively important greenhouse gases (e.g., the release of carbon dioxide during fuel combustion).

Endothermic: A chemical reaction that absorbs or requires energy (usually in the form of heat).

Energy: The ability to do work.

Energy crops: Crops grown specifically for their fuel value. They include food crops such as corn and sugarcane and nonfood crops such as poplar trees and switchgrass. Currently, low-energy crops are under development: short-rotation woody crops (fast-growing hardwood trees harvested in 5 to 8 years) and herbaceous energy crops, such as perennial grasses, which are harvested annually after taking two to three years to reach full productivity.

Enhanced geothermal systems (EGS): Engineered reservoirs that can extract economic amounts of heat from geothermal resources. Rock fracturing, water injection, and water circulation technologies sweep heat from the unproductive areas of existing geothermal fields or new fields lacking sufficient production capacity.

Enthalpy: A thermodynamic property of a substance, defined as the sum of its internal energy plus the pressure of the substance times its volume, divided by the mechanical equivalent of heat. The total heat content of air; the sum of enthalpies of dry air and water vapor, per unit weight of dry air, measured in Btu per pound (or calories per kilogram).

Environmental Impact Statement: A document created from a study of the expected environmental effects of a new development or installation.

Ethanol (CH_3–CH_2OH): Also known as ethyl alcohol or grain alcohol, a clear, colorless flammable oxygenated hydrocarbon with a boiling point of 173.5°F in the anhydrous state. It readily forms a binary azetrope with water, with a boiling point of 172.67°F at a composition of 95.57% by weight ethanol. It is used in the United States as a gasoline octane enhancer and oxygenate (maximum 10% concentration). Ethanol can be used in higher concentrations (E-85) in vehicles designed for such use. Ethanol is typically produced chemically from ethylene or biologically from the fermentation of various sugars from carbohydrates found in agricultural crops and cellulosic residues from crops or wood.

Evacuated tube: In a solar thermal collector, an absorber tube contained in an evacuated glass cylinder through which collector fluids flows.

Excess air: The amount of air required beyond the theoretical air requirements for complete combustion. This parameter is expressed as a percentage of the theoretical air required:

$$\text{Excess air} = \frac{(\text{Actual air rate} - \text{Theoretical air rate}) \times 100}{\text{Theoretical air rate}}$$

Exothermic: A chemical reaction that gives off heat.

Externality: A cost or benefit not accounted for in the price of goods and services. Often the term refers to the cost of pollution and other environmental impacts.

Fast pyrolysis: Thermal conversion of biomass by rapid heating to 450 to 600°C in the absence of oxygen.

Fault: A fracture in rock exhibiting relative movement between the adjoining surfaces.

Feller-buncher: A self-propelled machine that cuts trees with giant shears near ground level and then stacks the trees into piles to await skidding.

Feedstock: A product used as the basis for the manufacture of another product.

Fenestration: Windows, doors, and skylights; the whole-building design approach will determine what type of fenestration products should be used. It is desirable to select products with characteristics that accommodate the building's climate, which includes insulation, daylighting, heating and cooling, and natural ventilation needs.

Fermentation: Conversion of carbon-containing compounds by microorganisms for production of fuels and chemicals such as alcohols, acids, or energy-rich gases.

Fiber products: Products derived from fibers of herbaceous and woody plant material. Examples include pulp, composition board products, and wood chips.

Fine materials: Wood residues not suitable for chipping, such as planer shavings and sawdust.

Firebrick: Refractory brick made from fireclay.

Fischer–Tropsch fuels: Liquid hydrocarbon fuels produced by a process that combines carbon monoxide and hydrogen. The process is used to convert coal, natural gas, and low-value refinery products into high-value diesel substitutes.

Fixed grate: A grate without moving parts, also called a *stationary grate*.

Flash steam: Steam produced when the pressure on a geothermal liquid is reduced; also called *flashing*.

Flashpoint: Lowest temperature at which evaporation of a substance produces sufficient vapor to form an ignitable mixture with air near the surface of the liquid.

Flat plate pump: A medium-temperature solar thermal collector that typically consists of a metal frame, glazing, absorbers (usually metal), and insulation and that uses a pump liquid as the heat-transfer medium. Its predominant use is in water heating applications.

Flexible-fuel vehicle: A vehicle with a single fuel tank designed to run on varying blends of unleaded gasoline with either ethanol or methanol.

Flow: Volume of water, expressed as cubic feet or cubic meters per second, passing a point in a given amount of time.

Flow battery: An electrochemical energy storage device that utilizes tanks of rechargeable electrolyte to refresh the energy-producing reaction. Because its capacity is limited only by the size of its electrolyte tanks, it is useful for large-scale backup systems to supplement other forms of generation that may be intermittent in nature.

Flue gas: The products of combustion, including pollutants, emitted to the air after a production process or combustion takes place.

Flue gas desulfurization: The removal of sulfur oxides from exhaust gas streams of a combustion process.

Fluidized bed combustion: Oxidation of combustible material within a bed of solid, inert particles which under the action of vertical hot air flow will act as a fluid.

Fly ash: Airborne combustion residue from burning fuel.

Forced draft: The positive pressure created by the action of a fan or blower which supplies the primary or secondary combustion air in an incinerator.

Forebay: The structure that holds intake screening for the penstock.

Forest land: Land at least 10% stocked by forest trees of any size, including land that formerly had such tree cover and that will be naturally or artificially regenerated. Forest land includes transition zones, such as area between heavily forested and nonforested lands that are at least 10% stocked with forest trees and forest areas adjacent to urban and built-up lands. Also included are pinyon–juniper and chaparral areas in the West and afforested areas. The minimum area for classification of forest land is 1 acre. Roadside, streamside, and shelterbelt strips of trees must have a crown width of at least 120 feet to qualify as forest land. Unimproved roads and trails, streams, and clearings in the forest areas classified as forest if less than 120 feet wide.

Fouling: The impedance to the flow of fluid or heat that results when material accumulates in flow passages or on heat absorbing surfaces in an incinerator or other combustion chamber.

Fracture: Natural or induced break in rock.

Fracturing treatments: Treatments performed by pumping fluid into the subsurface at pressures above the fracture pressure of the reservoir formation to create a highly conductive flow path between the reservoir and the wellbore.

Fuel cell: One or more cells capable of generating an electrical current by converting the chemical energy of a fuel directly into electrical energy. Fuel cells differ from conventional electrical cells in that the active materials such as fuel and oxygen are not contained within the cell but are supplied from outside.

Fuel cell poisoning: The lowering of a fuel cell's efficiency due to impurities in the fuel binding to the catalyst.

Fuel cell stack: Individual fuel cells connected in a series. Fuel cells are stacked to increase voltage.

Fuel treatment evaluator (FTE): A strategic assessment tool capable of aiding the identification, evaluation, and prioritization of fuel treatment opportunities.

Fuel wood: Wood and wood products, possibly including coppices, scrubs, branches, etc., bought or gathered, and used by direct combustion.

Fugitive emissions: Emissions other than those from vents or stacks.

Full sun: The amount of power density in sunlight received at the Earth's surface at noon on a clear day (about 1000 W/m²).

Fumarole: A vent or hole in the Earth's surface, usually in a volcanic region, from which steam, gaseous vapors, or hot gases issue.

Furnace: A combustion chamber; an enclosed structure in which heat is produced.

Gallon: A volumetric measure equal to 4 quarts (231 cubic inches) used to measure fuel oil. One gallon equals 3785 liters; 1 barrel equals 42 gallons.

Gasification: A chemical or heat process to convert a solid fuel to a gaseous form.

Gasohol: A motor vehicle fuel that is a blend of 90% unleaded gasoline and 10% ethanol (by volume).

Gearbox: Device that increases the rpm of a low-speed shaft, transferring its energy to a high-speed shaft to provide enough speed to generate electricity.

Generation (electricity): The process of producing electric energy from other forms of energy; also, the amount of electric energy produced, expressed in watt-hours (Wh).

Geothermal: Of or relating to the Earth's interior heat.

Geothermal energy: The Earth's interior heat made available by extracting it from hot water or rocks. Hot water or steam extracted from geothermal reservoirs in the Earth's crust can be supplied to steam turbines at electric power plants that drive generators to produce electricity.

Geothermal gradient: The rate of temperature increase in the Earth as a function of depth. Temperature increases an average of 1°F for every 75 feet of descent.

Geothermal heat pump: Device that takes advantage of the relatively constant temperature of the Earth's interior, using it as a source and sink of heat for both heating and cooling. For cooling, heat is extracted from the space and dissipated into the ground; for heating, heat is extracted from the ground and pumped into the space.

Geothermal plant: A plant in which a turbine is driven by either hot water or natural steam that derives its energy from heat found in rocks or fluids at various depths beneath the surface of the Earth. The fluids are extracted by drilling and pumping.

Geothermal resources: The natural heat of the Earth that can be used for beneficial purposes when the heat is collected and transported to the surface.

Geyser: A spring that shoots jets of hot water and steam into the air.

Glazing: Transparent or translucent material (glass or plastic) used to admit light or to reduce heat loss; used for windows, skylights, or greenhouses or for covering the aperture of a solar collector.

Global positioning system (GPS): A navigation system using satellite signals to fix the location of a radio receiver on or above the Earth's surface.

Grassland pasture and range: All open land used primarily for pasture and grazing, including shrub and brush land types of pasture; grazing land with sagebrush and scattered mesquite; and all tame and native grasses, legumes, and other forage used for pasture or grazing. Because of the diversity in vegetative composition, grassland pasture and range are not always clearly distinguishable from other types of pasture and range. At one extreme, permanent grassland may merge with cropland pasture, or grassland may often be found in transitional areas with forested grazing land.

Gravimetry: The use of precisely measured gravitational force to determine mass differences that can be correlated to subsurface geology.

Green pricing/marketing: In the case of renewable electricity, green pricing represents a market solution to the various problems associated with regulatory valuation of the non-market benefits of renewables. Green pricing programs allow electricity customers to express their willingness to pay for renewable energy development through direct payments on their monthly utility bills.

Greenhouse gas: Gases that trap the heat of the sun in the Earth's atmosphere, producing the greenhouse effect. The two major greenhouse gases are water vapor and carbon dioxide. Other greenhouse gases include methane, ozone, chlorofluorocarbons, and nitrous oxide.

Grid: The layout of an electrical distribution system.

Gross generation: The total amount of electric energy produced by the generating units at a generation station or stations, measured at the generator terminals.

Growing stock: A classification of timber inventory that includes live trees of commercial species meeting specified standards of quality or vigor. Cull tress are excluded. When associated with volume, includes only trees 5.0 inches in dbh and larger.

HDR: Hot dry rock; subsurface geologic formations of abnormally high heat content that contain little or no water.

Hardwoods: Usually broad-leaved and deciduous trees.

Head: Vertical change in elevation, expressed in either feet or meters, between the head water level and the tailwater level.

Headwater: The water level above the powerhouse.

Heat exchanger: A device for transferring thermal energy from one fluid to another.

Heat flow: Movement of heat from within the Earth to the surface, where it is dissipated into the atmosphere, surface water, and space by radiation.

Heat pump: A year-round heating and air-conditioning system employing a refrigeration cycle. In a refrigeration cycle, a refrigerant is compressed (as a liquid) and expanded (as a vapor) to absorb and reject heat. The heat pump transfers heat to a space to be heated during the winter period and, by reversing the operation, extracts heat from the same space to be cooled during the summer period. The refrigerant within the heat pump in the heating mode absorbs the heat to be supplied to the space to be heated from an outside medium (air, ground, or groundwater); in the cooling mode, it absorbs heat from the space to be cooled to be rejected to the outside medium.

Heat pump, air-source: The most common type of heat pump. The heat pump absorbs heat from the outside air and transfers the heat to the space to be heated in the heating mode. In the cooling mode, the heat pump absorbs heat from the space to be cooled and ejects the heat to the outside air. In the heating mode, when the outside air approaches 32°F or less, air-source heat pumps lose efficiency and generally require a backup (resistance) heating system.

Heat pump efficiency: Directly related to range of temperatures in which it operates.

Heat pump, geothermal: A heat pump in which the refrigeration exchanges heat (in a heat exchanger) with a fluid circulating through an earth connection medium (ground or groundwater). The fluid is contained in a variety of loop (pipe) configurations depending on the temperature of the ground and the ground area available. Loops may be installed horizontally or vertically in the ground or submersed in a body of water.

Heating value: The maximum amount of energy available from burning a substance.

Heavy metals: Metallic elements having a high density (>5 g/cm^3).

Herbaceous: Non-woody type of vegetation, usually lacking permanent strong stems, such as grasses, cereals, and canola (rapeseed).

High-speed shaft: Transmits force from the gearbox to the generator.

High-temperature collector: A solar thermal collector designed to operate at a temperature of 180°F or higher.

Hub: The center part of the rotor assembly that connects the blades to the low-speed shaft.

Hydraulic stimulation: A stimulation technique performed using fluid.

Hydrocarbon: Any of a vast family of compounds containing carbon and hydrogen in various combinations, found especially in fossil fuels.

Hydrothermal: Pertaining to hot water.

Hydrothermal reservoir: An aquifer (subsurface body of water) that has sufficient heat, permeability, and water to be exploited without stimulation or enhancement.

Idle cropland: Land in cover and soil improvement crops, and cropland on which no crops were planted. Some cropland is idle each year for various physical and economic reasons. Acreage diverted from crops to soil-conserving uses (if not eligible for and used as cropland pasture) under federal farm programs is included in this component. Cropland enrolled in the Federal Conservation Reserve Program (CRP) is included in the idle cropland.

Ignitability: Liquid having a flash point of less than 140°F.

Incident light: Light that shines onto the face of a solar cell or module.

Incineration: An engineered process using controlled flame combustion to thermally degrade waste material.

Induced draft: The negative pressure created by the action of a fan, blower, or other gas-moving device located between an incinerator and a stack.

Induced seismicity: Refers to typically minor earthquakes and tremors caused by human activity that alters the stresses and strains on the Earth's crust. Most induced seismicity is of an extremely low magnitude, and in many cases human activity is merely the trigger for an earthquake that would have occurred naturally in any case.

Industrial wood: All commercial roundwood products except fuel wood.

Infiltration air: Air that leaks into the chambers of ducts of an incinerator.

Injection: The process of returning spent geothermal fluids to the subsurface; sometimes referred to as *reinjection*.

Interferometric synthetic aperture radar (InSar): A remote sensing technique that uses radar satellite images to determine movement of the surface of the Earth.

Internal collector storage: A solar thermal collector in which incident solar radiation is absorbed by the storage medium.

Inverter: Converts direct current (DC) into alternating current (AC).

Ionomer: A polyelectrolyte comprised of copolymers containing both electrically neutral repeating units and a fraction of ionized units.

Irradiance: The direct, diffuse, and reflected solar radiation that strikes a surface.

Joule: The basic energy unit for the metric system or, in a later more comprehensive formulation, the International System of Units (SI). It is ultimately defined in terms of the meter, kilogram, and second.

Kaplan turbine: A type of turbine that has two blades whose pitch is adjustable. The turbine may have gates to control the angle of the fluid flow into the blades.

Kilowatt (kW): One thousand watts of electricity.

Kilowatt-hour (kWh): One thousand watt-hours.

Landfill gas: Gas that is generated by decomposition of organic material at landfill disposal sites. Landfill gas is approximately 50% methane.

Langley (L): Unit of solar irradiance equal to 85.93 kWh/m^2 or 1 gram calorie per square centimeter.

Leeward: Away from the direction of the wind; opposite of windward.

Life-cycle analysis: Analysis focused on the environmental impact of a product during the entirety of its life cycle, from resource extraction to post-consumer waste disposal. It is a comprehensive approach to examining the environmental impacts of a product or package.

Lignin: Structural constituent of wood and (to a lesser extent) other plant tissues that encrusts the cell walls and cements the cells together.

Line shaft pump: Fluid pump that has the pumping mechanism in the wellbore and that is driven by a shaft connected to a motor on the surface.

Liner: (1) A casing string that does not extend to the top of wellbore but instead is anchored or suspended from inside the bottom of the previous casing string. (2) The material used on the inside of a furnace wall to ensure that a chamber is impervious to escaping gases.

Liquid collector: A medium-temperature solar thermal collector, employed predominately in water heating, which uses pumped liquid as the heat-transfer medium.

Lithology: The study and description of rocks, in terms of their color, texture, and mineral composition.

Live cull: A classification that includes live cull trees. When associated with volume, it is the net volume in live cull trees that are 5.0 inches in dbh and larger.

Load: The simultaneous demand of all customers required at any specified point in an electric power system.

Load balancing: Keeping the amount of electricity produced (supply) equal to the consumption (demand). This is one of the challenges of wind energy production, which produces energy on a less predictable schedule than other methods.

Local solar time: A system of astronomical time in which the sun crosses the true north–south meridian at 12 noon; it differs from local time according to longitude, time zone, and an equation of time.

Logging residues: The unused portions of growing stock and non-growing stock trees cut or killed by logging and left in the woods.

Lost circulation: Zones in a well that imbibe drilling fluid from the wellbore, thus causing a reduction in the flow of fluid returning to the surface. This loss causes drilled rock particles to build up in the well and can cause problems in cementing casing in place.

Low head: Heat of 66 feet or less.

Low-speed shaft: Connects the rotor to the gearbox.

Low-temperature collectors: Metallic or nonmetallic solar thermal collectors that generally operate at temperatures below 110°F and use pumped liquid or air as the heat-transfer medium. They usually contain no glazing and no insulation, and they are often made of plastic or rubber, although some are made of metal.

Magma: Molten rock within the Earth, from which igneous rock is formed by cooling.

Magnetic survey: Measurements of the Earth's magnetic field that are then mapped and used to determine subsurface geology.

Magnetotelluric: An electromagnetic method of determining structures below the Earth's surface using electrical currents and the magnetic field.

Mantle: The Earth's inner layer of molten rock, lying beneath the Earth's crust and above the Earth's core of liquid iron and nickel.

Matrix treatments: Treatment performance below the reservoir fracture pressure, generally designed to restore the natural permeability of the reservoir following damage to the near-wellbore area. Matrix treatments typically use hydrochloric or hydrofluoric acids to remove mineral material that reduces flow into the well.

Medium-temperature collectors: Solar thermal collectors designed to operate in the temperature range of 140 to 180°F, but that can also operate at a temperature as low as 110°F. The collector typically consists of a metal frame, metal absorption panel with integral flow channels (attached tubing for liquid collectors or integral ducting for air collectors), and glazing and insulation on the sides and back.

Megawatt (WM): One million watts of electricity.

Mesophilic bacteria: Grows best at moderate temperature, between 25 and 40°C.

Methane: A colorless, flammable, odorless hydrocarbon gas (CH_4) which is the major component of natural gas. It is also an important source of hydrogen in various industrial processes. Methane is a greenhouse gas.

Methanol: Also known as methyl alcohol or wood alcohol and having the chemical formula CH_3OH. Methanol is usually produced by chemical conversion at high temperature and pressure. Although usually produced from natural gas, methanol can be produced from gasified biomass.

Microseismicity: Small movements of the Earth causing fracturing and movement of rocks. Such seismic activity does not release sufficient energy for the events to be recognized except with sensitive instrumentation.

Mini-frac: A small fracturing treatment performed before the main hydraulic fracturing treatment to acquire stress data and to test presimulation permeability.

Modified Bardenpho process: Bardenpho process with the addition of an initial anaerobic zone; used to remove both total nitrogen and total phosphorus.

Modified Ludzack–Ettinger (MLE) process: Continuous-flow, suspended-growth process with an initial anoxic stage followed by an aerobic stage; used to remove total nitrogen.

Modified University of Cape Town (UCT) process: A^2/O process with a second anoxic stage where the internal nitrate recycle is returned; used to remove both total nitrogen and total phosphorus.

Moisture content: The amount of water per unit weight of biosolids. The moisture content is expressed as a percentage of the total weight of the wet biosolids. This parameter is equal to 100 minus the percent solids concentration, computed as follows:

$$\text{Moisture content} = \frac{(\text{Weight of solids}) - (\text{Weight of dry solids}) \times 100}{\text{Weight of wet solids}}$$

MSW (municipal solid waste): Residential solid waste and some nonhazardous commercial, institutional, and industrial wastes.

MTBE: Methyl tertiary butyl ether is a fuel oxygenate produced by reacting methanol wit isobutylene.

Multiplier effect: The multiplier effect is sometimes called the *ripple effect* because a single expenditure in an economy can have repercussions throughout the entire economy. The multiplier is a measure of how much additional economic activity is generated from an initial expenditure.

Nacelle (or cowling): Contains and protects the gearbox and generator; sometimes large enough for an engineer or technician to stand in while doing maintenance.

Natural draft: The negative pressure created by the height of a stack or chimney and the difference in temperature between flue gases and the atmosphere.

Net metering: Arrangement that permits a facility (using a meter that reads inflows and outflows of electricity) to sell any excess power it generates over its load requirement back to the electrical grid to offset consumption.

Net photovoltaic cell shipment: The difference between photovoltaic cell shipments and photovoltaic cell purchases.

Net photovoltaic module shipment: The difference between photovoltaic module shipments and photovoltaic module purchases.

Net summer capacity: The maximum output, commonly expressed in megawatts (MW), that generating equipment can supply to system load, as demonstrated by a multi-hour test at the time of summer peak demand. This output reflects a reduction in capacity due to electricity use for station service or auxiliaries.

Nocturnal cooling: The effect of cooling by the radiation of heat from a building to the night sky.

Nonforest land: Land that has never supported forests or land formerly forested where use of time management is precluded by development for other uses. Includes areas used for crops, improved pasture, residential areas, city parks, improved roads of any width and adjoining clearings, powerline clearings of any width, and 1- to 4.5-acre areas of water classified by the Bureau of the Census as land. If intermingled in forest areas, unimproved roads and nonforest strips must be more than 120 feet wide, and clearings must be more than 1 acre in area to qualify as nonforest land.

Nonindustrial private: An ownership class of private lands where the owner does not operate wood-using processing plants.

Nonutility generation: Electric generation by nonutility power producers to supply electric power for industrial, commercial, and military operations or sales to electric utilities.

Nonutility power producer: Corporation, person, agency, authority, or other legal entity or instrumentality that owns electric generating capacity and is not an electrical utility. Nonutility power producers include qualifying cogenerators, qualifying small power producers, and other nonutility generators without a designated, franchised service area that do not file forms listed in 18 CFR 141.

Occupied space: The space within a building or structure that is normally occupied by people and that may be conditioned (heated, cooled, or ventilated).

Opacity: Degree of obstruction of light; for example, a window has zero opacity, while a wall has 100% opacity.

Operations and maintenance (O&M): Activities related to the performance of routine, preventive, predictive, scheduled, and unscheduled actions aimed at preventing equipment failure or decline with the goal of increasing efficiency, reliability, and safety.

Other biomass: Includes agricultural byproducts and crops (e.g., straw); other biomass gas (digester waste alcohol); other biomass liquids (fish oil liquid acetonitrite, waste, tall oil, waste alcohol); other biomass solids (medical waste, solid byproducts, sludge waste, and tires).

Other forest land: Forest land other than timberland and reserved forest land. It includes available forest land that is incapable of annually producing 20 cubic feet per acre of industrial wood under natural conditions because of adverse site conditions such as sterile soils, dry climate, poor drainage, high elevation, steepness, or rockiness.

Other removals: Unutilized wood volume from cut or otherwise killed growing stock, from cultural operations such as precommercial thinnings, or from timberland clearing. Does not include volume removed from inventory through reclassification of timber land to productive reserved forest land.

Other sources: Sources of roundwood products that are not growing stock. These include available dead, rough, and rotten trees; trees of noncommercial species; trees less than 5.0 inches dbh; tops; and roundwood harvested from nonforest land (e.g., fence rows).

Oxidation ditch: Continuous-flow process using looped channels to create time-sequenced anoxic, aerobic, and anaerobic zones; used to remove both total nitrogen and total phosphorus.

Packer: Device that can be placed in the wellbore to block vertical fluid flow so as to isolate zones.

Paper pellets: Paper compressed and bound into uniform diameter pellets to be burned in a heating stove.

Parabolic dish: A high-temperature (above 180°F) solar thermal concentrator, generally bow shaped, with two-axis tracking.

Passive solar: A system in which solar energy alone is used for the transfer of thermal energy. Pumps, blowers, or other heat-transfer devices that use energy other than solar are not used.

Peak watt: A manufacturer's unit indicating the amount of power a photovoltaic cell or module will produce at standard test conditions (normally 1000 W/m^2 and 25°C).

Peaking plants: Electricity generating plants that are operated to meet the peak or maximum load on the system. The cost of energy from such plants is usually higher than from baseload plants.

Penstock: The pipe that brings water to the turbine.

Peat: Consists of partially decomposed plant debris. It is considered an early stage in the development of coal. Peat is distinguished from lignite by the presence of free cellulose and a high moisture content (exceeding 70%). The heat content of air-dried peat (about 50% moisture) is about 9 million Btu per ton. Most U.S. peat is used as a soil conditioner.

Penstock: A closed conduit or pipe for conducting water to the powerhouse.

Permeability: The ability of a rock to transmit fluid through its pores or fractures when subjected to a difference in pressure. Typically measured in darcies or millidarcies.

Photovoltaic (PV): A system that converts sunlight directly into electricity using cells made of silicon or other conductive materials. When sunlight hits the cells, a chemical reaction occurs, resulting in the release of electricity.

Photovoltaic (PV) cell: An electronic device consisting of layers of semiconductor materials fabricated to form a junction (adjacent layers of materials with different electronic characteristics) and electrical contacts and being capable of converting incident light directly into electricity (direct current).

Photovoltaic (PV) module: An integrated assembly of interconnected photovoltaic cells designed to deliver a selected level of working voltage and current at its output terminals; it is packaged for protection against environmental degradation and suited for incorporation in photovoltaic power systems.

Plate tectonics: A theory of global-scale dynamics involving the movement of many rigid plates of the Earth's crust. Tectonic activity is evident along the margins of the plates where buckling, grinding, faulting, and vulcanism occur as the plates are propelled by the forces of deep-seated mantle convection current. Geothermal resources are often associated with tectonic activity, as it allows groundwater to come in contact with deep subsurface heat sources.

Poletimber trees: Live trees at least 5.0 inches in dbh but smaller than sawtimber trees.

Polycrystalline diamond compact (PDC) drilling bit: A drilling bit that uses polycrystalline diamond compact inserts on the drill bit to drill by means of rotational shear of the rock face.

Population equivalent (PE) or unit loading factor: When it is impossible to conduct a wastewater characterization study and other data are unavailable, population equivalent or unit per capita loading factors are used to estimate the total waste loadings to be treated. If the BOD contribution of a discharger is known, the loading placed upon the wastewater treatment system in terms of equivalent number of people can be determined. The BOD contribution of a person is normally assumed to be 0.17 lb BOD per day.

$$\text{Population equivalent (PE)} = \frac{BOD_5 \text{ contribution (lb/day)}}{0.17 \text{ lb } BOD_5 \text{ per day per person}}$$

Porosity: The ratio of the aggregate volume of pore spaces in rock or soil to its total volume, usually stated as a percent.

Primary wood-using mill: A mill that converts roundwood products into other wood products. Common examples are sawmills that convert saw logs into lumber and pulp mills that convert pulpwood roundwood into wood pulp.

Process heating: The direct process end use in which energy is used to raise the temperature of substances involved in the manufacturing process.

Proppant: Sized particles mixed with fracturing fluid to hold fractures open after a hydraulic stimulation.

Protonic ceramic fuel cell: Based on a ceramic electrolyte material that exhibits high protonic conductivity at elevated temperatures.

Public Utility Regulatory Policies Act (PURPA) of 1978: Contains measures designed to encourage the conservation of energy, more efficient use of resources, and equitable rates, including suggested retail rate reforms and new incentives for production of electricity by cogenerators and uses of renewable resources.

Pulpwood: Roundwood, whole tree chips, or wood residues that are used for the production of wood pulp.

Pumped-storage hydroelectric plant: A plant that usually generates electric energy during peak load periods by using water previously pumped into an elevated storage reservoir during off-peak periods when excess generating capacity is available to do so. When additional generating capacity is needed, the water can be released from the reservoir through a conduit to turbine generators located in a power plant at a lower level.

Pyrolysis: The thermal decomposition of biomass at high temperatures (greater than 400°F, or 200°C) in the absence of air. The end product of pyrolysis is a mixture of solids (char), liquids (oxygenated oils), and gases (methane, carbon monoxide, and carbon dioxide), with the proportions determined by operating temperature, pressure, oxygen content, and other conditions.

Quadrillion Btu (quad): Equivalent to 10 to the 15th power Btu.

Qualifying facility (QF): A cogeneration or small power production facility that meets certain ownership, operating, and efficiency criteria established by the Federal Energy Regulatory Commission (FERC) pursuant to PURPA.

Rankine cycle: The thermodynamic cycle that is an ideal standard for comparing performance of heat engines, steam power plants, steam turbines, and heat pump systems that use a condensate vapor as the working fluid; efficiency is measured as work done divided by sensible heat supplied.

Recovery factor: The fraction of total resource that can be extracted for productive uses.

Renewable energy: Energy that is produced using resources that regenerate quickly or are inexhaustible. Wind energy is considered inexhaustible, because while it may blow intermittently it will never stop.

Renewable energy resources: Energy resources that are naturally replenishing but flow limited. They are virtually inexhaustible in duration but limited in the amount of energy that is available per unit of time. Renewable energy resources include biomass, hydro, geothermal, solar, wind, ocean thermal, wave action, and tidal action.

Renewable Portfolio Standard (RPS): A mandate requiring that renewable energy provide a certain percentage of total energy generation or consumption.

Reservoir: A natural underground container of liquids, such as water or steam (or, in the petroleum context, oil or gas).

Residues: Bark and woody materials that are generated in primary wood-using mills when roundwood products are converted to other products. Examples are slabs, edgings, trimmings, sawdust, shavings, veneer cores and clippings, and pulp screenings. Includes bark residues and wood residues (both coarse and fine materials) but excludes logging residues.

Resistivity survey: Measurement of the ability of a material to resist or inhibit the flow of an electrical current, measured in ohm-meters. Resistivity is measured by the voltage between two electrodes while an electrical current is generated between two other electrodes. Resistivity surveys can be used to delineate the boundaries of geothermal fields.

Resource base: All of a given material in the Earth's crust, whether its existence is known or unknown, and regardless of cost considerations.

Ribbon silicon: Single-crystal silicon derived by means of fabricating processes that produce sheets or ribbons of single-crystal silicon. These processes include edge-defined film-fed growth, dendritic web growth, and ribbon-to-ribbon growth.

Roller cone bit: Drill bit used to crush rock; studded rotating cones are attached to the bit.

Rotating biological contactor (RBC) process: Continuous-flow process using RBCs with sequential anoxic/aerobic stages; used to remove total nitrogen.

Rotten tree: A live tree of commercial species that does not contain a saw log now or prospectively primarily because of rot (that is, when rot accounts for more than 50% of the total cull volume).

Rough tree: (1) A live tree of commercial species that does not contain a saw log now or prospectively primarily because of roughness (i.e., when sound cull, due to such factors as poor form, splits, or cracks, accounts for more than 50% of the total cull volume). (2) A live tree of noncommercial species.

Roundwood: Wood cut specifically for use as a fuel.

Roundwood products: Logs and other round timber generated from harvesting trees for industrial or consumer use.

Runner: The rotating part of the turbine that converts the energy of falling water into mechanical energy.

Salvable dead tree: A downed or standing dead tree that is considered currently or potentially merchantable by regional standards.

Salinity: A measure of the quantity or concentration of dissolved salts in water.

Saplings: Live trees 1.0 inch through 4.9 inches in dbh.

Scroll case: A spiral-shaped steel intake guiding the flow into the wicket gates located just prior to the turbine.

Secondary wood processing mills: A mill that uses primary wood products in the manufacture of finished wood products, such as cabinets, moldings, and furniture.

Seismic: Pertaining to, of the nature of, or caused by an earthquake or earth vibration, natural or manmade.

Seismicity: Relative frequency, intensity, and distribution of earthquakes.

Seismometer: Electrical device that is used on the surface and within wellbores to measure the magnitude and direction of seismic events.

Self-potential: In geothermal systems, it is a measure of currents induced in the subsurface because of the flow of fluids.

Sequencing batch reactor (SBR) process: Suspended-growth batch process sequenced to simulate the four-stage process; used to remove total nitrogen (total phosphorus removal is inconsistent).

Silicon: A semiconductor material made from silica, purified for photovoltaic applications.

Single crystal silicon: An extremely pure form of crystalline silicon produced by the Czochralski method of dipping a single crystal seed into a pool of molten silicon under high vacuum conditions and slowly withdrawing a solidifying single crystal boule rod of silicon. The boule is sawed into thin wafers and fabricated into single-crystal photovoltaic cells.

Skylight: A window located on the roof of a structure to provide interior building spaces with natural daylight, warmth, and ventilation.

Slim hole: Drill holes that have a nominal inside diameter less than about 6 inches.

Slotted liner: Liner that has slots or holes in it to let fluid pass between the wellbore and surrounding rock.

Sludge: A dense, slushy, liquid-to-semifluid product that accumulates as an end result of an industrial or technological process designed to purify a substance. Industrial sludges are produced from the processing of energy-related raw materials, chemical products, water, mined ores, sewerage, and other natural and manmade products. Sludges can also form from natural processes, such as the runoff produced by rainfall, and accumulate on the bottom of bogs, streams, lakes, and tidelands.

Small hydro: Projects that produce 30 MW or less.

Smart tracer: Tracer that is useful in determining the flow path between a well injecting fluid into the subsurface and a well producing fluid from an adjacent well; it can also be used to determine temperature along the flow path, the surface area contacted by the tracer, the volume of rock that the tracer interacts with, and the relative velocities of separate phases (gas, oil, and water in petroleum fields; steam and liquid water in geothermal systems).

Solar energy: The radiant energy of the sun, which can be converted into other forms of energy, such as heat or electricity.

Solar heat gain coefficient (SHGC): Measures how well a product blocks heat caused by sunlight. The SHGC is expressed as a number between 0 and 1. The lower the SHGC, the less solar heat it transmits.

Solar thermal collector: A device designed to receive solar radiation and convert it into thermal energy. Normally, a solar thermal collector includes a frame, glazing, and an absorber, together with the appropriate insulation. The heat collected by the solar thermal collector may be used immediately or stored for later use.

Solar thermal collector, special: An evacuated tube collector or a concentrating (focusing) collector. Special collectors operate in the temperature range from just above ambient temperature (low concentration for pool heating) to several hundred degrees Fahrenheit (high concentration for air condition and specialized industrial processes).

Spent liquor: The liquid residue left after an industrial process; can be a component of waste materials used as fuel.

Spent sulfite liquor: End product of pulp and paper manufacturing processes that contains lignins and has a high moisture content; often reused in recovery boilers. Similar to black liquor.

Spinner survey: The use of a device with a small propeller that spins when fluid passes in order to measure fluid flow in a wellbore. The device is passed up and down the well continuously to measure flow to establish where and how much fluid enters or leaves the wellbore at various depths.

Spinning reserve: A reserve of generation capacity, where the generators are kept idle in anticipation of an unexpected increase in demand or decrease in supply.

Start-up speed: The wind speed at which a rotor begins to rotate.

Step feed process: Alternating anoxic and aerobic states. Influent flow is split to several feed locations and the recycle biosolids stream is sent to the beginning of the process; used to remove total nitrogen.

Stimulation: A treatment performed to restore or enhance the productivity of a well. Stimulation treatments fall into two main groups, hydraulic fracturing treatments and matrix treatments.

Stress: The forces acting on rock. In the subsurface the greatest force or stress is generally vertical caused by the weight of overlying rock.

Structural discontinuity: A discontinuity of the rock fabric that can be a fracture, fault, intrusion, or differing adjacent rock type.

Submersible sump: Pump with both the pumping mechanism and a driving electric motor suspended together at depth in the well.

Subsidence: A sinking of an area of the Earth's crust due to fluid withdrawal and pressure decline.

Syngas: A synthesis gas produced through gasification of biomass. Syngas is similar to natural gas and can be cleaned and conditioned to form a feedstock for production of methanol.

Tailrace: The channel that carries water away for a dam.

Tailwater: The water downstream of the powerhouse.

TDS: Total dissolved solids; used to describe the amount of solid materials in water.

Tall oil: The oily mixture of rosin acids, fatty acids, and other materials obtained by acid treatment of the alkaline liquors from the digesting (pulping) of pine wood.

Thermal drawdown: Decline in formation temperature due to geothermal production.

Thermal gradient: The rate of increase in temperature as a function of depth into the Earth's crust.

Thermosiphon system: A solar collector system for water heating in which circulation of the collection fluid through the storage loop is provided solely by the temperature and density difference between the hot and cold fluids.

Thin-film silicon: A technology in which amorphous or polycrystalline material is used to make photovoltaic (PV) cells.

Tiltmeter: Device able to measure extremely small changes in its rotation from horizontal. The "tilt" measured by an array of tiltmeters emplaced over a stimulation allows delineation of inflation and fracturing caused by the stimulation.

Tip-speed ratio (TSR): The ratio between the wind speed and the speed of the tips of the wind turbine blades.

Tipping fee: A fee for disposal of waste.

Tower: Steel structures that support the wind turbine assembly. Higher towers allow for longer blades and capture of the faster moving air at higher altitudes.

Tracer: A chemical injected into the flow stream of a production or injection well to determine fluid path and velocity.

Transformer: Used to step up or step down AC voltage or AC current.

Transmission line: Structures and conductors that carry bulk supplies of electrical energy from power-generating units.

Transmission system (electric): A set of conductors, insulators, supporting structures, and associated equipment used to move or transfer electric energy in bulk between points of supply and points at which it is transformed for delivery over the distribution system lines to consumers, or is delivered to other electrical systems.

Transportation sector: An energy-consuming sector that consists of all vehicles whose primary purpose is transporting people and goods from one physical location to another. Included are automobiles; trucks; busses; motorcycles; trains, subways, and other rail vehicles; aircraft; and ships, barges, and other waterborne vehicles. Vehicles whose primary purpose is not transportation (e.g., construction cranes and bulldozers, farming vehicles, warehouse tractors and forklifts) are classified in the sector of their primary use.

Turbine: A machine for generating rotary mechanical power from the energy of a stream of fluid (such as water, steam, or hot gas). Turbines convert the kinetic energy of fluids to mechanical energy through the principles of impulse and reaction, or a mixture of the two.

Turbulence: A state of being highly agitated. In turbulent fluid flow, the velocity of a given particle changes constantly both in magnitude and direction.

U-factor: Measures the rate of heat loss or how well a product prevents heat from escaping. It includes the thermal properties of the frame as well as the glazing. The insulating value is indicated by the R-value, which is the inverse of the U-factor. U-factor ratings generally fall between 0.20 and 1.20. The lower the U-factor, the greater a product's resistance to heat flow and the better its insulating value.

Ultra low head: Head of 10 feet or less.

Upwelling: A pattern of coastal and open water oceanic circulation. It is created by persistent winds blowing across the ocean surface. As winds move surface waters, they are replaced by deeper waters that are richer in nutrients and can support increased phytoplankton growth, which in turn supports higher populations of fish and other consumers.

Under reamer: A drilling device that can enlarge a drill hole. The device is placed about the drill bit and can be opened to drill and then closed to be brought back up through a smaller diameter hole or casing.

Upwind turbine: A turbine that faces into the wind and requires a wind vane and yaw drive to maintain proper orientation in relation to the wind.

Useful heat: Heat stored above room temperature (in a solar heating system).

Useful thermal output: The thermal energy made available for use in any industrial or commercial process or used in any heating or cooling application (i.e., total thermal energy made available for processes and applications other than electrical generation).

Vapor dominated: A geothermal reservoir system in which subsurface pressures are controlled by vapor rather than by liquid. Sometimes referred to as a *dry-steam reservoir*.

Variable-frequency drive (VFD): A specific type of adjustable-speed drive that controls the rotational speed of an alternating current (AC) electric motor by controlling the frequency of the electrical power supplied to the motor. VFDs are also know as *adjustable-frequency drives* (AFDs), *variable-speed drives* (VSDs), *AC drives*, or *inverter drives*.

Viewshed: The scenic characteristics of an area, when referred to as a resource.

Visible light transmittance: The amount of visible light that passes through the glazing material of a window, expressed as a percentage.

Voltage: The measure of electrical potential difference.

Volatile: Any substance that evaporates at low temperature.

Volatile organic compounds (VOCs): Carbon-containing chemical compounds that can vaporize easily and can play an important role in atmospheric chemistry. Based on their molecular structure, these organic species can be grouped in different classes of compounds, including aldehydes, alcohols, ketones, and acids, among others.

Volatility: The property of a substance or substances to convert into vapor or gas without chemical change.

Watt (electric): Electrical unit of power; the rate of energy transfer equivalent to 1 ampere of electric current flowing under a pressure of 1 volt at unity power factor.

Watt (thermal): A unit of power in the metric system, expressed in terms of energy per second, equal to the work done at a rate of 1 joule per second.

Watthour (Wh): The electrical energy unit of measure equal to 1 watt of power supplied to, or taken from, an electric circuit steadily for 1 hour.

Well log: Logging includes measurements of the diameter of the well and various electrical, mass, and nuclear properties of the rock which can be correlated with physical properties of the rock. The well log is a chart of the measurement relative to depth in the well.

Wetlands: Areas where water covers the soil or is present either at or near the surface of the soil all year or for varying periods of time during the year.

Wicket gates: Adjustable elements that control the flow of water to the turbine passage.

Wind energy: Energy present in wind motion that can be converted to mechanical energy for driving pumps, mills, and electric power generators. Wind pushes against sails, vanes, or blades radiating from a central rotating shaft.

Wind power plant: A group of wind turbines connected to a common utility system through a system of transformers, distribution lines, and (usually) one substation. Operation, control, and maintenance functions are often centralized through a network of computers monitoring systems, supplemented by visual inspection. This a term commonly used in the United States. In Europe, it is called a *generating station*.

Wind rose: A diagram that indicates the average percentage of time that the wind blows from different directions on a monthly or annual basis.

Wind turbine or windmill: A device for harnessing the kinetic energy of the wind and using it to do work, or generate electricity.

Wind vane: Wind direction measurement device, used to send data to the yaw drive.

Windward: Into or facing the direction of the wind; opposite of leeward.

Wood/wood waste: This category of biomass energy includes black liquor, wood/wood waste liquids (red liquor, sludge wood, spent sulfite liquor), and wood/wood waste solids (peat, paper pellets, railroad ties, utility poles, wood/wood waste).

Wood energy: Wood and wood products used as fuel, including roundwood (cord wood), limb wood, wood chips, bark, sawdust, forest residues, charcoal, pulp waste, and spent pulping liquor.

Wood pellets: Sawdust compressed into uniform-diameter pellets to be burned in a heating stove.

Yaw: The rotation of a horizontal axis wind turbine around its tower or vertical axis.

Yaw drive: Motor that keeps an upwind turbine facing into the wind.

Zonal isolation: Various methods to selectively partition portions of the wellbore for stimulation, testing, flow restriction, or other purpose.

Zone: An area within the interior space of a building, such as an individual room, to be cooled, heated, or ventilated. A zone has its own thermostat to control the flow of conditioned air into the space.

Index

A

acetogenesis, 179
acid esterification, 184, 185, 186
acid rain, 12
acidogenesis, 179
acidogenic bacteria, 180
AC motors, 53–57, 65, 66; *see also* motors
activated carbon, 12
activated sludge, 12, 20, 22, 75, 76, 77, 284, 286, 291, 306; *see also* return activated sludge solids, waste activated sludge solids
activated sludge process, 16, 18, 22, 285, 286, 291, 293, 295
active power, 277
adiabatic process, 115
adsorbents, 12
advanced wastewater treatment, 12, 38
advective winds, 248
aeration, 12, 76, 270
    systems, design and control of, 281–295
aerators, solar-powered, 240
aerobic conditions, 12
aerofoils, 253
Aerostrip®, 300
affinity laws, 67
agglomeration, 12
aging infrastructure, 6
air bearings, 127
air currents, 247–250
air density, 255, 256
    correction factors, 257
air flow control, 287–288
air–fuel mixture, 139
air gap, 13
air intake system, 282
air pollution, 161, 248
air pressure, 196
Airport Water Reclamation Facility (AWRF), 83–84
algae bloom, 13
algorithms, 292, 307
alpha ratio, 282
alternating current (AC), 54, 58, 59, 124, 125, 126, 127, 140, 195, 215, 263, 272, 273, 279
    generator, 55
    motors, 53–57, 65, 66
alum, 13

ambient conditions, 13
ammonia, 307
anaerobic conditions, 13
anaerobic digestion, 105, 161, 172, 176, 178, 179–181
anaerobic zone mixing, 307–311
anemometer, 253, 265
anhydrous HCL gas, 301
anoxic conditions, 13
anoxic zone mixing, 270, 301, 307–311
apparent power, 277
aquifer, 13, 16, 18
arc chutes, 57
area, velocity and, 205
armature, 54, 57
artesian water, 13
assets, useful life of, 24–26
Atlantic County Utilities Authority, 239–240, 266
auxiliary and supplemental power sources (ASPSs), 123, 161, 223, 245
auxiliary power source, 138, 161
average discharge limitations, 13
axial flow turbines, 125

B

backflow, 13
back-pressure turbine, 153
backwash, 13
bacteria, 13, 19, 176, 179, 180, 212, 293
    acidogenic, 180
    coliform, 14, 16
    methanogenic, 180
    nitrifying, 294
Baldor Electric Company, 277
barrel, 36
bar screen, 13
Bartlett Wastewater Treatment Plant, 289
baseline audit, 46–50
basin, groundwater, 13
batteries, deep-cycle, 263
battery vs. fuel cell, 161, 162
bearings
    air, 299
    magnetic, 299
    microturbine, 127
    pump, 281

Printed and bound by CPI Group (UK) Ltd, Croydon, CR0 4YY

21/10/2024

01777095-0018